Conference Proceedings of the Society for Experimental Mechanics Series

Series Editor
Tom Proulx
Society for Experimental Mechanics, Inc.,
Bethel, CT, USA

For further volumes:
http://www.springer.com/series/8922

Eann Patterson • David Backman • Gary Cloud
Editors

Composite Materials and Joining Technologies for Composites, Volume 7

Proceedings of the 2012 Annual Conference on Experimental and Applied Mechanics

Springer

Editors
Eann Patterson
University of Liverpool
UK

David Backman
National Research Council Canada
Ottawa, ON, Canada

Gary Cloud
Michigan State University
East Lansing, MI, USA

ISSN 2191-5644 ISSN 2191-5652 (electronic)
ISBN 978-1-4614-4552-4 ISBN 978-1-4614-4553-1 (eBook)
DOI 10.1007/978-1-4614-4553-1
Springer New York Heidelberg Dordrecht London

Library of Congress Control Number: 2012945402

Printed on acid-free paper

Springer is part of Springer Science+Business Media (www.springer.com)

Preface

Composite Materials and Joining Technologies for Composites, Volume 7: Proceedings of the 2012 Annual Conference on Experimental and Applied Mechanics represents one of seven volumes of technical papers presented at the Society for Experimental Mechanics SEM 12th International Congress & Exposition on Experimental and Applied Mechanics, held at Costa Mesa, California, June 11–14, 2012. The full set of proceedings also includes volumes on Dynamic Behavior of Materials, Challenges in Mechanics of Time-Dependent Materials and Processes in Conventional and Multifunctional Materials, Imaging Methods for Novel Materials and Challenging Applications, Experimental and Applied Mechanics, Mechanics of Biological Systems and Materials, and MEMS and Nanotechnology.

Each collection presents early findings from experimental and computational investigations on an important area within Experimental Mechanics. The Composite Materials and the first International Symposium on Joining Technologies for Composites conference track was organized by Eann Patterson, University of Liverpool; David Backman, National Research Council Canada; Gary Cloud, Michigan State University; and sponsored by the SEM Composite Materials Technical Division.

As composite materials have moved from smaller-scale applications to wider acceptance in larger-scale application areas such as automotive or aerospace structures, the need for improved joining of composites has become increasingly important. Composite joining technologies in the past have been widely grouped into mechanical joining or adhesive joining. Increasingly, joint optimization has required combinations of the two methods as well as introducing innovative new methods such as composite welding that provide high strength and light weight. Today, developments in composite joining technologies are being made at a rapid rate, driven by both technology and user requirements.

To provide a forum for an up-to-date account of the advances in the field of composite joining technologies and to promote an alliance between governmental, industrial, and academic practitioners, SEM has agreed to initiate a *Symposium Series on Joining Technologies for Composites*. The 2012 Symposium will be the first of a potential series and will address pertinent issues relating to design, analysis, fabrication, testing, optimization, reliability, and applications of composite joints, especially as these issues relate to experimental mechanics of both the macroscale and microscale structures.

Topics included in this volume are:

Composite Joining for Heavy Duty Applications
Advances in Fastening and Joining
Modeling and Validation of Composite Joints
Composite Joints in Aerospace Related Applications
Fatigue and Fracture of Composite Joints
Composite Characterization using Digital Image Correlation Techniques
Nanocomposites for Improved Composite Performance
Impact Behavior of Composites

The contribution of the organizing committee, session chairs, authors and keynote speakers is gratefully acknowledged along with the support from SEM staff.

The opinions expressed herein are those of the individual authors and not necessarily those of the Society for Experimental Mechanics, Inc.

Liverpool, UK — Eann Patterson
Ottawa, ON, Canada — David Backman
East Lansing, MI, USA — Gary Cloud

Contents

Chapter 1
Evaluating Bolted Joint Strength at High Strain Rates

Srinivasan Arjun Tekalur, Andy VanderKlok, Wei Zhang, and Abhishek Dutta

Abstract Experimental evaluation of bolted joint strength at high strain rates is a challenging but vital research. In the current study, a single lap bolted joint is studied using a Split Hopkinson Bar Pressure setup. The objective of the study is to understand the effect of variable geometrical parameters in the single lap bolted joint experimentally. Subsequently, only those aspects of these parameters that can confine to the dynamic experimental technique are selected. The challenges of the setup, the role of equilibrium and the occurrence of buckling are all studied critically to understand what is the best method to evaluate the joint strength under impact/high strain rate loading conditions.

Keywords Monolithic joint • Split hopkinson pressure bac • Equilibrium

1.1 Introduction

Bolted joints is a simple way to attach two structures together, and at the same time allows for a non-destructive means to separate the structures for repair or maintenance. This method of fastening has been extensively tested in the static to quasi-static regime with cross head speeds of 0.1–10 mm/s, but has been neglected at even higher rates of loading [1]. Static testing has been extensively carried out in both bearing and tensile configuration. Within the bearing configuration the primary modes of failure are considered to be out of plane buckling and hole elongation. Both failure modes are highly dependent upon geometrical parameters of the specimen. Results for static bearing testing show a trend for increasing diameter by thickness (d/t) yields lower ultimate bearing strength. Similarly, for edge by diameter (e/d) and width by diameter w/d show an asymptotic region where failure mode abruptly changes from shear at low stress to buckling at higher stresses. This asymptotic region of failure mode alteration is common in both the geometrical parameters e/d and w/d for bearing and tensile loading configuration, and occurs at a ratio of approximately 3.0 [2] [3]. This asymptotic behavior is has been extensively tested in the static regime where shear failure is abruptly changed to buckling for a certain e/d, w/d, and d/t ratio. The aim of this paper is to determine if any such behavior is present at high loading rates for monolithic specimens, along with subsequent determination of an optimal geometry for SHPB testing of bolted joints.

Most failures of bolted structures are predominant when loading acting on the structure is dynamic in nature thereby, motivating the need for further research on the behavior of bolted joints under impact rates of loading. In this paper we introduce the preliminary steps involved testing bolted joints in a split Hopkinson pressure bar at loading rates of approximately 500 MN/s.

This involves satisfying equilibrium and buckling conditions by testing a set of monolithic specimens with varying length and widths to determine optimum geometric parameters only. A finite element model had been constructed using ABAQUS to mimic the behavior of the experimentally tested monolithic specimens. This model was verified to match trends of monolithic specimen loading rate, equilibrium time, and equilibrium start time for a fixed d/t = 1, w/d = 2.83, and varying e/d = 1,2,3, and 4. Validation of the FEM against experimentally obtained results ensures the application of the model to

S.A. Tekalur (✉) • A. VanderKlok • W. Zhang • A. Dutta
Department of Mechanical Engineering, Michigan State University, 2727 Alliance Drive, Lansing, MI 48910, USA
e-mail: tsarjun@egr.msu.edu

E. Patterson et al. (eds.), *Composite Materials and Joining Technologies for Composites, Volume 7: Proceedings of the 2012 Annual Conference on Experimental and Applied Mechanics*, Conference Proceedings of the Society for Experimental Mechanics Series 44,
DOI 10.1007/978-1-4614-4553-1_1,

determine monolithic specimen behavior for other geometries that are not restrained by the diameter of the SHPB apparatus. Finally, an optimal geometry can be determined for ideal bolted joint testing where failure due to buckling and equilibrium times of specimen will be well established.

1.2 Materials and Methods

Testing of monolithic specimens was conducted using a Split Hopkinson Pressure Bar (SHPB) apparatus. The configuration utilizes 0.75 in. 6,061 Aluminum incidence and transmission bars with electro-resistive strain gages diametrically opposing centrally placed on each bar. The incident pulse was generated using a 5 in. striker powered by 30 psi of compressed Helium. For specimen equilibrium to be obtained, and to limit dispersive effects such as Pochhammer-Cree oscillations a pulse shaping material was employed between the striker and incidence bar [4]. Data acquisition system consists of an Ectron amplifier E513-2A with circuitry configured for half bridge and Lecroy Wave Jet 354A oscilloscope. Monolithic specimens that are geometrically similar to a uniaxial loaded bolted joint configuration were prepared with a fixed $d/t = 1$, $w/d = 2.83$, and varying $e/d = 1, 2, 3,$ and 4. For each e/d ratio five specimens were manufactured, and four of each was tested with one spare for any specimen that may have been lost during testing. Experimental results show 95% confidence error bars for 4 specimens. The FEM results show an average of 5 calculations of the same e/d configuration to reduce the bias of human calculation of takeoff and landing of incident, reflected, and transmission waves. The FEM model was constructed with ABAQUS (version 6.9–2) and contains an incident and transmission bar. An input pulse directly from experimental data was used for the system excitation. Using an element in the center of the incident and transmission bar within the model, the incident, reflected, and transmission waves were extracted for analysis of equilibrium times and loading rate similarly to the experiment (Fig. 1.1a).

Table 1.1 shows the geometrical parameters non-dimensionalized with respect to a hole diameter. It is important to note that the hole diameter is a fictitious value as there is no hole in the monolithic specimen. Each specimen was loaded into the SHPB setup where incidence, reflected, and transmission waves were recorded for computation of loading rate, total equilibrium time, and equilibrium start time. Parameters of w/d and e/d were fixed as the SHPB apparatus restrains from any larger specimens.

1.3 Results

Specimen geometries were altered according to Table 1.1 and incident, reflected, and transmitted waves recorded for each test. Using one dimensional wave theory and Eqs. 1.1 and 1.2 Figs. 1.2 and 1.3 were constructed to determine optimal specimen properties for later testing of bolted joints of similar dimensions. The force on the transmission face was calculated using Eq. 1.2 where the slope of the transmitted pulse was used to calculate the loading rate of the specimen. Five FEM simulations for e/d of 1 and 4 were run using ABAQUS and averaged to smooth any outliers in the data due to the human

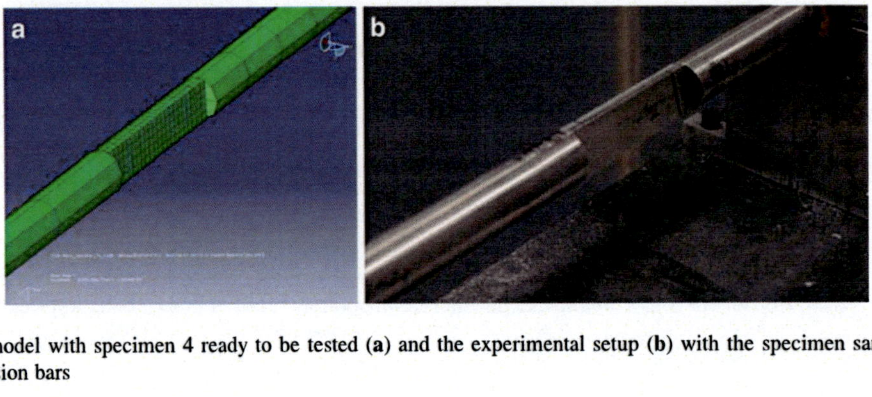

Fig. 1.1 ABAQUS model with specimen 4 ready to be tested (**a**) and the experimental setup (**b**) with the specimen sandwiched between the incident and transmission bars

Table 1.1 Geometric parameters for monolithic specimen testing

Specimen #	1	2	3	4
e/d	1.00	2.00	3.00	4.00
l/d	4.50	6.50	8.50	10.50

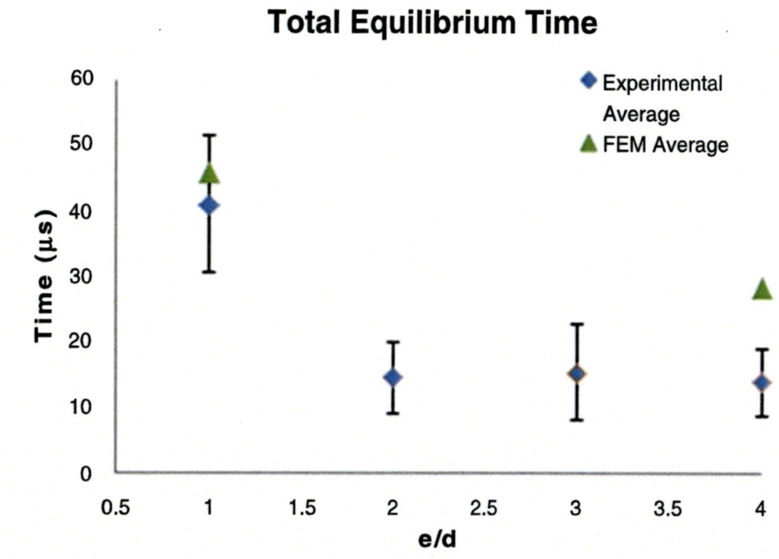

Fig. 1.2 Total amount of time the specimen is at equilibrium for varying e/d in compression. Experimental error bars are calculated with 95% confidence interval

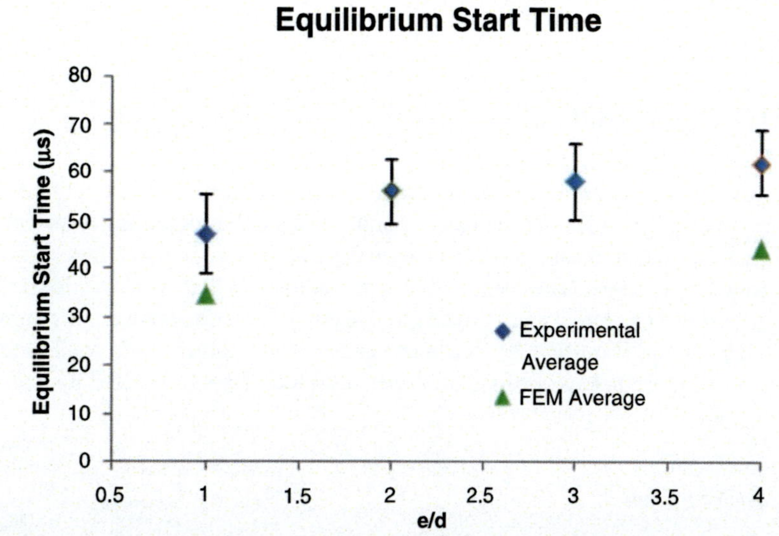

Fig. 1.3 The amount of time it takes the specimen to first reach equilibrium for varying e/d in compression. Experimental error bars are calculated with 95% confidence interval

interface between where to select where each pulse begins and ends. The general trend of the loading rate shows a slight increase from 450 MN/s to approximately 600MN/s as specimen e/d increase. Calculation of equilibrium times was conducted using both the force on the incident and transmission sides of the specimen. Plotting the incidence by transmission on the y-axis and time on the x-axis; equilibrium occurs when this force ratio is at unity. From this plot the total equilibrium time of the specimen can be obtained as well as when the specimen first reaches equilibrium as shown in Figs. 1.2 and 1.3 respectively.

1.4 Discussion

Using one dimensional wave propagation wave theory for elastic solids the following relations is established to determine loading rate, and equilibrium times [5].

$$F_{\text{inc.}}(t) = A_{\text{inc.bar}} E_{\text{bar}}(\varepsilon_{\text{i}}(t) + \varepsilon_r(t)) \tag{1.1}$$

$$F_{\text{trans.}}(t) = A_{\text{trans.bar}} E_{\text{bar}}(\varepsilon(t)) \tag{1.2}$$

Where $F_{inc.}$ and $F_{trans.}$ are the force on the incident and transmitted face of the SHPB. $A_{inc.bar}$ and $A_{trans.bar}$ is the cross-sectional area of the incident and transmission bars respectively. E_{bar} is the elastic modulus of the bars, and ε_i, ε_r, and ε_t are the recorded incident, reflected, and transmitted strain pulses from the adhesively bonded strain gages on the SHPB. Using split Hopkinson bar wave theory, specimen loading rate and equilibrium times can be calculated. This combined with visually inspecting each monolithic specimen for any mode of buckling will ensure proper specimen geometry for bolted joint testing in a SHPB setup.

Ideal specimen properties are high total equilibrium times, low equilibrium start times and low chances of buckling or bending. In the literature [4], it can be seen that equilibrium typically occurs at approximately 17–20 μs for sample lengths of 0.76–1.5 mm in length respectively. Length of the tested monolithic specimens varied from 1.125 to 2.875in (73.025 mm). For range of specimen length, it can be seen that the equilibrium start time is much higher, ranging from 45 to 63 μs due to inertia and non-uniform deformation due to wave propagation effects [4]. The increase in equilibrium start time is dependent on the length and inertia of the specimen as the longer specimen requires more time for the pulse to propagate through the specimen. As only a finite incidence pulse is generated tradeoff occurs between total time spent in equilibrium and equilibrium start times. This effect can be noticeably seen at e/d ratio of approximately 2.5. This behavior reflects nicely with the amount of specimens recorded as bent or deformed during testing with a decrease in total time spent in equilibrium. The FEM results follow closely to the actual experimental results in that general trends are still kept precise. However the accuracy of the model shows that some assumptions made in the model need to be refined. This error is mostly introduced between the interface between the specimen, incidence, and transmission bars has been modeled to be in perfect slip contact. This can cause the model to underestimate the actual equilibrium start time and over estimate the total time spent at equilibrium.

1.5 Conclusion

From the collected data it can be deduced that there does exist an optimum bolted joint geometry that can reach equilibrium and avoid local bending. Figures 1.2 and 1.3 show that increasing a specimen length will decrease the chance of buckling or local bending to occur, at the same time there exists a geometry where minimal local bending takes place with relatively high equilibrium times. The ideal bolted joint parameters that would yield the best result would be $e/d = 2.0$ or total length of 1.625 in. for minimal local bending and maximum time spent at equilibrium for a dynamic loading rate of approximately 500 MN/s. Upon identification of the critical e/d ratio, future work would include varying w/d and d/t parameters to identify where failure behavior of the joint would no longer be dominated by buckling.

References

1. Pearce GM, Johnson AF, Thomson R, Kelly DW (2010) Experimental investigation of dynamically loaded bolted joints in carbon fibre composite structures. Appl Compos Mater 17:271–291
2. Kretsis G, Matthews F (1985) The strength of bolted joints in glass fibre/epoxy laminates. Composites 16(2):92–102
3. Smith P, Pascoe K, Polak C, Stroud D (1986) The behaviour of single-lap bolted joints in CFRP laminates. Compos Struct 6:41–55
4. Wu X, Gorham D (1997) Stress equilibrium in the split Hopkinson pressure bar test. J Phys 7:C3-91–C3-96
5. Kolsky H (1949) An investigation of the mechanical properties of materials at very high rates of loading. Proc Phys Soc 62:C3-91–C3-96

Chapter 2
Fastening and Joining of Composite Materials

Sayed A. Nassar and Xianjie Yang

Abstract This paper presents some of the most recent advances in fastening and joining research on composite joints, which are often viewed as the weakest links that can significantly affect the safety and reliability of many mechanical and structural systems. A system approach is outlined for fastening and joining issues, which includes the effect of six groups of variables; namely, the joint, the joining element, tool used, process control, in-service loads, and the environmental variables. Challenging bolting issues are discussed as they apply to composite joints that are subjected to complex mechanical, thermo-mechanical, and environmental loading. This includes theoretical, FEA, and experimental models for investigating the effect of bolt holes, failure modes, vibration and impact-induced loosening of preloaded threaded fasteners, loss of clamp load due to mismatched thermal effects, and elastic interaction. Investigated joining variables in layered joints include normal and interfacial stresses, properties and thickness of the adhesive layers, additives, and damage modeling. Results from theoretical and experimental models are presented.

Keywords Composite materials • Composite joining • Bolted composites • Adhesive joined composites

2.1 Introduction

Composite materials have recently become more desirable for use in mechanical and structural components due to their relatively high specific strength and stiffness-to-weight ratios. Composites can also be designed and optimized to meet different strength and stiffness requirement in various directions as dictated by the design and performance of a structural component. Unlike metallic structures, the modeling and analysis of composite structures is a fairly challenging task. The load-carrying capacity of the composite material continues to be severely limited by the reduced load carrying capacity at the joint or attachment locations to the main structure. Due to anisotropy and inhomogeneity of composites, their response to loading is more complex, as compared to metallic structures. For a reliable design of composite structures, it is essential to thoroughly understand their behavior under static and dynamic mechanical and thermal loading under various environmental scenarios during the life of the component. The development of such design methods must be based on test data, analytical and computer modeling as well as numerical models using Finite Elements techniques.

The fastening and joining of composite materials mainly consist of the bonding and mechanical fastening. The main objective of bolted joints is to transfer applied load from one part of the joint structure to the other through fastener elements. However, bolt holes cause a stress concentration in the composite joint plates, which can severely reduce the mechanical strength and fatigue life of the joined structures. There are several possible joint failure modes in composites and three of the common ones are bearing, net tension and shear-out. Among them, bearing failure is often considered as the "desirable" mode because it usually gives a higher strength and the failure is less brittle. Other modes are often considered as "premature" failures which should be avoided through proper design of the joint geometry and the composite material itself. Referring to the geometric dimensions such as specimen width, hole diameter and hole-to-edge distance, Collings [1] proposed ultimate bearing strength, net ultimate tensile strength and ultimate shear strength based on average stresses at failure, and the actual failure mode and load are associated with the one with the lowest load value among them. Fatigue damage around bolt holes

S.A. Nassar (✉) • X. Yang
Department of Mechanical Engineering, Fastening and Joining Research Institute (FAJRI), Oakland University, Rochester, MI 48309, USA
e-mail: nassar@oakland.edu

E. Patterson et al. (eds.), *Composite Materials and Joining Technologies for Composites, Volume 7: Proceedings of the 2012 Annual Conference on Experimental and Applied Mechanics*, Conference Proceedings of the Society for Experimental Mechanics Series 44,
DOI 10.1007/978-1-4614-4553-1_2,

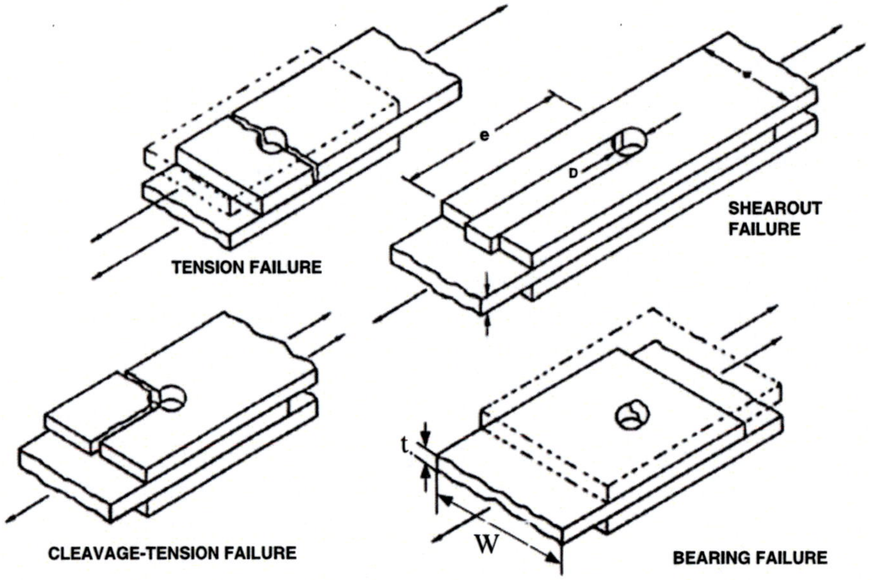

Fig. 2.1 Typical failure modes for bolted joints in composite laminate

consists of three types: hole wear, damage in the contact surface of the composite, and the growth of delamination around the bolt holes induced by the drilling. The hole wear is caused by the erosion of material around the bolt hole as a result of the friction forces. Damage at the contact surfaces is induced by bolt bending under loading. This would result in hole elongation during fatigue loading. It was found that the growth of delamination around bolt holes results in a decrease in the fatigue life of bolted joints. The failure load and pattern of composite bolted joints would obviously depend on the bolt preload level and distribution, orientation of layer reinforcing fibers, the ratio of bolt diameter to specimen width, bolt type, ratio of hole diameter to laminate thickness, number of bolts, and bolt arrangement.

Two problems should be addressed in the design of composite structures with joints that have multiple bolts. Firstly, it is necessary to adequately understand the behavior, strength, failure modes, and failure criterion of single-bolt joints. Secondly, it is necessary to accurately evaluate the loading magnitude and distribution. The strength prediction methods developed from single-bolt joints can also be utilized for determining the maximum failure load and the corresponding failure mode. The loading proportion is not generally equal for each fastener in the same joint. Therefore, determination of the ratio of loading proportion for each element becomes an important problem in the optimal design of joints with multiple threaded fasteners. Because this is mostly a highly statically indeterminate mechanical problem, FEA modeling is used for the determination of the loading ratio. Additionally, the FEA model can simulate the failure prediction based on the different failure criteria of the bearing, shear-out, and net-tension failure patterns.

An important consideration in joint testing and analysis is the selection of the type of test method with due attention to failure mode which is likely to lead to specific joint design requirements for a known composite system. The occurrence of a particular failure mode is primarily dependent on the joint geometry and laminate lay-up. Composite bolted joints may fail in various modes as shown in Fig. 2.1 [2]. The likelihood of a particular failure mode is influenced by bolt diameter D, laminate width w, edge distance e, and thickness t. The type of fastener used can also influence the occurrence of a particular failure mode.

For near-isotropic lay-ups in graphite/epoxy composite systems, net-section tensile/compressive failure occurs when the ratio of bolt diameter D to laminate width W is equal to or greater than one quarter (i.e. $D/w \geq 0.25$). It is characterized by failure of the plies in the primary load direction. Cleavage failure may initiate in the vicinity of the specimen end (rather than near the fastener); this mode of failure can be triggered by net-section failure. In some instances the bolt head may be pulled out through the laminate after the bolt is bent and/or deformed. Finally, it is important to note that for any given geometry, the failure mode may vary as a function of laminate lay-up and stacking sequence. Catastrophic failure modes such as tension, shear-out and cleavage-tension failures are avoidable through proper design of the joint geometry and the composite material itself. Most bolted composite structures are primarily designed to avoid bearing failure; investigating the effect of various joint parameters on bearing failure in a joint is of fundamental importance.

Adhesive joints increase structural efficiency and weight savings. They also minimize the potential for stress concentration within the joint, which cannot be achieved with mechanical fasteners. However, because of the lack of reliable, economical, and feasible inspection methods and due to the requirement for close dimensional tolerances in fabrication, designers have generally avoided bonded construction in primary structures. For bonded composite joints, the non-uniform stress distribution along the bonding surface should always be accounted for. The peak stress is mainly dependent on the bonding pattern of the joint, bonded length, adhesive thickness, joint geometry, adherend stiffness imbalance, ductile adhesive response, and the composite adherends.

The main purpose of an adhesively bonded joint is to transfer loads in a reliable fashion under various environmental conditions throughout the entire service life of the joint and the parent structural component. The joint interfacial stresses introduced by those loads must remain essential part of the design, analysis, test, and validation process. Much of the current methodologies used in the design and analysis of adhesive joints in composite structures is based on the approach developed by L. J. Hart-Smith in a series of NASA-Langley-sponsored contracts [3, 4] during the early 1970s. Some of the key principles on which that effort was based on includes: (1) the use of simple 1-dimensional stress analysis of generic composite joints wherever possible [5]; (2) the need to select the joint design so as to ensure failure in the adherend rather than the adhesive, so that the adhesive is never the week link; (3) a recognition that the ductility of adhesives is beneficial in reducing the stress peaks in the adhesive; (4) careful use of such factors as adherend tapering to reduce or eliminate peel stresses from the joint; and (5) recognition of slow cyclic loading, corresponding to such phenomena as a major factor controlling durability of adhesive joints, and the need to avoid the worst effects of this type of loading by providing sufficient overlap length. This ensures that some of the adhesive is so lightly loaded that creep was not significant under the most severe scenarios of extreme humidity and temperature over the component service life. After Volkersen's work [6] on the stress analysis of a single lap joint, many analytical models have been proposed for stress analysis of various adhesively bonded joints. Sample 3-D analytical models for interfacial stress are developed by Ma et al. [7].

Most recently, several investigational studies have been carried out by Nassar et al. at the Fastening and Joining Research Institute-Oakland University, on various fastening and joining aspects of advanced composite materials. Experimental and analytical techniques, as well as FEA simulation, are used to study bolted joint behavior and failure modes different loading and temperature conditions. That includes deformation, NDE and preload control, Tribology of threaded fasteners, vibration and impact-induced loosening of preloaded threaded fasteners, elastic interaction, loss of preload, creep relaxation modeling, and hole and washer variables. Summary of the results from the work by Nassar et al. is presented in this paper.

2.2 Composite Bolting

Mechanical fasteners (bolts and rivets) require drilling of holes in composite materials, which ruptures the composite fiber reinforcements. Those holes generally cause stress concentration, but equally or even more significant is the increased probability for micro-cracks and local damage to be introduced around the holes, which can cause structural instability [8]. In spite of those draw backs, mechanical fastening of composites can still be a viable and proven joining option.

2.2.1 *Effect of Geometry and Laminate Properties on the Bolt Bearing Behavior*

For a single-bolt composite joint, the bolt hole diameter with clearance, laminate thickness, material, and layer stack sequences, washer, and clamping pressure all have significant effect on the bolt bearing behavior. The bearing stress σ_{br} and bearing strain ε_{br} are defined as follows

$$\sigma_{br} = \frac{P}{Dh} \tag{2.1}$$

$$\varepsilon_{br} = \frac{\delta_{coupon,bearing}}{D} \tag{2.2}$$

where P is applied load, D is bolt hole diameter, h is thickness of the laminate, and $\delta_{coupon,bearing}$ represents the deformation in the bolt hole. The typical bearing stress–strain curve for composite bolted joint coupon have three distinct regions as shown in Fig. 2.2: initial sliding, linear bearing response prior to the damage, and a nonlinear post damage stress region. The bearing stiffness is determined from the initial linear part of the curve. Figure 2.2 shows the definition of the bearing stiffness, the 2% offset bearing strength and ultimate strength.

Other researchers did some work on the effect of geometry and laminate properties on the bolt bearing behavior. Oh et al. [9] worked on bolted joints for hybrid composites made of glass-epoxy and carbon-epoxy under tensile loading.

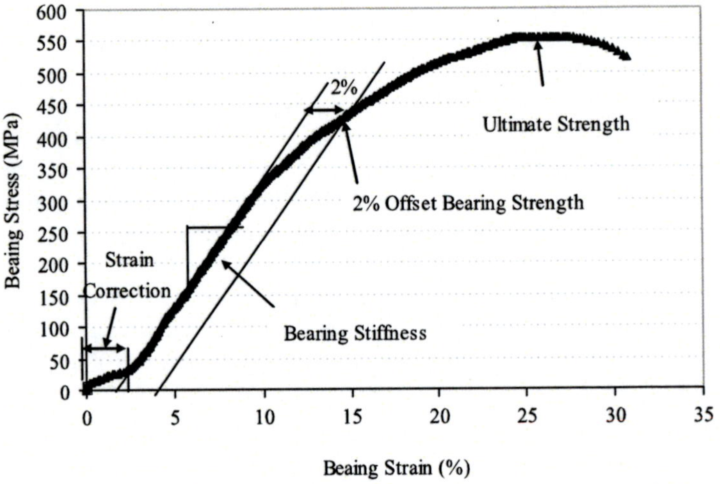

Fig. 2.2 Bearing stress–strain curve for bolted composite coupon

The design parameters investigated were laminate ply angle, stacking sequence, the ratio of glass-epoxy to carbon epoxy, the outer diameter of the washer and clamping pressure. Results showed that the peak load occurred before the maximum failure load due to the delamination of the laminate under the washer. The static test results of the hybrid composites with two kinds of stacking sequences revealed that the bearing strength increased as the $\pm 45°$ plies were distributed evenly along the thickness direction irrespective of the joint material and the stacking pattern. The bearing strength increased as the bolt clamping pressure increased to 71.1 MPa, thereafter the bearing strength saturated to a constant value. The failure mode changed from bearing failure to tension failure when a 20 mm diameter washer was used.

Aktas and Dirikolu [10] carried out an experimental and numerical study on the strength of a pinned-joint made of carbon epoxy composite. The experimental results showed that when the ratio of the edge distance to the pin diameter (e/D) is $e/D \geq 4$and the ratio of the specimen width to the pin diameter (W/D) is $W/D \geq 4$, bearing failure was dominant, where as when the ratios were below 4, net tension, shear-out and mixed model failure were observed. The [90/45/-45/0]s joint configuration showed 20% higher bearing strength than the [0/45-45/90]s configuration.

McCarthy et al. [11] investigated the effect of hole clearance on the strength and stiffness of single-lap, single-bolt composite joint. Hole clearance had a significant effect on the joint stiffness and ultimate strain and less effect on the joint strength. The clearance caused a delay in load take up and this was considered to be a significant factor for multiple-bolt joint applications.

2.2.2 *Effect of Tightening Torque and Clamping Pressure on Bearing Behavior*

Girard et al. [12] studied the effect of stacking sequence and clamping pressure on the carbon/epoxy bolted composite joints. Bearing stress vs. hole elongation curves and the bearing stress vs. strain curves showed significant effect of clamping pressure on the initial bearing stress and the maximum bearing stress. Tightening the bolt increased the initial bearing stress by 22% and the maximum bearing stress by 105%. The increase in clamping pressure increased the post-peak stiffness, where as the initial stiffness and the bolt-hole elongation decreased significantly. The bearing stress vs. hole elongation curves showed that the angle of ply lay-ups had the lowest initial stiffness and the cross-ply lay-up had the highest initial and post peak stiffness. Orienting the fibers at an angle of 45° improved the bearing behavior. The results from the rosette strain gage positioned on the bearing zone showed a linear behavior for the angle-ply laminate, where as a nonlinear behavior was observed for the cross-ply and the quasi-isotropic lay-ups. This nonlinear behavior was mainly due to the stresses corresponding to the initiation of damage due to local delamination around the hole.

Park [13] investigated the effect of stacking sequence and clamping force on delamination bearing strength and ultimate bearing strength of mechanically fastened carbon/epoxy composite joints using an acoustic emission (AE) and load–displacement technique. The stacking sequence and the clamping pressure had a significant effect on the delamination and ultimate bearing strength of the mechanically fastened composite joint. An increase in clamping pressure increased the ultimate bearing strength to saturation, whereas the delamination bearing strength increased progressively. The clamping pressure suppressed the delamination and the interlaminar cracks. The failure mode changed from catastrophic fracture to a progressive failure as the clamping pressure was increased.

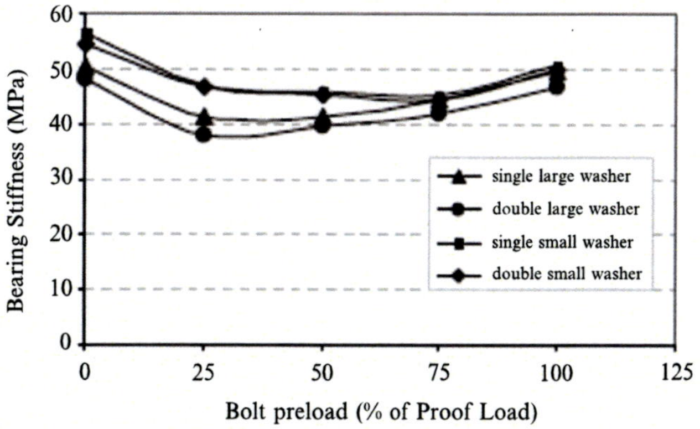

Fig. 2.3 Effect of bolt preload on joint bearing stiffness

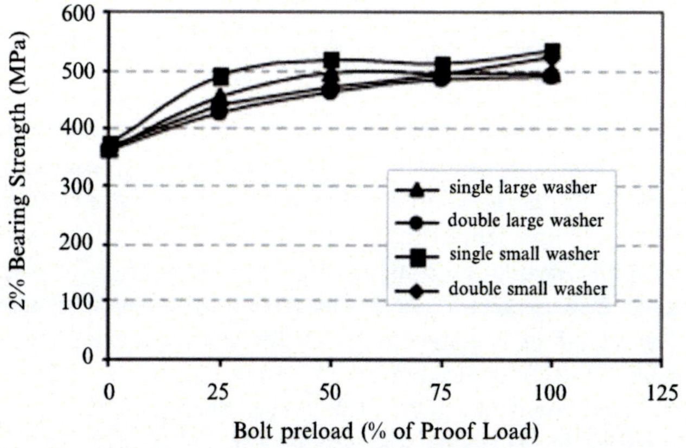

Fig. 2.4 Effect of bolt preload on offset bearing strength

Khashaba et al. [14] investigated the effect of bolt tightening torque and washer size on the bearing behavior of glass fiber reinforced epoxy composite ([0/±45/90]s) bolted joints. Damage analysis was carried out to understand the failure mechanism. Tightening torques of 0, 5, 10 and 15 Nm and the washer sizes of $D_{wo} = 14, 18, 22, 27$ mm were used in the study. The washer size of 18 mm and the tightening torque of 15 Nm produced the optimum clamping pressure. The composite bolted joints with 14 mm washers had higher clamping pressure but showed reduction in maximum bearing strength. The load–displacement curves of the finger tightened bolt joint showed least stiffness with non-linear behavior that indicated the unstable development of internal damage. Most of the tested specimens failed in a sequence, delamination, and net tension failure at 90° laminate, shear out failure at 0° layers and final failure which was nearly catastrophic due to the bearing failure of 45° layers.

2.2.3 Effect of Washers and Bolt Tension on the Behavior of Thick Composite Joints

Virupaksha and Nassar [15] studied the experimental characterization of thick composite bolted joints to explore the effect of washer size and bolt preload on bearing properties. S2-glass fabric-epoxy composite coupons [0/90; +45/−45@10sets] of 12.5 mm thickness were tested under double shear tensile loading. Two different washer sizes and thickness were used in the investigation. Five levels of bolt preload are investigated; namely, 0, 25%, 50%, 75% and 100% of the proof load of 1/2″−20 SAE Grade 5 fasteners. Figures 2.3, 2.4, and 2.5 show the effect of initial bolt load on the bearing joint stiffness, offset bearing strength, ultimate joint strength and joint strain for various joint configurations, respectively. The joint bearing stiffness was higher for the untightened bolted joint than that with much higher bolt preload (100% of proof load). The bearing stiffness was smallest for the joint with a preload equal to 25% of bolt proof load, and it increased with bolt preload. The offset bearing strength increased progressively with bolt preload. The ultimate joint strength was unaffected by increasing the bolt preload. Joint with small washers had higher bearing stiffness than those with large washers for initial bolt preload of 0%, 25% and 50%. Joints with small washers had higher offset bearing strength than the joints with large washers. The washer thickness had an insignificant effect on the ultimate joint strength and strain.

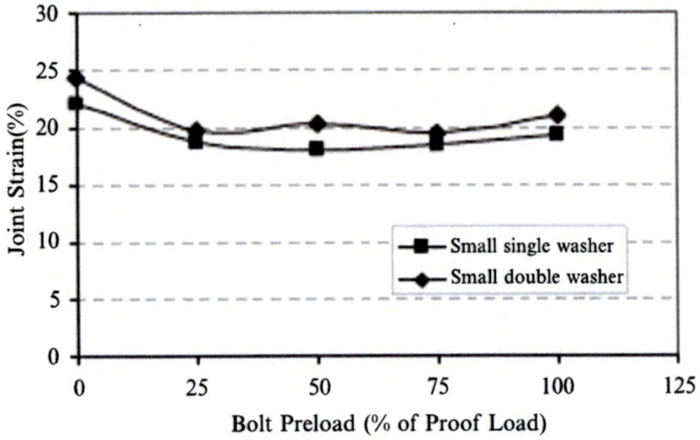

Fig. 2.5 Effect of bolt preload on joint strain

2.2.4 Vibration-Induced Loosening of Preloaded Threaded Fasteners

Threaded fasteners may have self-loosening when the joint is subjected to cyclic shear loads. The self-loosening leads to the partial loss or complete loss of the clamp load so that the function of the fastener will lose. When the clamp load loses completely, the bolt will sustain the whole separating force fluctuation under the cyclic separating force. This may result in the bolt fatigue failure much more easily. The shear force cannot be transferred between the two clamped joint members by using the friction force on the contact surfaces when the clamp load is zero. If the clamp load of the gasketed bolted joints loses, the leakage will occur. Therefore, the loss of the clamp load is one of the common failure modes of threaded fastener. There have been some studies on the self-loosening of threaded fasteners; most of them are experimental. Junker [16] studied the effect of the transverse vibrations on the self-loosening of threaded fasteners and showed that the loosening of threaded fasteners was far more severe when the joint was subjected to transverse cyclic loads. On the other hand, Junker designed a test machine for the self-loosening as shown in Fig. 2.6. In his work, he concluded that the self-loosening happens when slippage took place between engaged threads and/or under the bolt head/nut.

Based on the relative kinetic relationships as shown in Figs. 2.7 and 2.8 on the bearing surface and thread surface, Nassar and Yang [17] proposed the formulations for the sliding bearing friction torque and thread friction torque as shown in the following:

$$R_{Tb} = \left|\frac{T_b}{\mu_b q_{b0}}\right| = \int_{r_i}^{r_e} r^2 dr \int_0^{2\pi} \frac{(\eta_b \sin\theta + r)d\theta}{\sqrt{\eta_b{}^2 + 2\eta_b r \sin\theta + r^2}} \tag{2.3}$$

$$R_{Fb} = \left|\frac{F_{bs}}{\mu_b q_{b0}}\right| = -\int_{r_i}^{r_e} r dr \int_0^{2\pi} \frac{(\eta_b + r\sin\theta)d\theta}{\sqrt{\eta_b{}^2 + r^2 + 2\eta_b r \sin\theta}} \tag{2.4}$$

$$R_{Tt} = \left|\frac{T_t}{\mu_t q_{t0}}\right| = \sqrt{\sec^2\alpha + \tan^2\beta} \int_{r_{\min}}^{r_{maj}} r^2 dr \int_0^{2\pi} \frac{[r - \eta_t \sin\theta]d\theta}{\sqrt{\eta_t^2(1 + \tan^2\alpha\cos^2\theta) + r^2 - 2\eta_t r \sin\theta}} \tag{2.5}$$

$$R_{Ft} = \left|\frac{F_{ts}}{\mu_t q_{t0}}\right| = \sqrt{\sec^2\alpha + \tan^2\beta} \int_{r_{\min}}^{r_{maj}} r dr \int_0^{2\pi} \frac{(\eta_t - r\sin\theta)d\theta}{\sqrt{\eta_t^2(1 + \tan^2\alpha\cos^2\theta) + r^2 - 2\eta_t r \sin\theta}} \tag{2.6}$$

where T_b is the bearing frictional torque, F_{bs} is transverse bearing friction shear force, $\eta_b = v_{b1}/\omega_b$ is the bearing translation-rotational ratio, $v_{b1}$is the relative translation velocity along x direction, ω_b is the relative rotation angular velocity of the bolt underhead to the joint member, μ_b is bearing friction coefficient, and q_{b0} is the average bearing contact pressure. T_t is the thread friction torque, μ_t is the thread friction coefficient, and q_{t0} is the average thread contact pressure, and F_{ts} is transverse thread friction shear force, $\eta_t = v_{tx}/\omega_t$ is the thread translation-rotational ratio, $v_{tx}$is the relative thread translation velocity

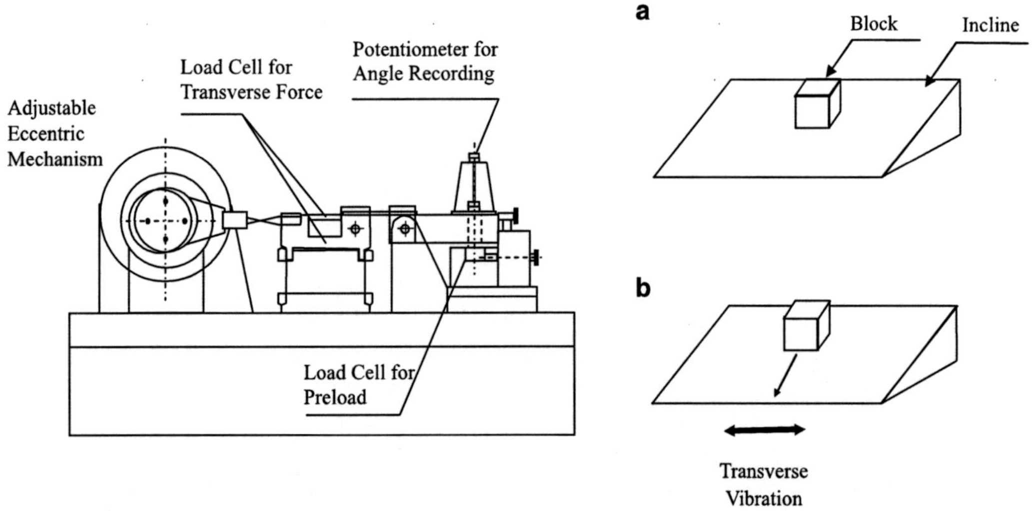

Fig. 2.6 Junker machine for self-loosening test; schematic of the block slippage due to vibration

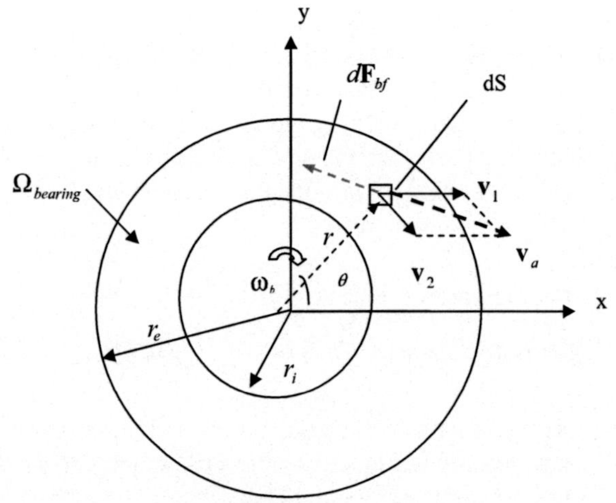

Fig. 2.7 The relative movement on the underhead bearing surface to the joint member

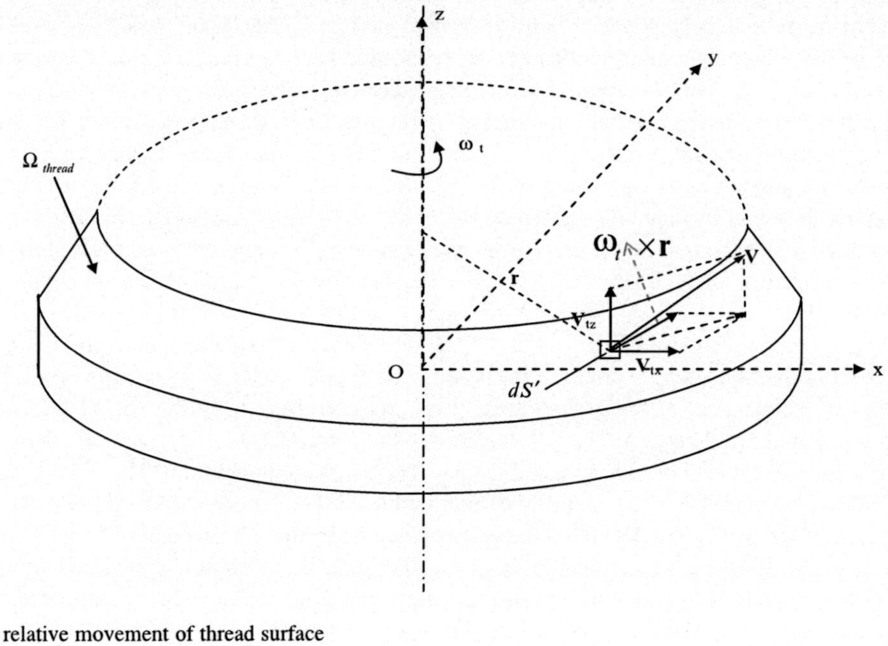

Fig. 2.8 Schematic relative movement of thread surface

along x direction, ω_t is the relative rotation angular velocity of the bolt thread to the joint member, and α is the half of the thread profile angle and β is lead helix angle.

From Eqs. 2.3, 2.4, 2.5, and 2.6, the sliding bearing and thread friction torque, the sliding bearing and thread shear force are dependent on the ratio of the relative transition movement to the rotation velocity of the bolt. When there is only the rotation velocity, the sliding friction torques are corresponding to the ones under bolt tightening and the resultant shear force is zero. When there is only relative transition movement, the sliding friction torques are zero.

When the pitch torque is larger than the sum of the sliding bearing friction torque and the sliding thread friction torque, the self-loosening occurs. Based on the dynamic movement equations of the bolted joint, the self-loosening under transverse cyclic loading can be predicted. Nassar and Housari [18, 19], Housari and Nassar [20] investigated the effect of thread pitch, initial bolt tension, hole clearance thread fit, bearing and thread friction, and amplitude of the cyclic transverse load on the loosening of threaded fasteners. They concluded that these parameters have a significant effect on the self-loosening of threaded fasteners. Yang and Nassar [21] developed a mathematical model for self-loosening of bolted joint under transverse cyclic loading, and Yang et al [22] proposed a criterion for preventing self-loosening under some specific conditions.

Following a different research route, Shoji and Sawa [23] proposed a 3-Dimensional finite element model to simulate the mechanism of the self-loosening caused by the relative slippage. Their simulation results show that during one cycle the nut rotates in both the tightening and the loosening directions with net rotation in the loosening direction. Independently, Jiang et al. [24] proposed a three dimensional elastic–plastic finite element model to investigate the early stage of self-loosening caused by the plastic ratcheting deformation of the thread roots, which does not include rotation of the bolt or the nut. After the bolt tightening, the micro-plastic deformation of the asperities on contact area occurs under the transverse cyclic loading. On the other side, the misalignment of the bolt assembly will be gradually adjusted in the initial cycles. Those lead to the initial clamp load loss. When the contact interfaces have relative movement, the wear of the contact area will decrease the bolt elongation and the joint compression, and it will change the surface roughness and the hardness conditions of the contact areas. The bolt elongation decrease results in the clamp load loss. The surface roughness and hardness change will make the friction coefficient on the contact area change with the transverse cyclic loading. As this phenomenon is affected by lots of parameters, quantitative results are still very scarce.

2.2.5 Elastic-Interaction of Multi-bolted Joints During Tightening

Nassar and Alkelani [25, 26] carried out the elastic-interaction of gasketed joints. Elastic interaction between the various fasteners in gasketed flanged joints is significantly influenced by the gasket material and thickness, bolt spacing, and by the tightening strategy and sequence. For the same tightening torque level, using simultaneous tightening of all bolts produces a higher and more uniform clamp load in the joint. Simultaneous tightening has also reduced the amount of tension drop off due to the combined effect of elastic interaction and gasket creep relaxation. The fastener grip length had insignificant effect on the clamp load loss for soft gaskets (e.g. red rubber). The residual clamp load level in gasketed joints was significantly affected by the gasket material used in the joint (i.e. soft versus hard gaskets). When a soft gasket was used in the joint, it was observed that increasing the tightening speed significantly has reduced the average residual clamp load in the joint; that was likely due to the continued gasket creep relaxation after the high-speed tightening process had been completed. Conversely, for a hard gasket (e.g. flexible graphite), it was observed that increasing the tightening speed increased the residual clamp load in joint. The study showed that simultaneous tightening of all bolts produced both higher initial and higher residual clamp loads, as compared to a star pattern tightening of the individual bolts. Using a second tightening pass increased the residual clamp load more for the individual bolt tightening as compared to simultaneous tightening of all bolts in the joint. Elastic interaction between preloaded bolts in the same joint was significantly increased by increasing the thickness of a soft gasket (styrene butadiene rubber), while the thickness of a hard gasket (flexible graphite) has not significantly affected the elastic interaction.

Alkelani et al. [27, 28] proposed a closed form mathematical modeling of elastic interaction and creep relaxation for the clamp load as a function of the time elapsed after the initial tightening of the joint as shown in Figs. 2.9 and 2.10. Gasket constants are found to be independent of the gasket stress level, but are affected by the gasket thickness for the gasket material considered in this study (red rubber). The bolt tension loss as a percentage of bolt preload does not depend on the fastener preload level. Fastener preload level has insignificant effect on elastic interaction. The clamp load formulation has been successfully used to determine the required initial clamp load level that is necessary to provide the desired level of a steady state residual clamp load in the joint, by taking the gasket creep relaxation into account. The good agreement between the mathematical model results and the experimental data suggests that the proposed model can be used to accurately describe the gasket behavior and the clamp load loss due to gasket creep relaxation. The amount of elastic interaction is

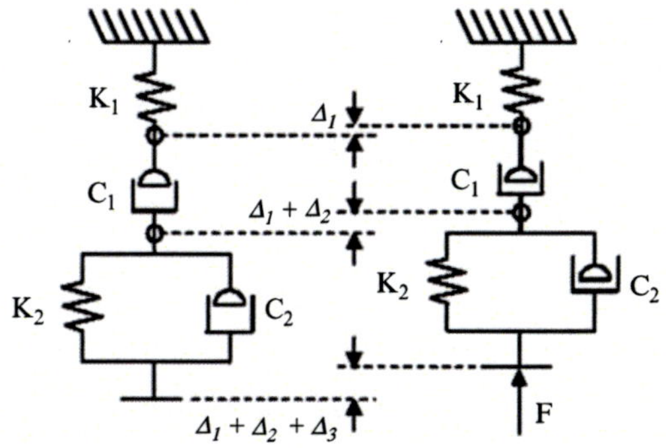

Fig. 2.9 Mechanical model of a typical gasket

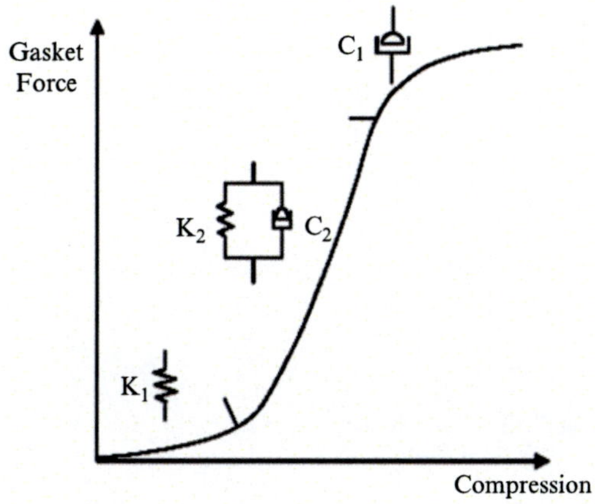

Fig. 2.10 Gasket Force-Compression

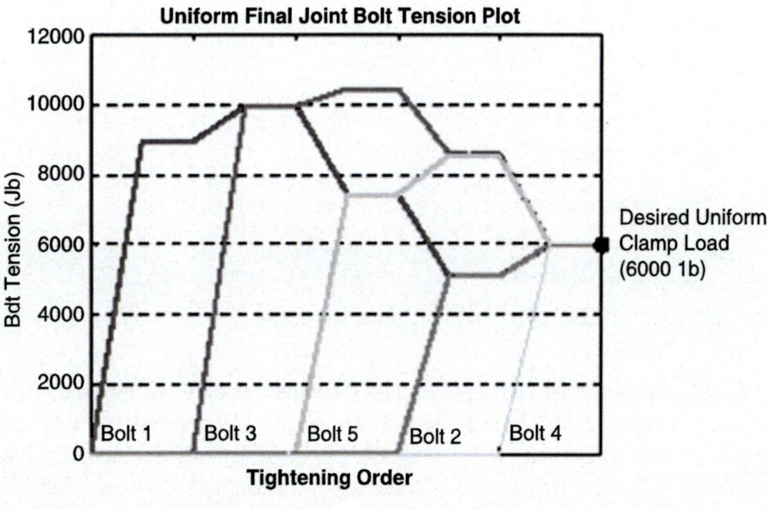

Fig. 2.11 Model prediction of a desired uniform clamp load

increased by increasing the gasket thickness. Reducing bolt spacing causes more loss of bolt tension due to increased elastic interaction. The final clamp load level and uniformity is significantly influenced by the tightening pattern. Good agreement exists between the proposed model and the experimental results as shown in Figs. 2.11 and 2.12. Hence, the model can reliably be used to predict and/or achieve a uniform clamp load distribution in flat-faced flanged joint.

Nassar et al. [29] did FEA simulation to Study the Effect of Multi-Pass Tightening of Gasketed Bolt Joint. The FEA model is shown in Figs. 2.13 and 2.14 shows the experimental and FEA results for multi-pass star tightening. For gasketed joint, the finite element simulation results have a good agreement with the experimental ones. The finite element simulation methodology can be used to develop bolting procedures for any bolted joint assembly. Four-pass sequential and star tightening operation is enough to achieve uniform preloads for the 5-bolt gasketed joint model. The experimental and the FE simulated results show that the multi-pass tightening strategy is very effective for the uniformity of the bolt tensions of

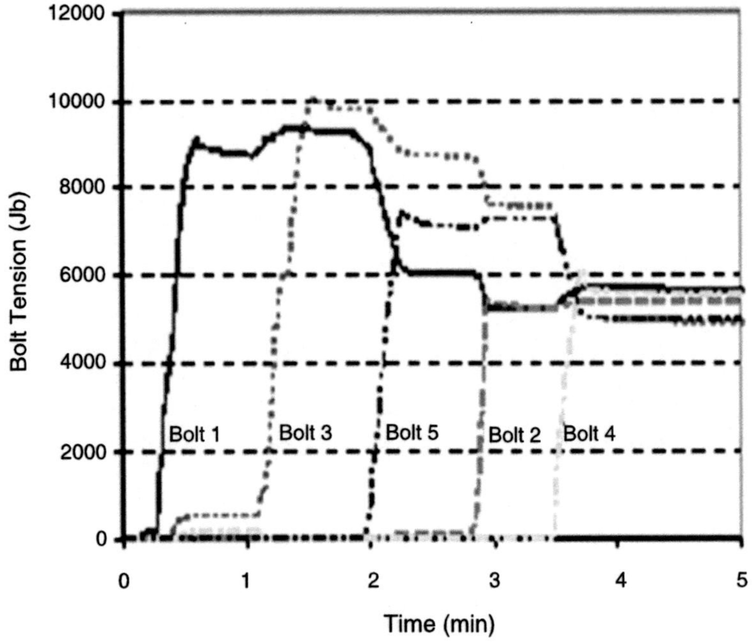

Fig. 2.12 Experimental verification of model prediction of a desired uniform clamp load

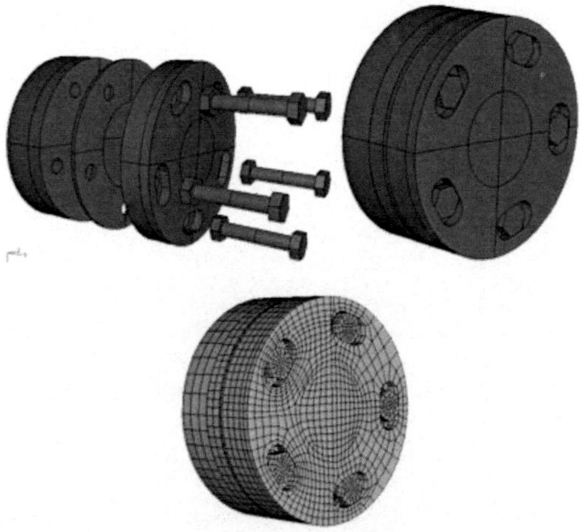

Fig. 2.13 Finite element models

the gasketed joint. The scatter of bolt load for the star pattern tightening is much higher than that for the sequential tightening with the same target preload at the end of each pass. For both of tightening patterns, the effect of the tightening history on the bolt tension variation of the subsequential tightening very significantly even though the previous tightening was completely removed before the subsequential tightening. It is easier to achieve uniform preload for the lower level of target preload with multi-pass sequential tightening procedure for the flexible graphite gasketed joint. The single pass tightening approach with the linear elastic gasket for the uniformity of bolt tensions was proposed, and the approach has been verified by FEA simulation.

2.3 Adhesively Bonded Composite Joints

This section highlights some of the recent analytical and experimental investigations of the behavior of bonded laminated joints under mechanical and thermo-mechanical loads.

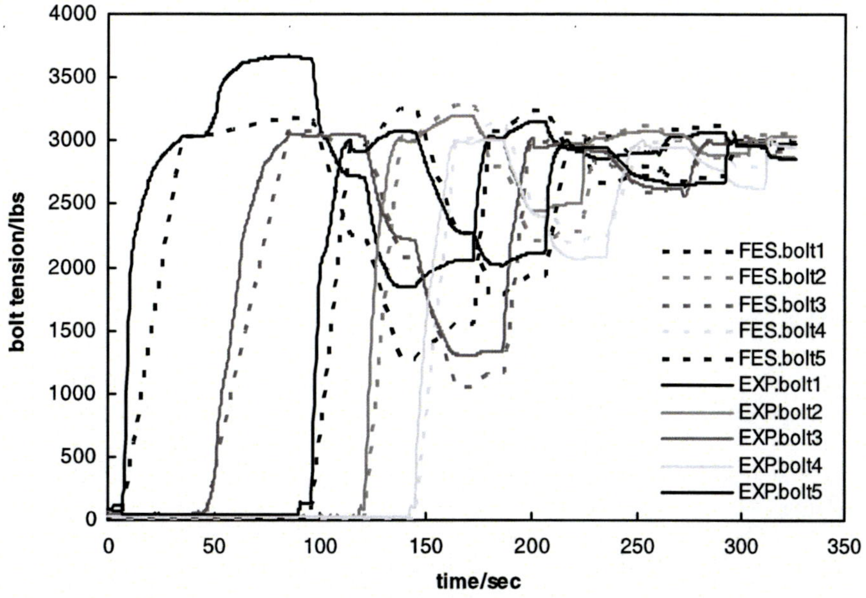

Fig. 2.14 The experimental and simulated bolt tension variations of gasketed joint in the Multi-pass star tightening with the preload of 3,000 lb

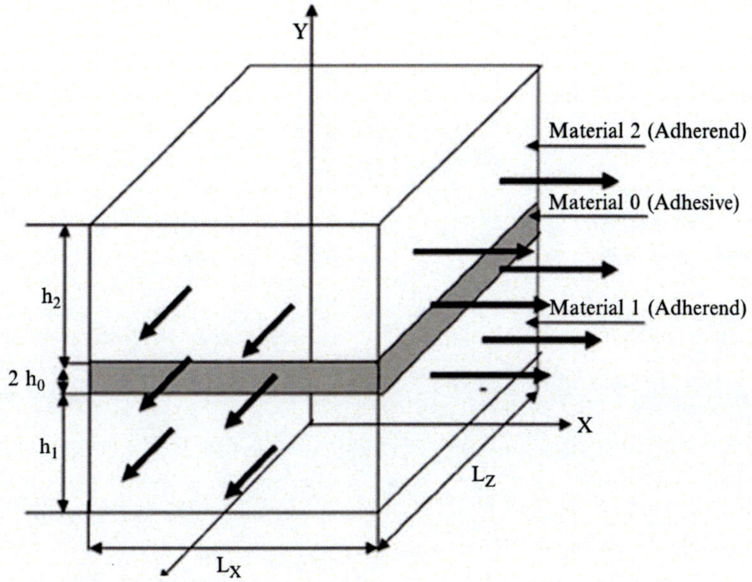

Fig. 2.15 Complete model of adhesive bonded joint

2.3.1 Interfacial Stresses in Adhesively Bonded Joints

Nassar and Virupaksha [30] introduced a linear continuum mixture model of periodically-stacked laminates with adhesive bonding. They derived a set of partial differential equations for interfacial shear stresses on both sides of a representative adhesive layer using the 3-D multi-layer model shown in Fig. 2.15. Analytical expressions were obtained for the interfacial shear stresses under thermo-mechanical loading. The numerical results on the model show that increasing the thickness of the adhesive causes a significant increase in the interfacial shear stresses; the larger difference in the elastic and thermal properties between the adhesive and the layered adherend leads to the higher corresponding interfacial shear stress.

2.3.2 *Damage and Fracture Analysis of Adhesively Bonded Composite Laminates*

The analysis of adhesive joint is usually based on through-thickness average stresses and strains; the corresponding basic variables are the peel stress σ, the shear stress τ, the elongation w, and the shear deformation v of the adhesive layer. To study the two stress components σ and τ separately, two loading scenarios are considered, namely, adhesive bonded joint loaded in shear and peel [31, 32].

2.3.2.1 Damage Model of Single Lap Joint

In this scenario, single lap joint is used, the joint is made of S2 glass/SC-15 epoxy resin composite, and bonded by 3 M Scotch-Weld Epoxy Adhesive DP405 Black [31]. Damage couple constitutive model is developed for this joint as follows

1. **Adhesive material**
   (a) Stress–strain relation

$$\sigma_{ij} = (1 - D_A)\bar{\sigma}_{ij} \tag{2.7}$$

$$\bar{\sigma}_{eq} = \begin{cases} E\bar{\varepsilon}_{eq} & before \quad yielding \\ K_A\bar{\varepsilon}_{eq}^{n} & after \quad yielding \end{cases} \tag{2.8}$$

   (b) Damage initiation

$$w = \int \frac{d\bar{\varepsilon}^p}{\bar{\varepsilon}_D} \geq 1 \tag{2.9}$$

   (c) Damage propagation

$$\dot{D}_A = \dot{u}_p^m \tag{2.10}$$

2. **Adherend material**
   (a) Stress–strain relation

$$\sigma_{ij} = M_{ik}\bar{\sigma}_{kj} = \bar{C}_{ik}\varepsilon_{kj} \tag{2.11}$$

   (b) Damage initiation

$$F_f = \begin{cases} \left(\frac{\bar{\sigma}_{11}}{X^T}\right)^2 + \left(\frac{\bar{\tau}_{12}}{S^L}\right)^2 & \bar{\sigma}_{11} \geq 0 \\ \left(\frac{\bar{\sigma}_{11}}{X^C}\right)^2 & \bar{\sigma}_{11} < 0 \end{cases} \tag{2.12}$$

$$F_m = \begin{cases} \left(\frac{\bar{\sigma}_{22}}{Y^T}\right)^2 + \left(\frac{\bar{\tau}_{12}}{S^L}\right)^2 & \bar{\sigma}_{11} \geq 0 \\ \left(\frac{\bar{\sigma}_{22}}{2S^T}\right)^2 + \left[\left(\frac{Y^C}{2S^T}\right)^2 - 1\right]\frac{\bar{\sigma}_{22}}{Y^C} + \left(\frac{\bar{\tau}_{12}}{S^L}\right)^2 & \bar{\sigma}_{11} < 0 \end{cases} \tag{2.13}$$

(c) Damage propagation

$$d_i = \left[\frac{\delta^f_{eq}}{\delta^f_{eq} - \delta^0_{eq}}\left(1 - \frac{\delta^0_{eq}}{\delta_{eq}}\right)\right]_i \tag{2.14}$$

3. **Interface**

(a) Stress–strain relation

$$t_i = \begin{cases} K_i\delta_i & w_I \le 1 \\ K_i\delta_i(1 - D_I) & w_I > 1 \end{cases} \tag{2.15}$$

(b) Damage initiation

$$w_I = \left(\frac{t_n}{X}\right)^2 + \left(\frac{t_s}{S}\right)^2 \tag{2.16}$$

(c) Damage propagation

$$d_i = \left[\frac{\delta^f_{eq}}{\delta^f_{eq} - \delta^0_{eq}}\left(1 - \frac{\delta^0_{eq}}{\delta_{eq}}\right)\right]_i \tag{2.17}$$

The results compared with experimental ones are shown in Figs. 2.16 and 2.17. The correlation between experiment and prediction is acceptable, the standard deviation is 0.75 MPa in Fig. 2.16 and 1.4 MPa in Fig. 2.17, and with the increase of adhesive thickness or overlap length, the prediction tends to be improved as shown in Figs. 2.16 and 2.17. Figures 2.18, 2.19, and 2.20 give the shear stress distribution in the joint. From the figures, the overlap length affects the distribution of the interfacial stress; the adhesive thickness does not significantly affect the interfacial stresses except at locations near the edges; increased plastic deformation of the adhesive makes the interfacial stress distribution significantly more uniform except at locations near the edges, plastic deformation also reduces the difference between the maximum and minimum stress values.

2.3.2.2 Fracture Analysis of Double Cantilever Beam

Double Cantilever Beam (DCB) is used in this scenario. Unlike classical beam theory and El-Zein and Reifsnider [32, 33], the crack tip rotation is considered. Also, the material properties and geometry of adhesive layer are taken into account in the model. Another difference compared with existing theories is the parameter c in the model, a nonlinear equation as list below is derived to solve this parameter

$$\frac{1}{c^2(l-a)} + \frac{E_x^{(1)} I h_a}{2E_y^2}\frac{c^2}{b(l-a)} + \frac{\frac{1}{2}(l-a) - \frac{1}{c}}{1 - e^{c(a-l)}} = \frac{1}{2}(l+a) \tag{2.18}$$

Model for determining the compliance and strain energy release rate of adhesive bonded composite joint is proposed as

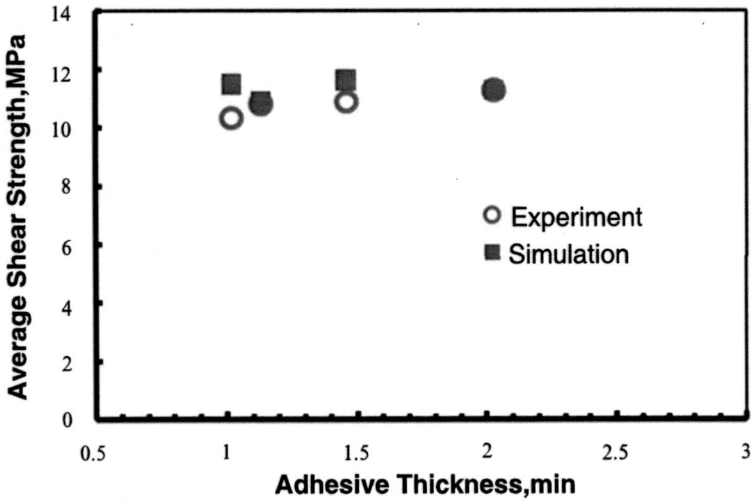

Fig. 2.16 Average shear strength versus adhesive thickness

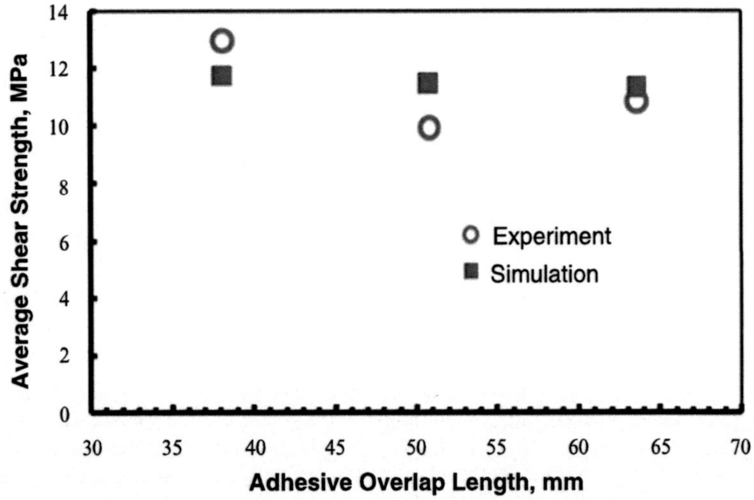

Fig. 2.17 Average shear strength versus adhesive overlap length

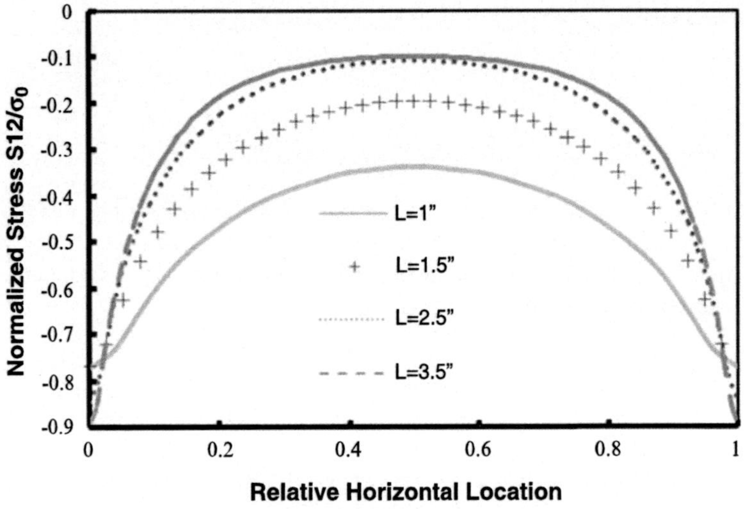

Fig. 2.18 Shear stress S12 distribution for various overlap length

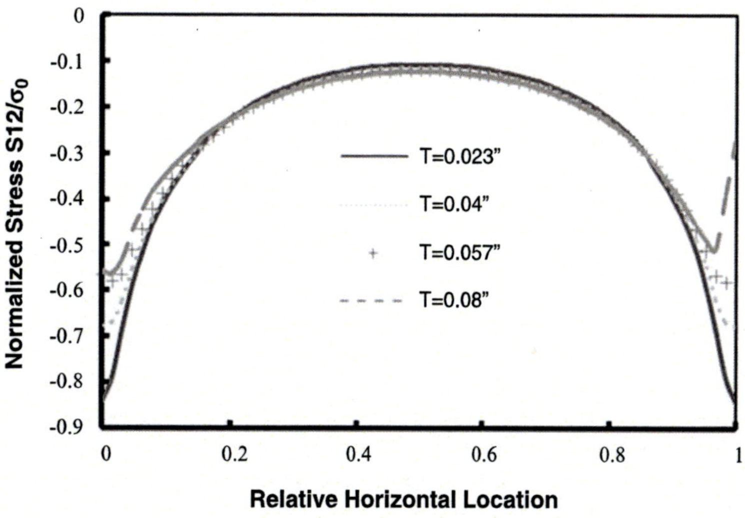

Fig. 2.19 Shear stress S12 distribution for various adhesive thickness

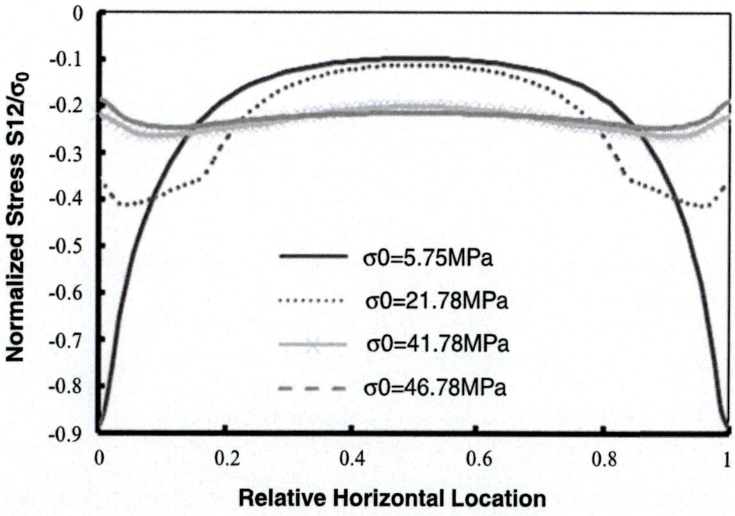

Fig. 2.20 Shear stress S12 distribution for various loading condition

$$C = \frac{2}{E_x^{(1)}I}\left[\frac{E_x^{(1)}Ih_a c(ca+1)(l-a)}{2E_y^{(2)}b[1-e^{c(a-l)}]lk} + \frac{1}{3}l^3 + (l-a)^2 l - l^2(l-a) - \frac{1}{3}(l-a)^3\right] \tag{2.19}$$

$$G_I = \frac{P^2}{2b}\frac{dC}{da} = \frac{P^2}{E_x^{(1)}Ib}\left\{\frac{E_x^{(1)}Ih_a c^2(l-a)}{2E_y^{(2)}blk}\frac{1+cae^{c(a-l)}}{[1-e^{c(a-l)}]^2} + a^2\right\} \tag{2.20}$$

Four existing theories are used to compare with the current proposed model, as well as the experimental result given by Meo and Thieulot [34].

From Table 2.1, the predicted critical load is the one closest to experiment, in regards to strain energy release rate, current developed theory is close to Pradeep's theory [35] and beam theory. Note that the adhesive layer is relatively thick, the Penado's theory [36] gives very poor prediction.

The effect of various parameters on the model prediction is discussed in detail below.

1. **Effect of Young's modulus**

   Effect of Young's modulus varies depends on the crack length. When the crack length is relatively small, the influence of Young's modulus ratio between adherend and adhesive is obvious, with the increase of the ratio, strain energy release decreases a lot. However, when the crack length is relatively larger, the increase of the ratio has almost no effect on the strain energy release rate (Fig. 2.21).

Table 2.1 APC-2/AS4-CFRP JOINT

	Critical load, N	G_I, 10^3 J/m^2
EXP.	65	
Current theory	65.02813	0.813683716
Penado 1993 [36]	27.39576	0.342422089
Pradeep et al. 2010 [35]	65.10282	0.813726064
El-Zein and Reifsnider 1988 [33]	65.09943	0.967903038
Beam theory	65.09943	0.812440909

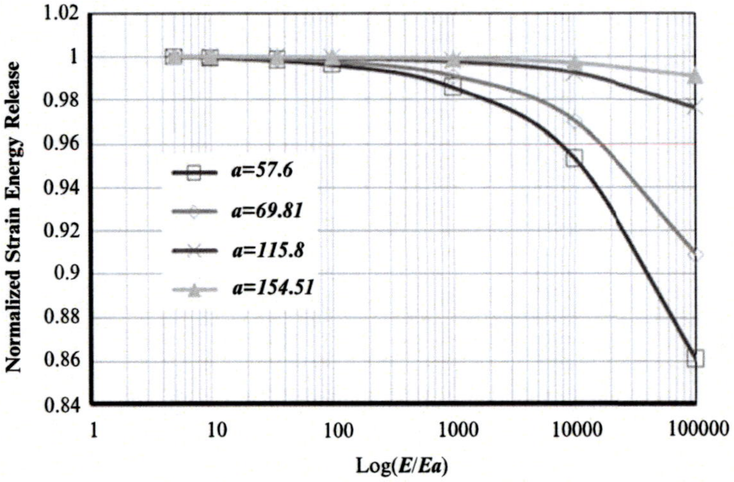

Fig. 2.21 Effect of Young's modulus

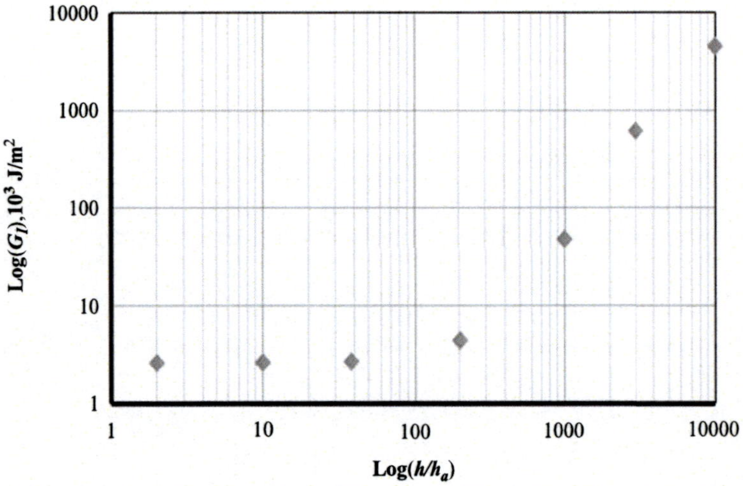

Fig. 2.22 Effect of adhesive thickness at small crack length

2. **Effect of adhesive thickness**
   Effect of adhesive thickness is shown in Fig. 2.22. The change of adhesive thickness h_a has almost no influence on critical SERR when the thickness ratio of adherend to adhesive is below about 100, beyond which, however, the influence becomes significant, decrease of h_a greatly increases critical SERR. In real application, the thickness ratio of adherend to adhesive is rare beyond 100, current study provides a theoretical support for other theories which neglect the effect of adhesive thickness.
3. **Effect of initial crack length**
   Joint compliance C versus crack length a is shown in Fig. 2.23. It is illustrated in the figure that joint compliance increases nonlinearly with the extension of crack length, and the increase rate becomes larger as crack length increase. This indicates that the load is going to decrease as crack propagates.

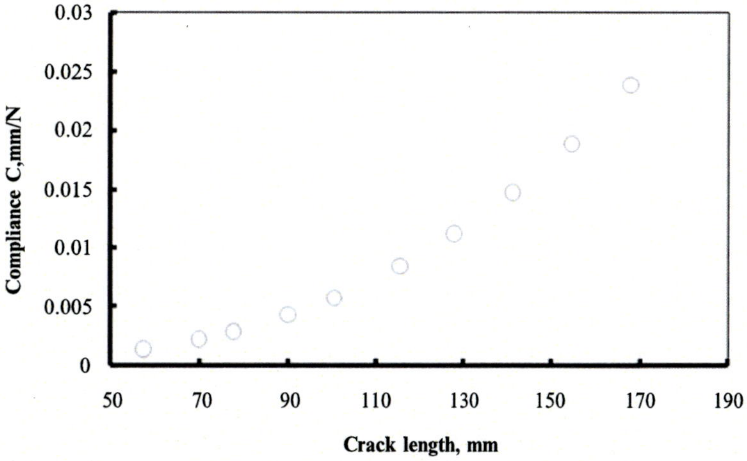

Fig. 2.23 Effect of initial crack length

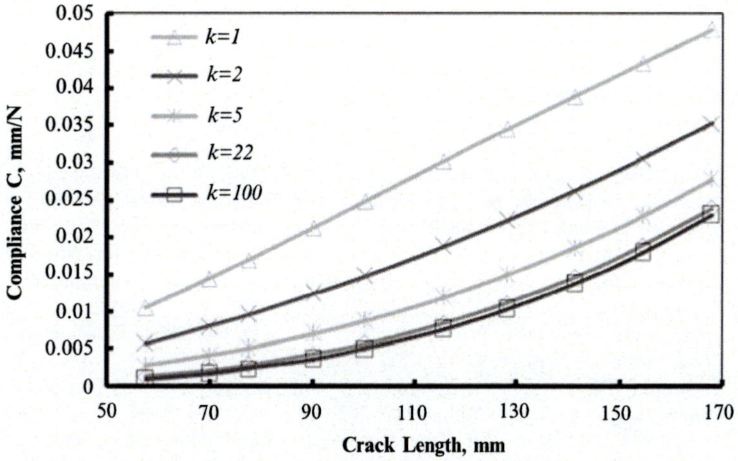

Fig. 2.24 Effect of *k* in correction factor on joint compliance

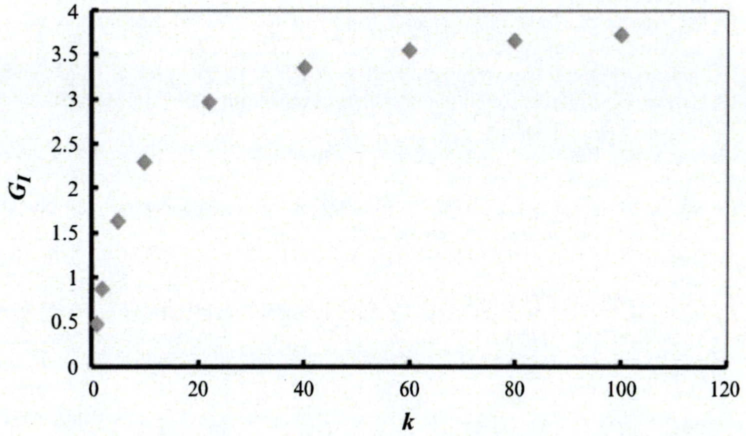

Fig. 2.25 Effect of k in correction factor on strain energy release rate

4. **Effect of correction factor**

   Figures 2.24 and 2.25 show, respectively, the effect of k in correction factor on the joint compliance and critical SERR. From these figures, the joint compliance is in inverse proportion to k, increase of k decreases joint compliance at a given crack length, and results in an increased nonlinearity compliance versus crack length curve, meanwhile, the curves becomes much closer. On the contrary, the critical SERR is in direct proportion to k, increase of k increases critical SERR, while the rate of increase is decreased as the tangent of the curve decrease.

2.4 Conclusions

This paper highlights some of recent research advances in bolting and adhesive joining of composite materials. Significant findings are summarized regarding joint modeling, testing, behavior, failure modes, vibration-induced loosening, gasket creep relaxation, bolt elastic interaction, and bearing strength. Novel criteria for preventing vibration loosening of preloaded threaded fasteners are discussed. For adhesively bonded joints, studies on peel strength and shear stress distributions under mechanical and thermo-mechanical loading, damage coupled modeling, and fracture analysis are outlined.

References

1. Collings TA (1982) On the bearing strengths of CFRP laminates. Composites 1982:241–252
2. Collings TA (1987) Chapter 2: experimentally determined strength of mechanically fastened joints. In: Matthews FL (ed) Joining fibre-reinforced plastics, Elsevier Applied Science, New York
3. Hart-Smith LJ (1973) Analysis and design of advanced composite bonded joints, NASA Langley Contractor Report NASA CR-2218
4. Hart-Smith LJ (1974) Advances in the analysis and design of adhesive-bonded joints in composite aerospace structures, SAMPE process engineering series, vol 19. SAMPE, Azusa, pp 722–737
5. The composite materials handbook MIL17 (2002) Polymer matrix composites: materials, usage, design, and analysis, vol 3. ASTM International, West Conshohocken
6. Volkersen O (1938) Die Nietkrftverteilung in Zugbeanspruchten Nietverbindungen mit Konstanden Laschenquerschnitten. Luftfahrtforschung 15:4–47
7. Ma Weijian, Gomatam R, Fong RD, Sancaktar E (2001) A novel mathematical procedure to evaluate the effect of surface topography on the interfacial state of stress: part I: verification of the method for flat surfaces. J Adhes Sci Technol 15(13):1533–1558
8. Walker SP (2001) Thermal effects on the bearing behavior of composite joints. Ph.D. dissertation, University of Virginia, May 2001
9. Oh JH, Kim YG, Lee DG, Lee DG (1997) Optimum bolted joints for hybrid composite materials. Compos Struct 38(1–4):329–341
10. Aktas A, Dirikolu MH (2004) An experimental and numerical investigation of strength characteristics of carbon-epoxy pinned-joint plates. Compos Sci Technol 64:1605–1611
11. McCarthy MA, Lawlor VP, Stanley WF, McCarthy CT (2002) Bolt-hole clearance effects and strength criteria in single-bolt, single-lap, composite bolted joints. Compos Sci Technol 62:1415–1431
12. Girard C, Dano ML, Picard A, Gendron G (2003) Bearing behavior of mechanically fastened joints in composite laminates-part I: strength and local strains. Mech Adv Mater Struct 10:1–21
13. Park HJ (2001) Effect of stacking sequence and clamping force on the bearing strengths of mechanical fastened joints in composite laminates. Compos Struct 53:213–221
14. Khashaba UA, Sallam HEM, Al-Ahorbagy AE, Seif MA (2006) Effect of washer size and tightening torque on the performance of bolted joints in composite strcutures. Compos Struct 73:310–317
15. Virupaksha VL, Nassar SA (2008) Effect of washers and bolt tension on the behavior of double-lap S2-glass fabric-epoxy composite joints. In: Proceedings of PVP2008, PVP2008-61727, Chicago, 27–31 July 2008
16. Junker GH (1969) New criteria for self-loosening of fasteners under vibration. SAE Trans 78:314–335
17. Nassar SA, Yang X (2009) A mathematical model for vibration induced loosening of preloaded threaded fasteners. ASME J Vib Acoust 131:021009-1–021009-13
18. Nassar SA, Housari BA (2006) Effect of thread pitch on self-loosening of threaded fasteners due to cyclic transverse loads. ASME J Press Vessel Technol 128:590–598
19. Nassar SA, Housari BA (2006) Study of the effect of hole clearance and thread fit on the self-loosening of threaded fasteners due to cyclic transverse loads. ASME J Mech Des 128:586–594
20. Housari BA, Nassar SA (2007) Effect of thread and bearing friction coefficients on the vibration-induced loosening of threaded fasteners. ASME J Vib Acoust 129:484–494
21. Yang XJ, Nassar SA (2011) Analytical and experimental investigation of self-loosening of preloaded cap screw fasteners. ASME J Vib Acoust 133:0310071–0310078
22. Yang XJ, Nassar SA, Wu ZJ (2011) Criterion for preventing self-loosening of preloaded cap screws under transverse cyclic excitation. ASME J Vib Acoust 133:041013-1–041013-11
23. Shoji Y, Sawa T (2005) Analytical research on mechanism of bolt loosening due to lateral loads. In: Proceedings of the ASME pressure vessels and piping conference, vol 2, PVP2005-71333, Denver, 17–21 July 2005, pp 59–65
24. Jiang Y, Zhang M, Lee CH (2003) A study of early stage self-loosening of bolted joints. ASME J Mech Des 125:518–526
25. Nassar Sayed A, Alkelani AA (2006) Clamp load loss due to elastic interaction and gasket creep relaxation in bolted joints. ASME J Press Vessel Technol 128:394–401
26. Nassar SA, Alkelani AA (2006) Effect of tightening speed on clamp load distribution in gasketed joints. SAE Transactions Journal of Materials and Manufacturing, pp 811–825
27. Alkelani Ali A, Housari BA, Nassar SA (2008) A proposed model for creep relaxation of soft gaskets in bolted joints at room temperature. ASME J Press Vessel Technol 130(1):0112111–0112116
28. Alkelani Ali A, Nassar SA, Housari BA (2009) Formulation of elastic interaction between bolts during the tightening of flat-face gasketed joints. ASME J Mech Des 131(2):0210041–0210049

29. Nassar SA, Wu ZJ, Yang Xianjie (2010) Achieving uniform clamp load in gasketed bolted joints using a nonlinear finite element model. J Press Vessel Technol 132(3):031205-1–031205-10
30. Nassar SA, Virupaksha VL (2009) Effect of adhesive thickness and properties on the biaxial interfacial shear stresses in bonded joints using a continuum mixture model. ASME J Eng Mater Technol 131:0210151–0210159
31. Nassar SA, Mao JH, Yang XJ, Templeton D (2011) Effect of adhesives on the mechanical behavior of thick composite joints. In: Proceedings of the ASME 2011 pressure vessels and piping division conference, PVP2011-57692, Baltimore 17–21 July 2011
32. Nassar SA, Mao JH, Yang XJ (2012) A model for fracture characterization of adhesive bonded joint. In: First international symposium on joining technology for composites, 2012 international congress on experimental mechanics, Costa Mesa, 11–14 June 2012
33. El-Zein M, Reifsnider K (1988) Evaluation of GIC of a DCB specimen using an anisotropic solution. J Compos Technol Res 10:151–155
34. Meo M, Thieulot E (2005) Delamination modelling in a double cantilever beam. Compos Struct 71:429–434
35. Pradeep K, Rao B, Sivakumar S, Balasubramanium K (2010) Interface fracture assessment on sandwich DCB specimens. J Reinf Plast Compos 29(13):1963–1977
36. Penado FA (1993) Closed form solution for the energy release rate of the double cantilever beam specimen with an adhesive layer. J Compos Mater 27:383–407

Chapter 3
Inter-cellular Joining for Amorphous Honeycombs

Balaji Jayakumar, Masoud Allahkarami, and Jay C. Hanan

Abstract The discovery of metallic glass ribbons with excellent Yield strength (σ_{ys}) and Elastic Modulus (E_s) has lead to several new technologies and products otherwise impractical or even impossible. However, the alloys are generally high density and brittle, limiting their use in low density, high toughness applications. To reduce density, cellular solids offer promise as a mechanism to employ high strength base materials. Recent research has led to the discovery of amorphous metal honeycombs. Using novel manufacturing approaches, even cell sizes as small as 1 mm have been achieved. Such honeycombs offer a mechanical performance exceeding other similar low density materials. However, to be practical as a future solution for applications ranging from spacecraft to construction, manufacture of the material must be scalable so that significant volumes are produced quickly and cost effectively. The inter-cellular bonding method is critical due to its influence on mechanical strength and weight of the honeycomb. The feasibility and effectiveness of an adhesive and the selected welding techniques for inter-cellular bonding in amorphous $Fe_{45}Ni_{45}Mo_7B_3$ honeycombs was investigated. Results from lap-joint tests using adhesive bonding show a wide range of performance. Diffraction results indicated crystallization-induced embrittlement in the ribbons from welding. The embrittled amorphous metal showed significantly inferior mechanical strength. Due to embrittlement of this alloy, adhesives provide the most effective path for production of viable high specific strength materials. However, the weight advantage of welding drives future work to overcome the challenge.

Keywords Amorphous metal honeycomb • Adhesive bonding • Welding • Shear strength

3.1 Introduction

Metallic glasses lack microscopic defects such as grain boundaries and exhibit higher yield strengths, elastic modulus and elastic strains, in contrast to crystalline materials [1, 2]. They also exhibit good corrosion and wear resistance. These unique mechanical properties have attracted interest in using them in structural, biomedical, and MEMS applications [3–5]. However, the limited plasticity and the lack of large-scale manufacturability have restricted their use in bulk form for large scale structural applications. Owing to their unique properties, metallic glasses have also been produced in cellular form to make materials with a high strength-to-weight ratio. The first cellular glass was made in the form of a Pd foam with an 83% porosity and a compression strength of 30 MPa [6]. Analytical models indicate that, for the same base material, honeycombs offer a 16X strength advantage in comparison with foams [7]. The high elastic strains (2%) and the lack of global plasticity had limited amorphous material fabrication as honeycombs, produced using the conventional expansion method with a hexagonal cellular structure. The first Amorphous Metal Honeycomb (AMH) using an amorphous $Fe_{45}Ni_{45}Mo_7B_3$ alloy with a nominal density of 0.29 g/cc was produced with a "teardrop" cellular structure using a bottom-up manufacturing approach [8, 9]. Analytical models reveal that the attainable axial strengths exceed any existing honeycomb. The influences of defects and limited cell sizes have restricted the achievable strengths of AMH, and inter-cellular joining plays a critical role.

The selection of a bonding process is critical due to the weight added to the cellular structure, and connectivity of the cellular network. From a manufacturing standpoint, the cost, complexity of the application process and their mechanical

B. Jayakumar • M. Allahkarami • J.C. Hanan (✉)
Mechanical and Aerospace Engineering, Oklahoma State University, Tulsa, OK 74106, USA
e-mail: jay.hanan@okstate.edu

E. Patterson et al. (eds.), *Composite Materials and Joining Technologies for Composites, Volume 7: Proceedings of the 2012 Annual Conference on Experimental and Applied Mechanics*, Conference Proceedings of the Society for Experimental Mechanics Series 44,
DOI 10.1007/978-1-4614-4553-1_3,

properties are important. The majority of commercially available metallic honeycombs made using base materials such as Aluminum, Stainless steel and Titanium are either adhesively bonded or welded to the neighboring cells [10]. The most commonly used honeycomb base material is Aluminum. Crystalline base materials allow for weldability and have been tested with compatible adhesives that provide the required strength in the case of expanded honeycombs. On the other hand, AMH need to be produced using a suitable joining method, to provide highest net specific strength while not affecting the base material properties. In earlier work, joining methods such as laser welding of metallic glasses indicated amorphous-crystalline phase changes, indicating embrittlement and surface cracking [11–13]. It is therefore necessary to study the influence of bonding method on the honeycomb structure and evaluate the which methods can be used in the bottom-up manufacturing technology of AMH production. This work outlines the study of adhesive bonding resistance spot welding, and laser welding techniques on amorphous $Fe_{45}Ni_{45}Mo_7B_3$ foils, the base material of AMH. Lap joint shear strength testing was used to evaluate the strength of bonding. XRD was used to study crystallization due to welding and the influence of welding on amorphous $Fe_{45}Ni_{45}Mo_7B_3$ ribbons.

3.2 Background

3.2.1 Welding Metallic Glasses

Metallic glasses in ribbon form are produced using slip casting at a critical cooling rate of 10^5 to $10^6 Ks^{-1}$. One of the limitations of amorphous alloys is their tendency to crystallize on heating. Based on which, joining methods for amorphous alloys can be grouped into two categories [14] dependent on the crystallization temperature. Bonding methods under conditions lower than the crystallization temperature; typical examples include adhesive bonding, brazing or soldering, cold pressure welding, explosive welding, and ultrasonic welding. The others involve temperatures greater than the crystallization temperature and at short time intervals; typical examples include resistance welding, electron beam welding, and laser beam welding. In consideration of Bulk Metallic Glasses (BMG) as structural materials for several applications, welding techniques have been previously examined.

Welding of amorphous alloys refers to a non-equilibrium process of reheating and cooling a disordered microstructure with small solidification shrinkage. The cooling process is a non-equilibrium condition leading to Coring, a formation of higher melting temperature element(s) in the external layers. It is important to suppress grain growth which may cause centre cracking while avoiding deviation from the glass forming composition. The amorphous state of the weld interface needs to be retained for a successful bond. When the interface temperature is less than T_g (Glass Transition temperature), the presence of surface oxide film prevents bonding. Subsequently, when the temperature is equaled or exceeded T_g, there is super cooled liquid state; where the surface oxides need be either broken or prevented from formation in order to form successful bonding. Further, a surface contact can be created at the atomic scale through fresh surfaces to form an amorphous weld [15]. Other work [11–13] has shown evidence that crystallization during the weld deteriorates properties with cracks formed due to surface oxidation.

Focuses on achieving higher length-scales of BMGs have attracted interest on welding. Pulse-current, Friction [15–17], and Electron beam welding techniques [18–21] have been reported to work on BMG-BMG and BMG-crystalline material combinations. Other welding studies reveal that Zr based and Pd based bulk glassy alloys can be welded together by the use of Joule heating. Work [13] on Zr-based metallic glass indicated that high power laser welding was suitable for welding amorphous Zr-based BMG with no crystallization around the heat affected zone (HAZ). In welding studies for amorphous foils, Capacitor discharge welding has been reported to successfully weld Co and Fe based amorphous foils with no measurable oxide formation or crystallization [22]. A Ni-based glassy alloy was reported weldable using electron beam welding with no crystallization [23]. Ultrasonic welding and explosive welding have proven to work on $Fe_{40}Ni_{40}P_{14}B_6$ amorphous foils [24].

3.2.2 Laser Welding and Resistance Spot Welding

Laser welding has been identified most successful in welding metallic glasses, due to the relatively smaller melting volume of base material [11, 25–27]. Successful welding has been achieved for a $Ni_{53}Nb_{20}Ti_{10}Zr_8Co_6Cu_3$ metallic glass foil with a 25 μm thickness using high power fiber laser welding [28]. The same work also indicates that weld rate affects Crystallinity. Also, high quality welds have been achieved on $(CoFe)_{70}(MoSiB)_3$ foils using Nd:YAG laser welding, but limited to a narrow window of voltage, pulse duration, and focus position, with tensile shear strength data of these amorphous foils supporting these claims [14]. Fe-based and Co-based thin amorphous ribbons were tested for weldability using the spot welding method [12]. The welding results of the $Fe_{40}Ni_{40}Mo_4B_{16}$ showed the embrittlement of spot welded regions, and crystallization induced embrittlement which caused deterioration in joint strength. Results from this work indicate that the spot weld-ability of the amorphous ribbon largely depends upon the chemistry of the foil.

3.2.3 Adhesive Bonding

In the expansion route of honeycomb manufacturing, thermally activated adhesives are pre-printed to accommodate cellular bonding in both metallic and non-metallic honeycombs. The handling time, curing time, viscosity, cost and complexity of use are some practical consideration involved in the choice of an adhesive used in high volume manufacturing [10]. With the advent of several welding techniques successful in welding amorphous base materials, it is natural to imagine welded amorphous honeycombs consisting of a chemically homogenous cellular network with a high strength-to-weight ratio. However, the cost and availability of amorphous base material, in addition to manufacturing process compatibility limits some techniques and base materials. This work focuses on the study of adhesive bonding, laser welding and resistance spot welding technique on amorphous $Fe_{45}Ni_{45}Mo_7B_3$ substrates for use as an inter-cellular joining method for AMH.

3.3 Materials and Methods

The amorphous base material under study is a $Fe_{45}Ni_{45}Mo_7B_3$ alloy, in ribbon form with an 8 mm width and 28 μm thickness. The same alloy was used in producing AMH with a cell size as small as 1 mm. Several other cell sizes were also made and tested [9]. Different adhesives were chosen to test for adhesion strength using lap joint shear strength testing. Epoxy based adhesives commercially known as 3 M-DP110, 3 M-DP 420, Gorilla (bisphenol-A type), 3 M78 aerosol spray type insulating adhesive, and Ad tech. 271, a hot melt adhesive were chosen. Adhesives were chosen based on compatibility of use with the production method of AMH, adhesive handling time, curing time, viscosity, cost and ease of applicability. Lap-joints were formed between $Fe_{45}Ni_{45}Mo_7B_3$ ribbons (Fig. 3.1).

Two sets of samples, one with the as-is condition and the other set with surface roughened (using 380 grit sandpaper) conditions were made. ASTM D 1002, "Standard Test Method for Apparent Shear Strength of Single Lap-Joint Adhesively Bonded Metal Specimens by Tension Loading (metal–metal)" was followed for tension tests. The lap joints were allowed to cure and tensile shear strength was measured using a 5967 Instron UTM.

While forming the "teardrop" cellular structure [8], the high elastic limit (2%) of the ribbon allows for the ribbon to spring back to its original state, the adhesive force varies as a function of the honeycomb cell size. The lower the cell size, higher the adhesive force required to form the teardrop cell. A simple bending formula was used to calculate the adhesive force required for inter-cellular bonding, and experimentally tested using a ribbon-bending experiment between flat platens.

Amorphous $Fe_{45}Ni_{45}Mo_7B_3$ ribbon samples were also welded using resistance spot welding and laser welding methods. The weld conditions are presented in Table 3.1

In the laser welding process carried out in EWI, only the top side of the weld was exposed to an argon atmosphere. Laser welding was done using a 600 W fiber laser at 10% power and a rate of 285 mm/s. Sample sizes were chosen to accommodate for shear strength tests under tension (Fig. 3.2).

Microscopic observations were made before subjecting the samples for destructive testing. Non-destructive testing on the welded samples was done using XRD. Diffraction was performed with Cu–Kα radiation at tube parameters of 40 kV/40 mA using a Bruker D8 Discover XRD^2 micro-diffractometer equipped with the General Area Diffraction Detection System (GADDS) and Hi-Star 2D area detector. The detector distance to the center of diffraction was kept at 30 cm which covers approximately the area of 20° in 2θ and 20° in χ with 0.02° resolution. A motorized five axis (X, Y, Z (translation), χ (tilt), φ (rotation)) stage was used to move the measurement spot to the instrument center within 12.5 μm position accuracy. Sample positioning was controlled by video-laser positioning system before each exposure to ensure diffraction patterns comes from welded region of ribbon. After XRD measurements, the samples were tested following ASTM D1002 for lap joint shear strength in tension.

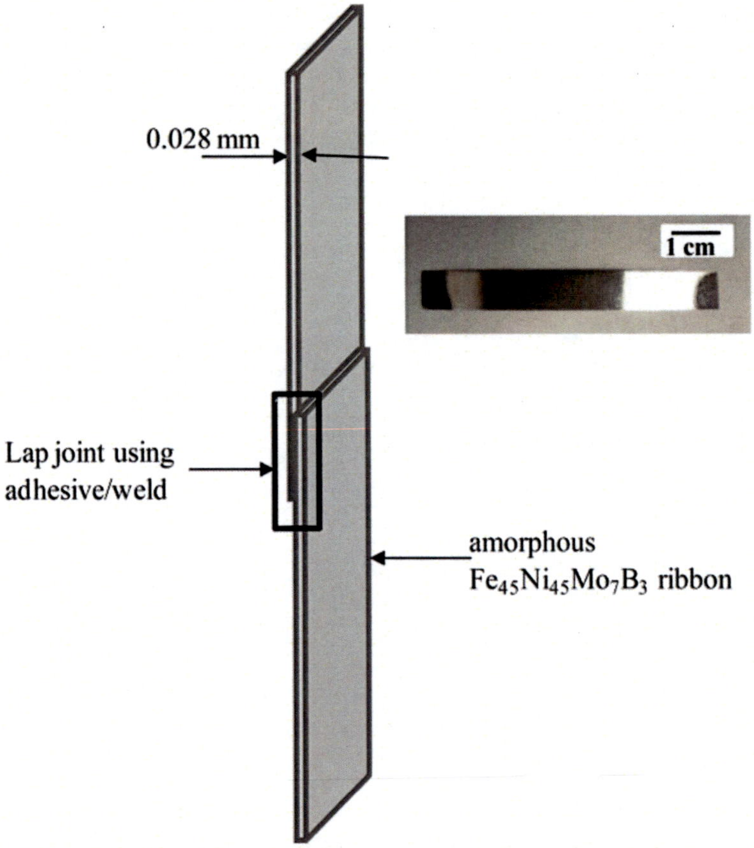

Fig. 3.1 Schematic of a lap joint on ribbon, *not drawn to scale*: Inset shows a $Fe_{45}Ni_{45}Mo_7B_3$ ribbon

Table 3.1 Welding conditions

Type/place of welding	Miyachi Unitek (Set A)		EWI (Set B)		
Resistance welding	*Weld force (N)*	*Hold time (ms)*	*Weld force (N)*		*Hold time (ms)*
	13.3–17.8	150	62		30
Fiber laser welding	*Spot size (mm)*	*Power (KW)*	*Type*	*Spot size (mm)*	*Power (KW)*
	0.3	0.2	Pulse	0.006 s pulse	0.06
			Cont.	0.009	0.06

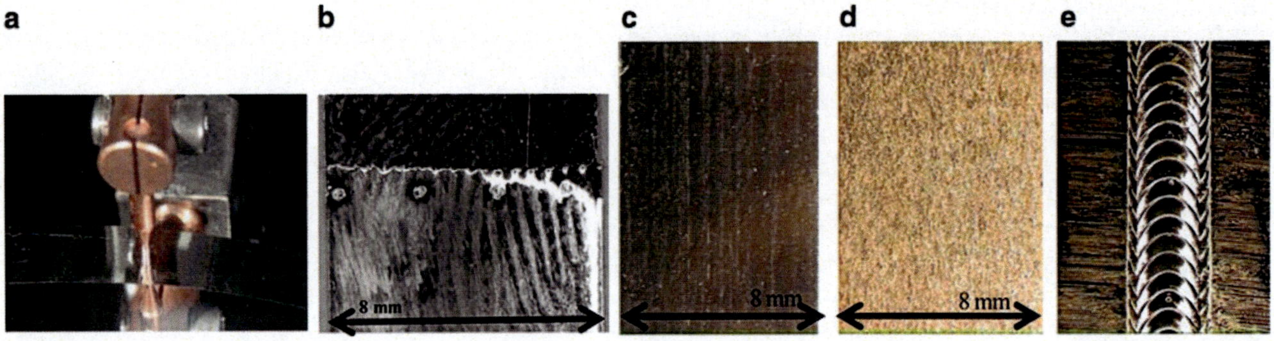

Fig. 3.2 (**a**) Resistance spot welding of amorphous ribbons between copper electrodes, (**b**) Welded spots (4) across the 8 mm ribbon width, (**c**) Matte side of the amorphous $Fe_{45}Ni_{45}Mo_7B_3$ ribbon, (**d**) Matte side of the amorphous $Fe_{45}Ni_{45}Mo_7B_3$ ribbon, (**e**) Top side of the laser weld

3.4 Results and Discussion

The failure load for each of the adhesively bonded lap joint samples was normalized with the bonding area. The tensile shear strengths of the adhesives are presented in Fig. 3.3. Clearly, bisphenol-A based epoxies have better adhesion to the amorphous $Fe_{45}Ni_{45}Mo_7B_3$ substrate. The aerosol based spray epoxy had the least strength. The influence of surface treatment is not clearly observed. There is 10–12% difference in the shear strength values of the surface treated and untreated samples; however for a substrate as thin as 28 μm, increasing the surface roughness may cause surface damage, with a possibility of increasing the surface roughness of the ribbon to increase the adhesion properties. Amorphous ribbons manufactured using the slip casting method exhibit different surface characteristics on the two faces. Since the manufacturing method involves a high cooling rate of 10^6 K/s, the reflective and smoother side is the surface in contact with the copper disc, while the other side has a darker matte finish. Earlier work on the measurement of surface roughness indicate that the mean roughness is higher for the matte side than for the reflective side [14]. The bonding surfaces were noted for each of the tested samples, but there was no clear trend for the influence of surface roughness on the sample. Increasing the number of samples would be necessary for statistical significance to determine the role of surface roughness and surface treatment. The present results reveal adhesion properties on a comparative scale. In addition to these, the handling time, cost and adhesive viscosity can help in selection of the most suitable adhesive.

While forming a "teardrop" cellular structure, a minimal adhesive force is required to form the cell. This force was calculated using the simple bending formula and was experimentally validated. The bending formula is given by

$$F = \frac{EI}{Rd} \tag{3.1}$$

E is the Young's modulus of the material, I the resistance to bending, R the radius of curvature, and d the teardrop cell dimension. In a teardrop cellular structure, the elastic limit of the ribbon allows for storing elastic energy in the form of residual stress and remains elastic up to a stress limit. Beyond this limit, plastic deformation occurs in the amorphous ribbon as an onset of shear bands. Figure 3.4 shows that a decrease in the honeycomb cell size, (cell size <1 mm), requires a higher adhesive force for inter-cellular joining. A corresponding trend for the axial strength is also presented in the secondary y-axis. This is however not the case for honeycombs with crystalline base materials. Their low elastic limit (about 10X lower) and ability to plastically deform, enables the formation of a hexagonal structure where no force is required to suppress elastic relaxation.

Inter-cellular bonding in defective honeycombs due to broken or missing cell walls have shown to reduce the strength and stiffness by up to 30% [7]. Therefore it is critical to select the appropriate bonding method. This becomes an even more difficult constraint with the formation of smaller cell sizes. Manufacturing limitations for smaller cell sizes can also be

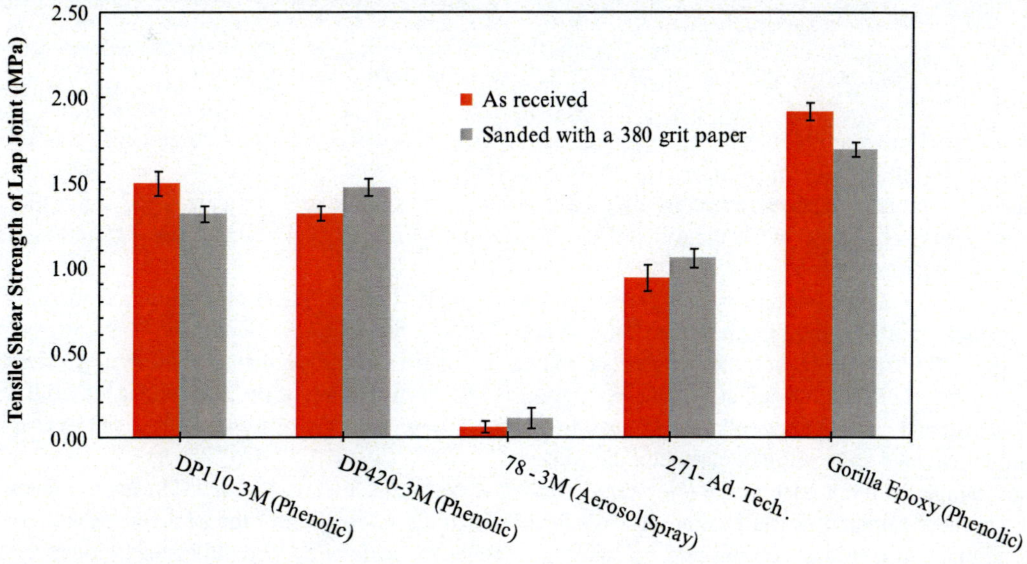

Fig. 3.3 Adhesive lap joint shear strength comparison

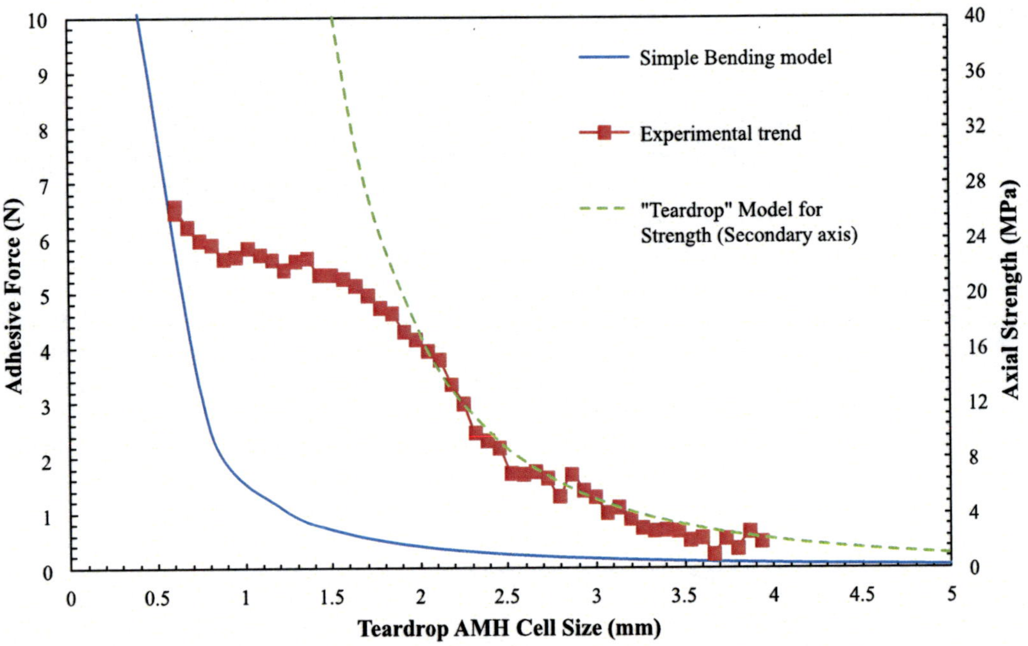

Fig. 3.4 Adhesive force required to from a teardrop cellular structure

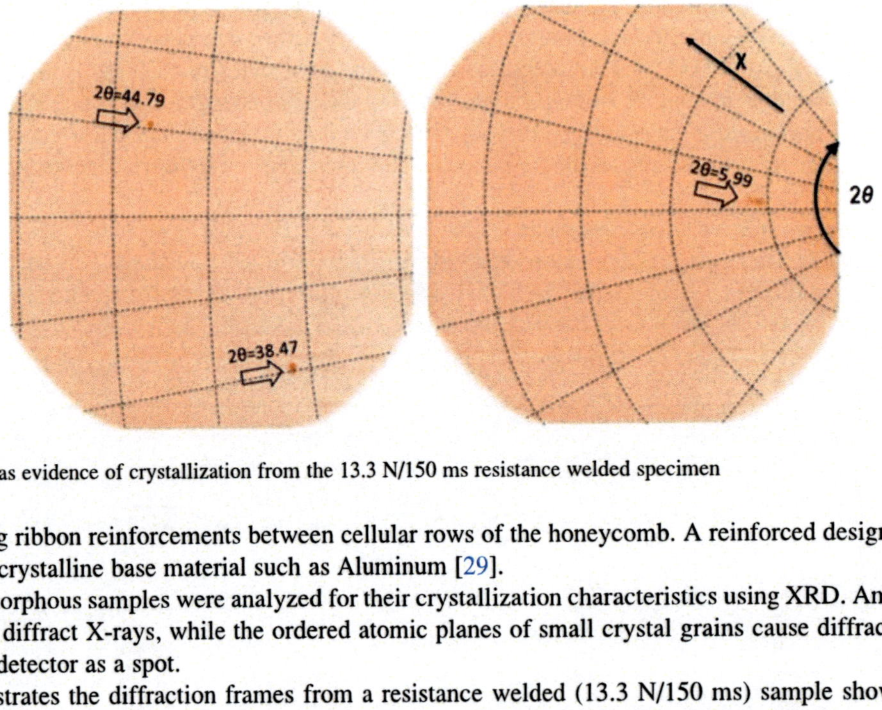

Fig. 3.5 XRD spots as evidence of crystallization from the 13.3 N/150 ms resistance welded specimen

overcome by using ribbon reinforcements between cellular rows of the honeycomb. A reinforced design is already used in the industry for a crystalline base material such as Aluminum [29].

The welded amorphous samples were analyzed for their crystallization characteristics using XRD. Amorphous structures do not coherently diffract X-rays, while the ordered atomic planes of small crystal grains cause diffraction. This is easily detected by a 2D detector as a spot.

Figure 3.5 illustrates the diffraction frames from a resistance welded (13.3 N/150 ms) sample showing isolated spots reflected from single crystal grains. The number of the observed spots depends on the beam size, detector area, and number of crystal grains in the irradiation volume. In practice only a few grains in the irradiated volume may orient properly to satisfy Bragg's law and diffract. Observations of these spots reveal crystallization of the amorphous material at and around the weld due to heat and consequent cooling at an insufficient rate. These spots are observed only at the Heat Affected Zone (HAZ) around the welded spots and not elsewhere on ribbon.

The welded samples did not possess the minimum handling strength for loading on the UTM for tensile lap joint shear strength tests. In these samples, crystallization of was caused due to the interaction of the heat source with the amorphous substrate. In addition to a slower cooling rate (in resistance welding), surface melting and re-solidification could be the governing process leading to crystallization. Earlier work on spot welding of the same alloy revealed successful welds with a

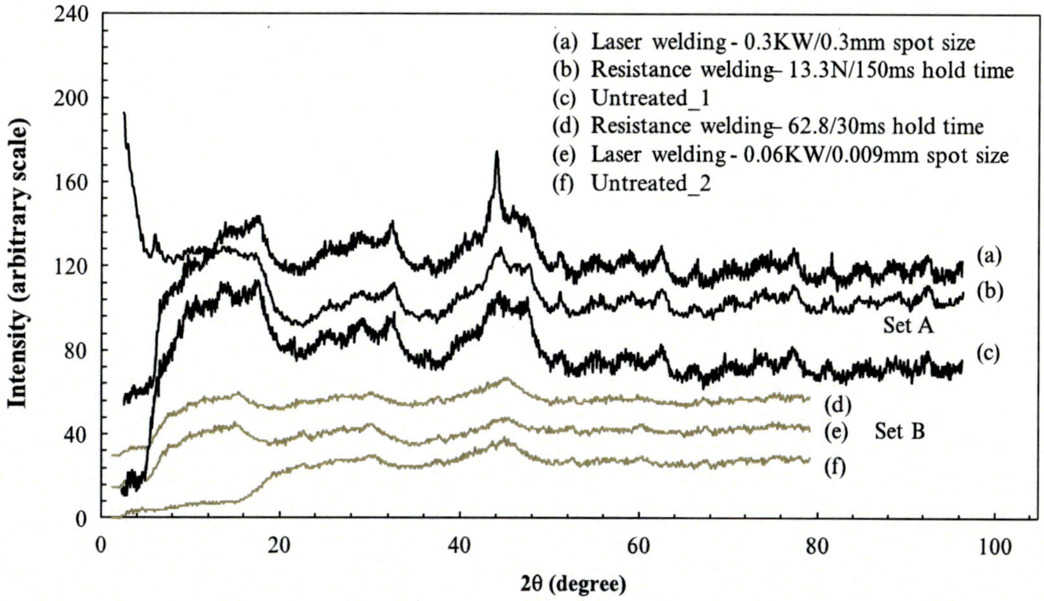

Fig. 3.6 XRD results for resistance and laser welded amorphous $Fe_{45}Ni_{45}Mo_7B_3$ ribbons

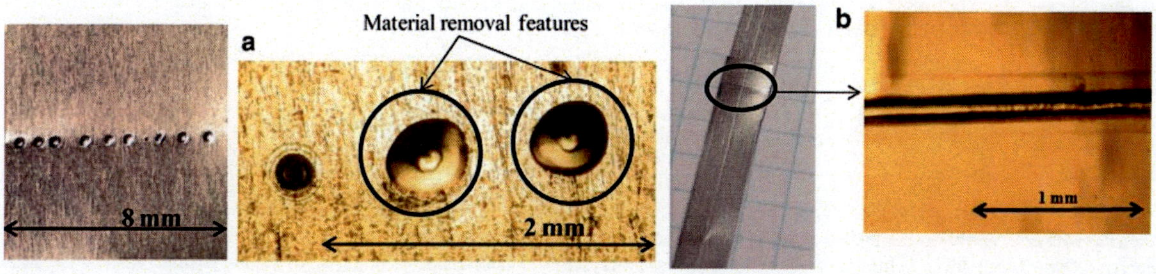

Fig. 3.7 Microscopic observations: (**a**) pulsed laser weld across the amorphous $Fe_{45}Ni_{45}Mo_7B_3$ ribbon, (**b**) Continuous laser weld on the ribbon-ribbon interface

lap joint shear force of 30 N/spot. Multiple spots increased the strength of weld, while the tensile strength was also dependent of the electrode force [12]. XRD results from the same work also reveal crystallization as a result of spot welding. Nuggets were reported to brittle off the welded sample and the XRD measurements of the nugget and the HAZ on the ribbon matched, substantiating the deterioration in joint strength due to crystallization. A similar result was observed while testing the spot welded $Fe_{45}Ni_{45}Mo_7B_3$ sample, but with no handling strength on the lap joint.

With laser welding techniques, a cooling rate of 10^5 Ks^{-1} is common during re-solidification of melted metals. It is seemingly easy to retain an amorphous state, with a rapid cooling rate to accommodate solidification shrinkage, without crystallization. However, Crystallization of many materials under laser melting has been reported [23, 26, 30, 31] (Fig. 3.6).

In contrast to the welding results discussed earlier, the resistance welded samples under a 62 N force and a 30 ms hold time and the laser welded samples using a 0.06 KW power laser on a 0.009 mm spot size did not reveal signs of crystallization. XRD plots indicate no sharp peaks revealing a retained amorphous state. Resistance welded samples and both the pulsed laser weld and the continuous laser weld samples showed no crystallization. When tested for their tensile lap joint shear strength, the samples exhibited a brittle failure in comparison with adhesives. The failure loads of resistance welds were 3.4 N for 4 spots, and the highest strength for a Laser welded sample was 52.8 N over a 2.4 mm^2 area. Laser welded lap joints exhibited the highest strength of 22 MPa in tension. However, the sample failed in a brittle fashion under bending. Micrographs of the laser welded regions of sample set-A are presented in Fig. 3.7. In the pulsed laser welded sample shown in Fig. 3.7a, there are traces of visible thermal ablation. This process occurs as a result of surface vaporization caused by a localized increase in temperature. Smaller thermal diffusivities lead to efficient ablations. The thermal diffusivities of amorphous materials are in the order of 0.2 cm^2s^{-1} [32]. In the pulsed laser welding case, a long (6 ms) pulse duration was used, making thermal ablation an unlikely reason [30]. However the visible material removal features from the micrographs remain suggestive.

The continuous wave laser welded sample in Fig. 3.7b revealed a straight line, which can be treated as a laser irradiated track which also failed in a brittle fashion. The tensile lap joint shear strength was measured at 52.8 N; however there were signs of embrittlement. Embrittlement effects have been postulated to occur due to structural relaxation involving short-range order accompanied by a reduced quenched-in free volume [33, 34]. Visible "ripple-like" formation was observed initiating from the laser track. DSC measurements and Microhardness testing are ongoing work for further confirmation. For structural applications involving inter-cellular joining of amorphous metal honeycombs it is unfavorable to have a weak brittle bond between the cells.

3.5 Conclusions

Adhesive bonding, resistance spot welding, and laser welding techniques were attempted on a $Fe_{45}Ni_{45}Mo_7B_3$ ribbon investigated for use in inter-cellular joining of amorphous metal honeycombs. Several adhesives were chosen and tested for their tensile lap joint shear strength with an amorphous substrate. Bisphenol-A based epoxies showed better bonding strength compared with aerosol based and hot melt adhesives. Future work and further testing of samples could reveal the influence of surface roughness on the bonding properties. Under limited welding conditions, laser welding and resistance spot welding were attempted and results reveal embrittlement of the amorphous substrate, deeming the bond weak and brittle. Due to the effect of embrittlement, adhesives currently provide the more effective solution for inter-cellular joining of amorphous metal honeycombs. However, the weight advantage offered by welding drives ongoing and future work.

Acknowledgements Miyachi Unitek, CA and EWI, OH for laser welding and Resistance spot welding. Sudheer Bandla for XRD measurements. Partial funding also provided by ONR Grant No. N00173-071-G001.

References

1. Demetriou MD et al (2011) A damage-tolerant glass. Nat Mater 10:123–128
2. Chen M (2008) Mechanical behavior of metallic glasses: microscopic understanding of strength and ductility. Annu Rev Mater Res 38:445–469
3. Ashby MF, Greer AL (2006) Metallic glasses as structural materials. Scr Mater 54:321–326
4. Greer AL (2011) Metallic glasses: damage tolerance at a price. Nat Mater 10:88–89
5. Hanan JC et al (2009) Mechanical properties of amorphous metal honeycombs for ballistic applications. ASME Conference Proceedings 2009:73–74
6. Demetriou MD et al (2008) Stochastic metallic-glass cellular structures exhibiting benchmark strength. Phys Rev Lett 101:145702
7. Gibson LJ, Ashby MF (1999) Cellular solids: structure and properties, vol 2. Cambridge University Press, Cambridge
8. Hanan JC (2011) Teardrop lattice structure for high specific strength materials, US, WO/2011/056659 Patent
9. Jayakumar B, Hanan JC (2012) Modeling the axial response of amorphous $Fe_{45}Ni_{45}Mo_7B_3$ honeycombs. Metall Mater Trans A 43(8):2669–2675
10. Bitzer T (1997) Honeycomb technology: materials, design, manufacturing, applications and testing. Chapman and Hall, London
11. Kim JH et al (2007) Pulsed Nd: YAG laser welding of $Cu_{54}Ni_6Zr_{22}Ti_{18}$ bulk metallic glass. Mat Sci Eng A-Struct Mat Properties Microstruct Process 449:872–875
12. Fukushima S (1991) Spot welding of amorphous alloys. Weld Int 5:654–659, 1991/01/01 1991
13. Wang HS et al (2010) Combination of a Nd:YAG laser and a liquid cooling device to (Zr53Cu30Ni9Al8)Si0.5 bulk metallic glass welding. Mater Sci Eng, A 528:338–341
14. Runcev D, Dorn L (2004) Welding of cobalt-iron-based amorphous alloy foils with Nd: YAG laser, pp 449–455
15. Kawamura Y, Ohno Y (2001) Superplastic bonding of bulk metallic glasses using friction. Scr Mater 45:279–285
16. Shoji T et al (2003) Joining of $Zr_{41}Be_{23}Ti_{14}Cu_{12}Ni_{10}$ bulk metallic glasses by a friction welding method. Mater Trans 44:1809–1816
17. Shoji T et al (2004) Friction welding of bulk metallic glasses to different ones. Mater Sci Eng, A 375–377:394–398
18. Kawamura Y, Ohno Y (2001) Successful electron-beam welding of bulk metallic glass. Mater Trans 42:2476–2478
19. Kawamura Y, Ohno Y (2001) Spark welding of $Zr_{55}Al_{10}Ni_5Cu_{30}$ bulk metallic glasses. Scr Mater 45:127–132
20. Yokoyama Y et al (2002) Electron beam welding of $Zr_{50}Cu_{30}Ni_{10}Al_{10}$ bulk glassy alloys. Mater Trans 43:2509–2515
21. Kawamura Y et al (2001) Electron beam welding of Zr-based bulk metallic glass to crystalline Zr metal. Mater Trans 42:2649–2651
22. Kim Y et al (1988) Consolidation of metallic glass ribbons using electric discharge welding. Metall Mater Trans, A 19:1634–1638
23. Louzguine-Luzgin DV et al (2008) Structural investigation of Ni–Nb–Ti–Zr–Co–Cu glassy samples prepared by different welding techniques. Mater Sci Eng, B 148:88–91
24. Kawamura Y, Ohno Y (2001) Metallurgical bonding of bulk metallic glasses. Mater Trans 42:717–719
25. Li B et al (2006) Laser welding of $Zr_{45}Cu_{48}Al_7$ bulk glassy alloy. J Alloys Compd 413:118–121

26. Kim J et al (2006) Phase evolution in $Cu_{54}Ni_6Zr_{22}Ti_{18}$ bulk glass Nd: YAG laser weld. Mat Sci Eng A-Struct Mat Properties Microstruct Process 434:194–201
27. Wang HS et al (2010) Combination of a Nd:YAG laser and a liquid cooling device to (Zr(53)Cu(30)Ni(9)Al(8))Si(0.5) bulk metallic glass welding. Mat Sci Eng A-Struct Mat Properties Microstruct Process 528:338–341
28. Tsumura T, Nakata K (2011) Laser welding of Ni-based metallic glass foil. Weld Int 25:491–496, 2011/07/01
29. Hexcel (1986) Mechanical properties of HexCel honeycomb material, HexCel Corporation, Pleasanton
30. Harimkar SP et al (2011) Periodically laser patterned Fe-B-Si amorphous ribbons: phase evolution and mechanical behavior. Adv Eng Mater 13:955–960
31. Katakam S et al (2012) Laser-induced thermal and spatial nanocrystallization of amorphous Fe–Si–B alloy. Scr Mater 66:538–541
32. Chen HS et al (1980) Preparation of glassy metals. Annu Rev Mater Sci 10:363–391
33. Egami T (1978) Structural relaxation in amorphous alloys – compositional short range ordering. Mater Res Bull 13:557–562
34. Takahara Y, Matsuda H (1994) Reversible structural relaxation in Fe-B-Si amorphous alloys. Mater Sci Eng, A 179–180, Part 1

Chapter 4
Milled Glass Reinforced Polyurea Composites: The Effect of Surface Treatment

Zhanzhan Jia, Kristin Holzworth, and Sia Nemat-Nasser

Abstract Polyurea is commonly utilized in blast-mitigating applications due to its excellent thermo-mechanical properties. In this work, we seek to develop polyurea-based composite materials capable of enhanced blast-induced stress-wave management through material design. Typically the matrix, the filler, and the interfacial surface chemistry comprise the basic structure of a composite material. Here we evaluate the effect of the matrix-filler interfacial properties through the integration of surface treated milled glass fibers into polyurea. The milled glass fibers are connected to the matrix via weak (van der Waals force), intermediate (hydrogen bonding), or strong (covalent bonding) interactions using a variety of surface treatments. Although a complete understanding of the interfacial relationships is exceedingly complex, experimental studies are essential in providing a basic understanding to support simulations and ultimately guide the optimal design of the polyurea-based composites. The properties of the resultant composite materials are thermo-mechanically characterized using dynamic mechanical analysis. Additionally, the surface treatments are also applied to glass slides in order to allow for water droplet contact angle measurements to assess the hydrophobicity/hydrophilicity. Furthermore, interfacial adhesion tests are conducted on samples fabricated by casting polyurea on surface treated glass slides. These efforts are part of an ongoing initiative to develop elastomeric composites with optimal properties to manage blast-induced stress-wave energy.

Keywords Polyurea elastomer • Interface • Milled glass • Surface treatment • Dynamic mechanical properties

4.1 Introduction

Polyurea is the generic name of the type of elastomer derived from the reaction of an isocyanate component and a resin component. In this study, polyurea is synthesized using the diamine component Versalink P-1000 [1] and the isocyanate component Isonate 143 L [2]. The resulting polyurea elastomer is a block copolymer, which consists of hard domains dispersed throughout a soft matrix. The hard domains reinforce the soft matrix and create a cross-linked backbone. Polyurea has excellent thermo-mechanical properties, and it is an attractive choice for developing composites with optimal stress-wave energy management capabilities.

Milled glass is a type of finely powdered glass fiber used as reinforcement in composite material systems. The milled glass fibers utilized in this study are cylindrical-shaped micro-glass with an aspect ratio of 16. The glass surface is grafted with various functional groups such that different types of interfacial bonds form between the polyurea matrix and the milled glass.

With the polyurea matrix and the surface treated milled glass filler, composite materials with different interfacial bonding strengths are fabricated. Interfacial properties are critical in composite material systems. In theory, the onset of damage is controlled by the interfacial strength [3]. However, the manner in which the interface affects the dynamic properties of the composite material is not fully understood. Thus, the milled glass reinforced polyurea composite materials with different interfacial bonding strengths are characterized using dynamic mechanical analysis (DMA).

Z. Jia (✉) • K. Holzworth • S. Nemat-Nasser
Department of Mechanical and Aerospace Engineering, Center of Excellence for Advanced Materials, University of California, San Diego, 9500 Gilman Drive, La Jolla, CA 92093-0416, USA
e-mail: jiazhanzhan@gmail.com

E. Patterson et al. (eds.), *Composite Materials and Joining Technologies for Composites, Volume 7: Proceedings of the 2012 Annual Conference on Experimental and Applied Mechanics*, Conference Proceedings of the Society for Experimental Mechanics Series 44, DOI 10.1007/978-1-4614-4553-1_4, © The Society for Experimental Mechanics, Inc. 2013

Additionally, the same surface treatments are applied to glass slides in order to allow for evaluation of hydrophobicity/ hydrophilcity through water droplet contact angle measurements. Lastly, interfacial adhesion test samples are made by casting polyurea on the surface treated glass slides in order to evaluate the surface treatment effect on bond strength.

4.2 Surface Treatments

Three types of silanes are used to treat the glass surfaces including: a silane with an aliphatic chain, a silane with a urea group, and a silane with an amine group, all purchased from Sigma-Aldrich. These three surface treatments graft three types of functional groups on the glass surfaces that connect the glass surface with the polyurea matrix via van de Waals force, hydrogen bonding, and covalent bonding respectively. Prior to the silane surface treatment, all glass surfaces are washed using piranha solution, a 3:1 mixture of sulfuric acid and hydrogen peroxide.

4.3 Composite Materials

Composite materials are fabricated by mechanically mixing the surface treated milled glass fibers with the polyurea matrix. For each type of surface treatment, samples containing 10%, 15%, and 20% fiber volume fraction are created. SEM images of the fracture surface for the 10% fiber volume fraction samples of each surface treatment are shown in Fig. 4.1. The fracture surface of the amine sample is noticeably different than the other three surfaces; the attached polyurea at the surface of the milled glass fiber is indicative of strong bonding (i.e. this surface treatment is effective). Since the chemical mechanisms of the three types of surface treatments are very similar, it is expected that the other two types of surface treatments are also effective. The adhesion test results, which are discussed later in this paper, support this hypothesis.

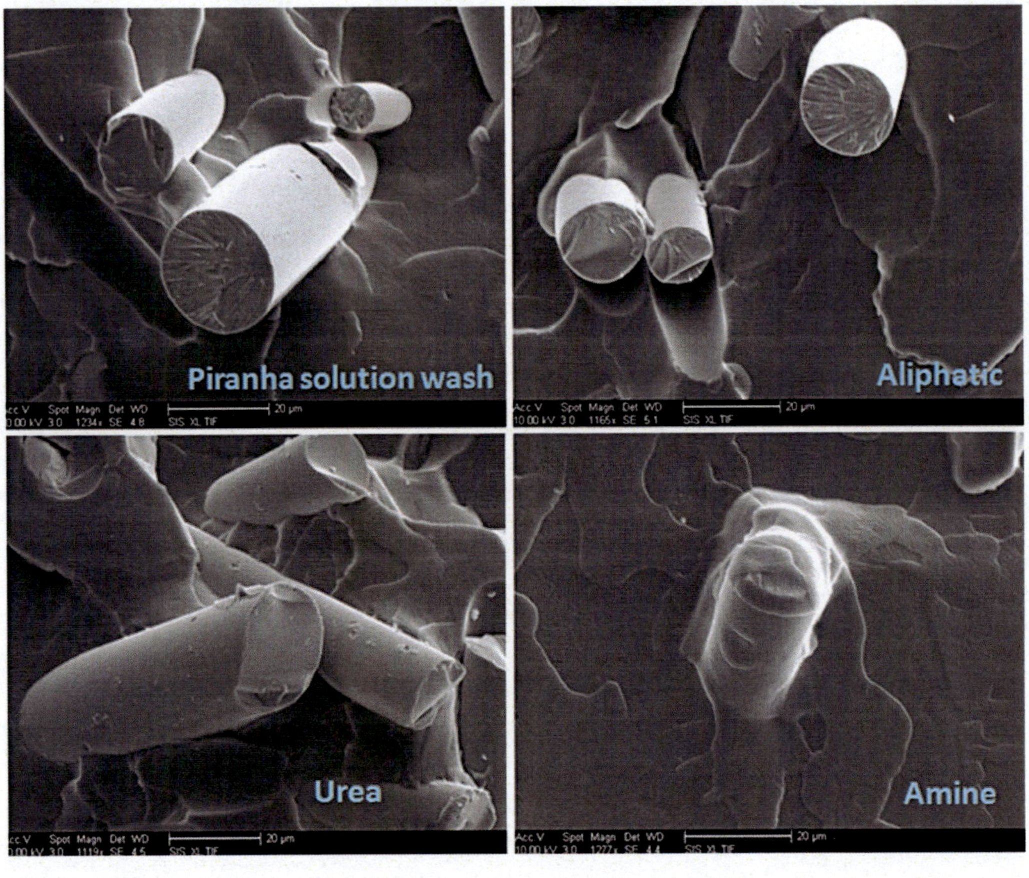

Fig. 4.1 SEM images of 10% fiber volume fraction milled glass reinforced polyurea composites

4.4 DMA Characterization

DMA characterization is conducted for each milled glass reinforced polyurea composite configuration. The DMA results indicate that the fiber volume fraction is the dominant contributing factor to changes in the storage modulus and loss modulus, while the effect of the surface treatment is inconclusive. The results demonstrate that the 10%, 15% and 20% fiber volume fraction milled glass reinforced polyurea composites increase both the storage and loss moduli by more than 50%, 100% and 200% respectively in comparison to pure polyurea.

4.5 Surface Treatments on Glass Slides

The three surface treatments are also applied to glass slides, and static water droplet contact angles are measured for each type of surface treatment. Figure 4.2 shows the water droplet contact angles for the three surface treatments as compared to the glass surface only washed by piranha solution. The functional group grafted to the glass surface has a minimal effect on the static water droplet contact angle.

Lastly, adhesion test samples are made by casting polyurea on the surface treated glass slides. Two columns of polyurea are aligned and cast on opposing sides of the surface treated glass slides. The opposing polyurea columns are loaded in tension at a ramping rate of 0.5 N/s until failure occurs. The amine surface treatment, which establishes covalent bonds at the interface, results in very strong interfacial bonding, such that the polyurea column starts to yield prior to failure at the interface. The adhesion test qualitatively shows that different surface treatments create interfaces with various bonding strengths. The strength of the interface with amine surface treatment is at least 50% stronger than that with urea surface treatment, and 600% stronger than that with aliphatic surface treatment. The development of a refined adhesion test is currently in progress.

4.6 Discussion

Based on the contact angle measurement, the adhesion test, and SEM imaging, the effectiveness of the surface treatments is evident. However, different interfacial bonding strengths has a negligible effect on the dynamic mechanical properties of the surface treated milled glass reinforced polyurea composites.

Different functional groups on the glass surface have different polarity. Yet, we did not observe a significant difference in the water droplet contact angle measurements for the various surface treatments. Literature shows the contact angle is sensitive to both the reaction condition and also to the molecular topology [4]. It is a complex phenomenon affected by many factors and not directly related to the interfacial bonding properties, which is of greatest interest here. Nevertheless, it is a useful way to qualitatively evaluate the repeatability of surface treatment.

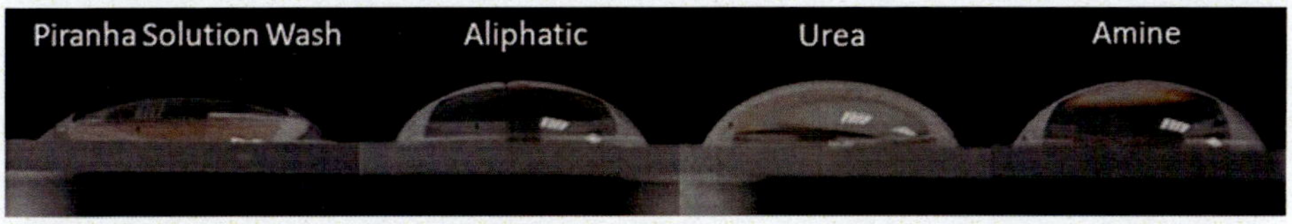

Fig. 4.2 Water droplet contact angles on the surface treated glass slides

Acknowledgements This research has been conducted at the Center of Excellence for Advanced Materials (CEAM) at the University of California, San Diego. This work has been supported by the Office of Naval Research (ONR) grant N00014-09-1-1126 to the University of California, San Diego, under Explosion Resistant Coating Joint Enhanced Explosion Resistant Coating Exploitation (JEERCE) Advanced Concept Technology Demonstration (ACTD) research program which was directed by Dr. Roshdy Barsoum.

References

1. Air Product Chemicals, Inc., (2003) Polyurethance specialty products (Air Products and Chemicals, Allentown)
2. The Dow Chemical Company (2001) Isonate 143L, Modified MDI (Dow Chemical, Midland, MI)
3. Dibenedetto AT, Lex PJ (1989) Evaluation of surface treatments for glass fibers in composite materials. Polymer Eng Sci 29(8):543–555
4. Fadeev AY, McCarthy TJ (1999) Trialkylsilane monolayers covalently attached to silicon surfaces: wettability studies indicating that molecular topography contributes to contact angle hysteresis. Langmuir 15(24):3759–3766

Chapter 5
Experimental and Numerical Characterization of Relaxation in Bolted Composite Joints

Ronald F. Gibson and Srinivasa D. Thoppul

Abstract This paper reports on experimental and numerical studies of the effects of bolt preloads, viscoelasticity, and external applied static and dynamic loads on bolt load relaxation in a unidirectional carbon/epoxy composite bolted joint. Experimental measurements of bolt-connected joints in three-point bending specimens were employed in the studies, and relaxation was observed to depend on the initial preload and external dynamic applied loads. It was observed that for any magnitude of external load the bolt load relaxation decreases with increasing initial preload. These findings emphasize the importance of the magnitude of the preload. It was concluded that only about 1/3 of the bolt force relaxation in the composite joints could be attributed to viscoelastic behavior of the polymer matrix in the composite, and the remaining 2/3 of the relaxation is likely caused by other mechanisms such as bolt thread slip, plasticity and/or external excitation. This paper also briefly reviews some relevant relaxation studies found in the literature for mechanically fastened composite and hybrid joints, as well as the effects of environmental conditions such as temperature and moisture on joint relaxation, and points out some gaps where more research needs to be carried out to understand the behavior of such joints.

Keywords Composites • Bolted joints • Viscoelastic • Relaxation

5.1 Introduction

In today's economy, aircraft and automotive vehicle manufacturers are scrambling to improve fuel efficiency by improving the engine technology and/or reducing weight by using light weight composite materials, which have high strength-to-weight ratio. Commercial airliner and business jet makers such as Boeing, Airbus and Gulfstream Aerospace Corporation are making greater use of composite materials in their airframes and primary structures than ever before. Advanced composites on the Boeing 787 account for 50% of its structural weight mainly in the wings and fuselage [1]. Gulfstream Aerospace makes modest use of composite materials on its new and innovative business jet G650 from tail assembly to floor panels and furnishings [2]. The usage of composite materials as primary structures in aircrafts presents challenges in terms of manufacturing and joining the parts. In composite structures, three types of joints are commonly used, namely, mechanically fastened joints, adhesively bonded joints, and hybrid combinations of mechanical and bonded joints. Combinations of bonded and mechanically fastened joints are primarily used in the aerospace industry to join two or more composite parts, or to join a composite part to a metallic structure (hybrid joint). Mechanically fastened joints usually consist of either solid bolts with collars or nuts, rivets, screws and blind bolts (pull-type). Some common advantages of fastened joints are: simple joint configuration, and ease of assembly, manufacturing and inspection. But when joining composite structures one has to consider stress concentration, poor bearing strength of composites, joint fatigue and environmental exposure. Any joint in a composite structure, if not designed properly, may act as damage initiation point and may lead to failure of the component at that location. For fiber-reinforced composites joined mechanically through bolts with an initial through-the-thickness

R.F. Gibson (✉)
Department of Mechanical Engineering, University of Nevada, Reno, Reno, NV 89557, USA
e-mail: ronaldgibson@unr.edu

S.D. Thoppul
Gulfstream Aerospace Corporation, Savannah, GA 31402, USA
e-mail: srinithoppul@gmail.com

E. Patterson et al. (eds.), *Composite Materials and Joining Technologies for Composites, Volume 7: Proceedings of the 2012 Annual Conference on Experimental and Applied Mechanics,* Conference Proceedings of the Society for Experimental Mechanics Series 44, DOI 10.1007/978-1-4614-4553-1_5, © The Society for Experimental Mechanics, Inc. 2013 preload, viscoelastic relaxation has a direct bearing on how much of the initial preload will be retained in the joint over time, and how much the preload will be affected by external mechanical loads and environmental conditions. The question becomes hard to answer when the composite material is fastened to metallic structures to form hybrid joints.

The procedure for designing mechanically fastened composite or hybrid joints relies mostly on experimental data. An overview of design requirements, different approaches, and industrial codes of practice used for the design and analysis of bolted and bonded joints is given by Broughton, et al. [3]. In the most recent review on mechanics of mechanically fastened joints in polymer composite structures [4], standard methods available to determine joint design allowables such as bearing strength, open hole tension (OHT), open hole compression (OHC), filled hole tension (FHT) filled hole compression (FHC), and bearing by-pass are given. The article also discusses the effects of geometric parameters (width-to-diameter (w/d), edge distance-to-diameter (e/d), washer size, and bolt type), hygrothermal effects, and bolt clamp-up and relaxation (or preload) on bearing strength, failure load and fatigue life in composite joints.

In both of the above articles [3, 4], it is stressed that for composite laminates, the through-the-thickness (TTT) force (preload) exerted by the fastener on the joined members is critical to the performance of the joint. Bolt preload induces compressive load in the washer and material being joined, and tensile load in the bolt shank. This serves two purposes: in terms of carrying load by friction and suppressing delamination driven failure modes, while increasing bearing strength. However, for composite structures, if the preload is not sufficient to begin with, some of the preload may be lost in the joint due to viscoelastic behavior. It is well known that for polymeric composites, viscoelastic effects are predominant in the matrix dominated TTT direction, and this viscoelastic behavior is magnified when the composite is exposed to elevated moisture and/or temperature levels. The relevant previous studies found in the literature on clamp-up force relaxation in composite-to-composite joints and composite-to-metal or hybrid joints, are reviewed in the next two subsections.

5.1.1 Composite-to-Composite Joints

As mentioned earlier, it is important to select an optimum preload for the joint or torque applied to the fastener. Some designers use the existing MSFC-STD-486B design code as a guideline from NASA Marshall Space Flight Center (MSFC) [5–7], which was developed originally for metallic structures. A process has been developed for recommending the torque range to apply for fasteners as per MSFC-STD-486B to join composite structures [8]. This process calls for the use of a torque versus tension test, and monitoring of the joints using acoustic emission during the test. It is also recommended that nondestructive thermal images and thermographs be taken after the tests. Some studies also recommend using instrumented bolts to measure the pre-load in composite joints while applying torque to the fastener [9]. Earlier experiments in 1980s showed the clamp-up force relaxation in graphite/epoxy (T300/5208) laminates [10, 11], under three steady state environments: room temperature ambient (RTA) with 0.46% moisture content by weight, room temperature dry (RTD), and elevated temperature (66°C) dry (ETD). Relaxation in clamp-up force of 12% for RTD and 14% for RTA for a 100 day period was reported for a double-lap joint configuration tightened with an initial torque of 5.65 N-m, which clearly indicates that increasing moisture content increases the relaxation.

Fastener type, different torque values and also the material may have different effects on bolt load relaxation. Horn and Schmitt [12, 13] showed that bolt load relaxation for graphite/thermoplastic composite materials (Dupont's IM6/KII) is 3.8% more than that of the ICI-Fiberite's IM8/APC(HTA) in a single-lap shear configuration for a 1,000 h period, with initial torques of 65 and 100 in-lbs. It was also found that the relaxation rate increased by 4.7% for hot (250°F)/dry (0.0% moisture) conditions. To understand the effect of bolt-load relaxation on the bearing strength, the composite specimens were tested to failure after the end of 1,000 h relaxation tests. On comparing the bearing strength of relaxed joints with those of joints tested to failure immediately after tightening, it was concluded that the bolt-load relaxation did not significantly affect the bearing strength. Zhao and Gibson [14] showed that compressive clamping stress relaxed by 18% and 15% in E-glass/epoxy beams with and without polymeric interleaves, respectively, for a period of 50 h, whereas the corresponding relaxation in an aluminum beam was negligible.

It is clear from the previous experiments that exposure to high temperature and moisture increases the clamp-up force relaxation in composite joints. But the effect of hygrothermal cycling on clamp-up force relaxation is unclear as observed by Chen [15, 16]. Experiments on IM6/3501-6 $[45/90/-45/0_3/\pm45/0_3/\pm45]_s$ showed a 5.5% and 7.6% reduction in clamp-up force over a period of 4 months at RTA and 98°F/98% Room humidity (RH) 12 h on-off cycling, respectively, but at 120°F/98% RH cycling the results showed an increase in the clamp-up force by about 55%. For IM7/8552 $[45/0]_{10}$,21.7% reduction in clamp-up force was observed at RTA, whereas there was no change in the clamp-up force at 120°F/98% RH cycling. It was concluded that material swelling under humid room conditions and some possible nonlinearity of the instrumented bolt characteristics were responsible for this increase.

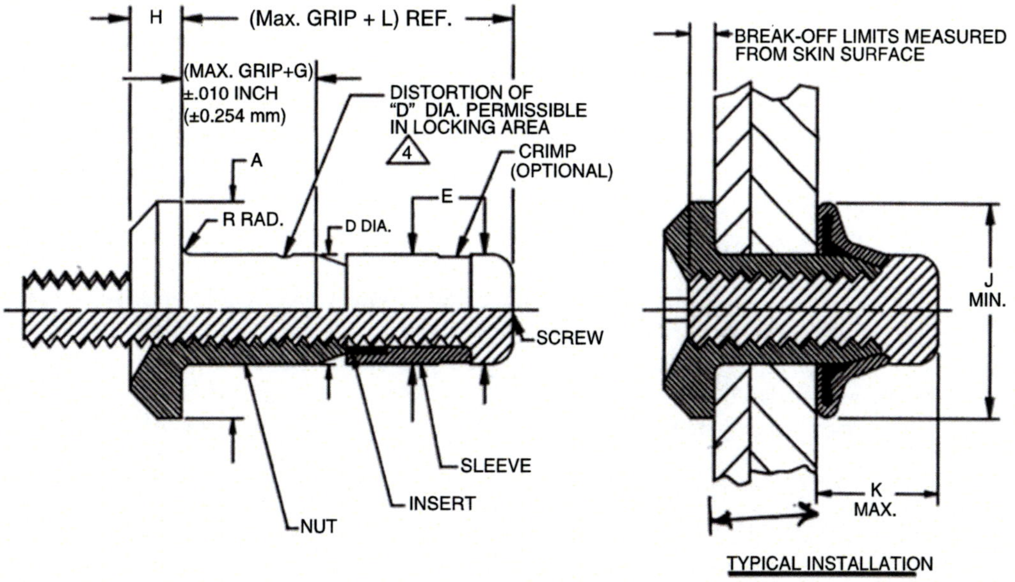

Fig. 5.1 Blind Fastener protruding head MBF 2010 [18]

It is suggested that titanium alloy fasteners be used to prevent galvanic corrosion and tension heads be used to avoid pull-through for composite assemblies in aircraft primary structures [17]. It is very common practice to use blind or one-sided fasteners in primary structures where access to both sides of the assembly is not possible or practical (e.g., general aviation wing covers, empennage surfaces, and control surfaces). Monogram Aerospace Fastener is one of the leaders in blind bolt technology, its patented Composi-Lok® is approved for use on most of the composite airframe programs [18]. A typical blind fastener joint is shown in Fig. 5.1. The blind fastener consists of a stem and a sleeve which forms the fastener head. The stem extends through the sleeve head in the form of a threaded shank. The other end of the stem has a diameter slightly larger than the sleeve. The sleeve rests against the part being fastened and the stem is pulled out, flaring the sleeve and thus forming the bulb which serves as a nut. The quality of the joint, the clamp-up force and the strength of the joint have been shown to depend on the ratio of the part thickness to the manufacturer specified grip length ratios (part thickness to recommended grip length) [19]. Clamp-up force relaxation in graphite/epoxy ([±45/0]s) composite laminate joints using blind fasteners (NAS1919-M06) has been investigated experimentally [20]. The clamp-up force was measured for different ratios of part thickness to recommended grip length. For a 21 day monitoring period, the relaxation was observed to be 3–8%, and it was also found that the rate of relaxation increased with increasing grip ratio.

5.1.2 Hybrid Joints

It is very difficult to avoid composite-to-metal connections in aircraft structures, and the design of these connections presents its own challenges as additional loads are generated due to thermal expansion differences, and the behavior is different when the joint is subjected to elevated or cold temperature conditions. Comprehensive experimental studies have shown that the clamp-up force relaxation in composite-aluminum (C/Al) and composite-steel (C/St) bolted joints is significantly different and highly variable [21–23]. The experimental study involved testing C/Al and C/St joints using both protruding head and countersunk head bolts over a 3 month period under ambient and elevated temperature (62°C). The C/Al and C/St joints showed 45% and 30% clamp-up force relaxation, respectively for 2,000 h, while the composite-to-composite joint had 55% relaxation. It was concluded that using a countersunk bolt did not affect the relaxation, when compared to a protruding head bolt. Significant thermal response was observed from the above joints as expected, because of the coefficient of thermal expansion mismatch between the materials. Structural health monitoring methods which come under the umbrella of structural vibration analysis have been successfully used to detect the clamp-up force relaxation in a large hybrid (C/St) structure [24]. Three different methods for assessing the change included: (1) fundamental model properties; (2) transfer functions; and (3) transmittance function using a piezoelectric actuator bonded to the composite panel to deliver the controlled vibration. Experiments involved detecting the loosening of bolts using the above three methods on a 622 mm square composite plate joined to a steel plate with 16 bolts at its perimeter in two ways: first, by reducing torque of only one bolt, and second where the torques in all the bolts were reduced by the same amount. It was concluded that the transmittance function was the most promising method of the three in reliably detecting a single bolt loosening.

Most of the above studies were carried out without external static or dynamic loads acting on the joint. Bickford [25] suggests that external vibrations on any bolted joint will increase the clamp-up force relaxation because of wear and hammering. After sufficient pre-load is lost, friction forces drop below a critical level and the nut actually starts to back off and shake loose. With higher initial pre-load, longer or more severe vibration is required to reduce pre-load to the critical point at which back-off occurs. In fact, in some circumstances, if the pre-load is high enough to start with, nut back-off will never take place. It is not known if these observations apply to composite bolted joints. Hence, an effort has been made to determine the clamp-up force relaxation in a single composite bolted lap joint subjected to static and dynamic loads. The focus of the current investigation was to characterize the effects of various clamping loads (bolt preloads), external static loads and dynamic beam loads on relaxation in a composite bolted joint under combined bending and shear loads (3 point bending). An attempt was also made to back out the effect of viscoelasticity on relaxation by comparing experimental relaxation curves for a composite bolted joint with those of a steel joint, which does not exhibit viscoelastic behavior at room temperature.

5.2 Experimental Set-up

The composite specimen used for the test was made out of unidirectional P2254-20-305 T800 carbon/epoxy prepreg tape from Toray Composites America, Inc. A total of 48 layers of prepreg were laid on top of each other in the mold along with bleeder cloth and release fabric. The mold assembly was placed in the molding chamber of a TMP autoclave-style vacuum press. After curing, the thickness of the sample was 7.75 mm. There was no post curing cycle. After removing the laminate from the mold assembly, it was cut into two-piece rectangular beams of 269.87 mm × 25.4 mm and machined with a diamond cutting tool to the required dimensions as shown in Fig. 5.2. The joint was clamped together by a standard hexagonal head cap screw steel bolt with built in strain-gages (Strainsert®, 9.525 mm or 3/8 in. in diameter, model: SXS-FB) and washers (19 mm diameter). The joint was subjected to 3-point bending using an Enduratec servo-pneumatic testing machine. The design of the specimen support was selected in such a way to maintain combined bending and shear loading.

The joints were clamped with three different bolt preloads, 4,200 N, 5,050 N and 7,850 N, which correspond to 12.5%, 15% and 23.5%, respectively, of the maximum manufacturer recommended tensile load in the instrumented bolt. The corresponding through-the-thickness compressive stresses generated in the composite specimen were approximately 25%, 29% and 45%, respectively of the estimated transverse compressive strength of the composite. After applying an initial preload to the composite bolted joint, the bolt load was monitored for a period of 30 h under the following conditions: (a) Preload only condition without application of external load, (b) preload and a static applied load of 250 N (combination of ramp and dwell) in the 3 point bend setup (Fig. 5.2) and (c) preload and a 250 N amplitude dynamic load in the 3 point bend setup at frequencies of 1 Hz, 2 Hz and 5 Hz. External static and dynamic load experiments were carried out using the 3-point bend fixture.

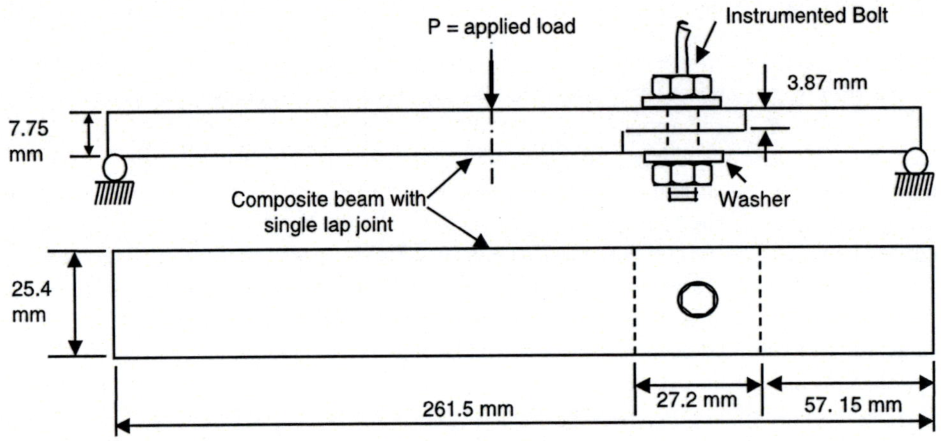

Fig. 5.2 Specimen configuration and three point bend set-up

Two separate data acquisition systems were used; one for the instrumented bolt and another to control the Enduratec servo-pneumatic testing machine and also to record the beam load and displacement values. A stability check on the data acquisition system was carried out by monitoring the signal from the instrumented bolt for a period of 30 h without the application of load to the bolts. The drift in the signal for the above period was less than 1%. Also, the drift in the signal was taken into account when the percentage relaxation was calculated. Preload only experiments were also conducted on a steel joint having the same dimensions as those of the composite joint. The bolt load was monitored for a period of 30 h using the same 3-point bend set-up and the instrumented bolt for preloads mentioned above.

5.3 Experimental Results

5.3.1 Effect of Preload Only Conditions on Relaxation in Composite and Steel Joints

Experimental bolt load relaxation results for composite bolted joints with initial preloads of 5,050 N and 7,850 N for preload only condition (i.e. no external load) are shown in Fig. 5.3, where the normalized bolt load is the instantaneous load divided by the initial load. It is clear from the results that for both steel and composite bolted joints, the magnitude of relaxation decreases with increasing bolt preload. However, for steel joints, the magnitude of relaxation is less than that of the composite joints. By comparison with the steel joints, some additional relaxation due to viscoelastic behavior in the composite joint was expected. Relaxation of approximately 3.0% and 2.5% was observed after 30 h in the composite joints preloaded to 5,050 N and 7,850 N, respectively, whereas the corresponding relaxation in the steel joint was approximately 1.9% and 1.75%, respectively. Numerical results from finite element analysis (FEA) will be discussed later.

5.3.2 Effect of Static External Loading Conditions on Relaxation in Composites Joints

Figure 5.4 shows the effect of a 250 N static beam load on bolt load relaxation in the composite bolted joint under the 3 point bend setup. It is observed that the magnitude of relaxation decreases with increasing bolt preload. However, when compared with the preload only condition, the bolt load relaxation increases for lower preloads (5,050 N) and the bolt load relaxation decreases for higher preloads (7,850 N) by approximately 0.5%, and the result for the lower preload (4,200 N) seems inconsistent with that trend.

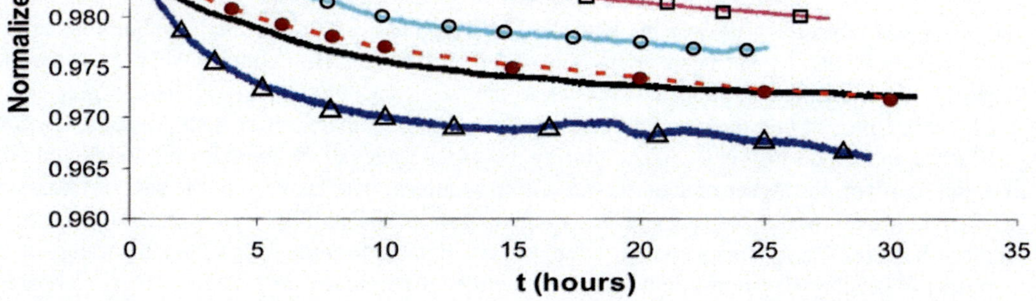

Fig. 5.3 Bolt preload relaxation in composite and steel joints for preload only condition without any external beam load (Note: S = Steel joint, C = Composite joint)

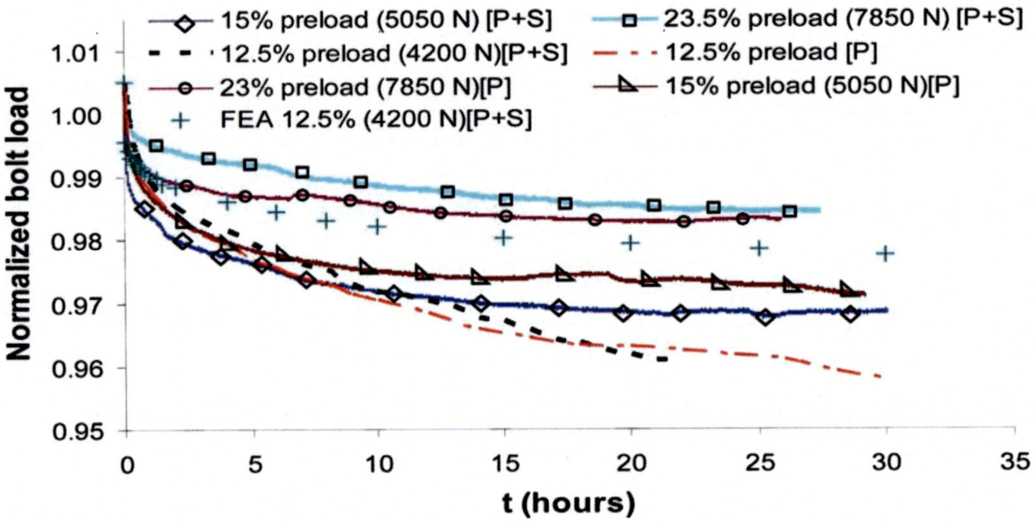

Fig. 5.4 Bolt preload relaxation in composite joints for preload only and preload with a static external beam load of 250 N (Note: P = preload only condition, $P + S$ = preload + static 250 N beam load)

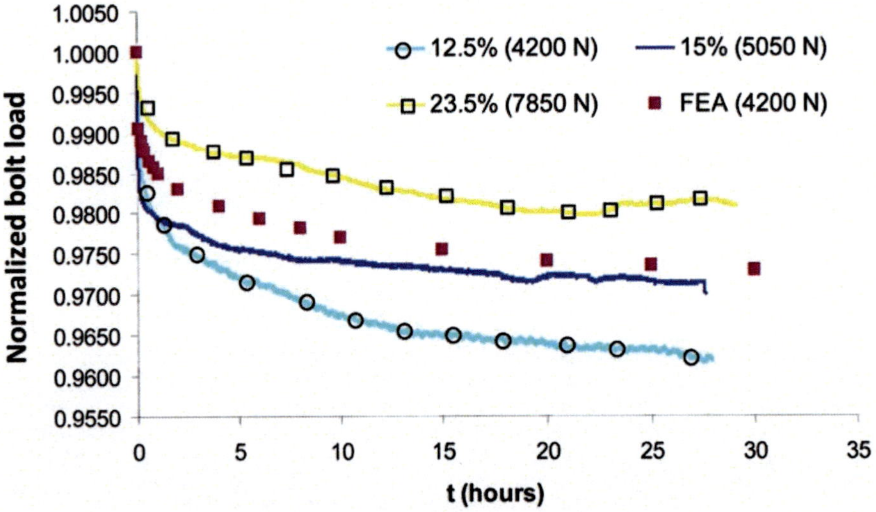

Fig. 5.5 Bolt preload relaxation in composite joints for preload with a dynamic external beam load of 250 N at 1 Hz

5.3.3 Effect of Dynamic Loading Conditions on Relaxation in Composites

Figures 5.5 and 5.6a, b show the effects of dynamic load at 1 Hz, 2 Hz and 5 Hz, respectively, on bolt load relaxation in the composite bolted joint under the 3 point bend setup. It is observed that at 1 Hz frequency (Fig. 5.5), the magnitude of relaxation decreases with increasing bolt preload. However, as the frequency of external beam load is increased to 2 Hz and 5 Hz (Fig. 5.6a, b), it is observed that the magnitude of relaxation tends to increase for both preloads. For lower preload (5,050 N) the reduction in the bolt load is significant during the initial stages of the experiment, and the rate of relaxation decreases with time, however, for higher preload the relaxation is higher. The increase in the rate of relaxation at higher frequencies of excitation may be due to the increase in temperature at the joint interfaces due to frequency-dependent frictional sliding and heating. The specimen configuration was not suitable for monitoring the temperature at the interface during the experiment. When the relaxation in bolt load due to applied dynamic load at 2 Hz and 5 Hz is compared with those due to applied static loads, it is observed that for lower preload (5,050 N) the applied static load increases the bolt load relaxation whereas for higher bolt preload (7,800 N) the bolt load relaxation decreases.

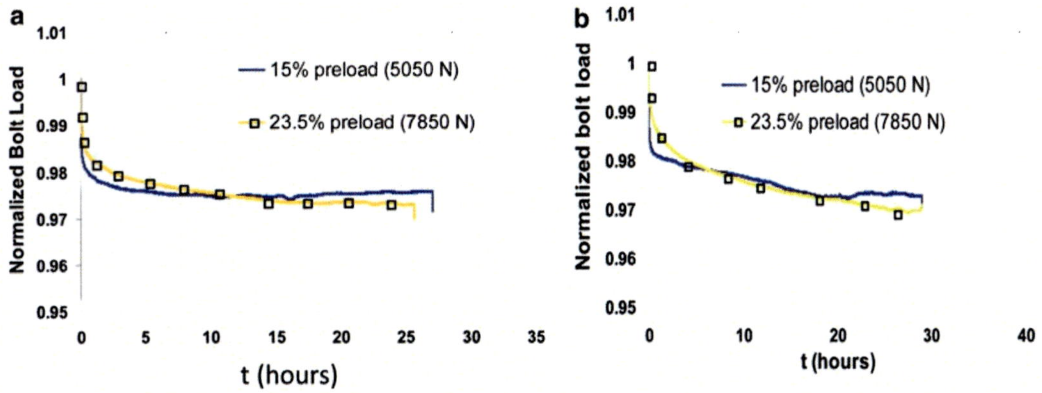

Fig. 5.6 Bolt preload relaxation in composite joints for preload with a dynamic external beam load of 250 N at (**a**) 2 Hz (**b**) 5 Hz

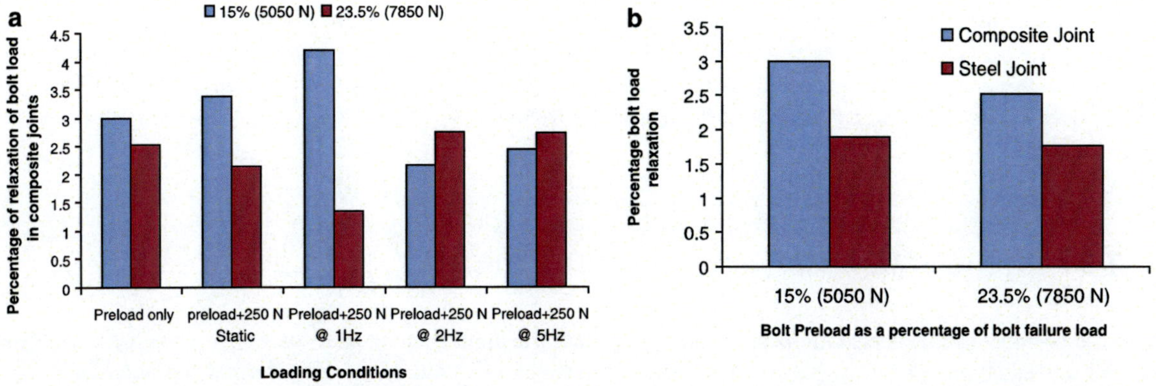

Fig. 5.7 (**a**) Effect of external beam load on bolt load relaxation in composite bolted joints for a period of 30 h (**b**) comparison of relaxation in composite and steel joints for a period of 30 h under preload only condition

A comparison of the bolt load relaxation in composite joints at the end of the 30 h duration is shown in Fig. 5.7a for two preloads with different external loading conditions. It is observed that for any external loading condition, the bolt load relaxation decreases with increasing initial bolt preload. These findings emphasize the importance of preload selection. For higher preload (7,850 N) the bolt load relaxation increases with increasing frequency of excitation, which could be partly due to an increase in the temperature at the interface due to frictional heating. Comparing the bolt load relaxation in steel and composite joints for the duration of 30 h (Fig. 5.7b), it is observed that only about 1/3 of the total relaxation in the composite joint is due to viscoelastic behavior of the polymer matrix material in the composite, while the remaining 2/3 is apparently due to the other relaxation mechanisms such as plasticity and/or slip in the bolt threads, which probably occur in steel joints as well.

5.4 Numerical Studies and Results

In this section, a brief summary of finite element analysis (FEA) simulations is given. For more detailed explanations of material characterization and complete FEA results, see [26]. The purpose of FEA is to develop predictive numerical models, and to promote a more meaningful interpretation of the experimental results. Hypermesh® 5.0 [27] was used to develop a three-dimensional (3D) FEA model of the composite bolted joint (see Fig. 5.8), and post-processing the results. ABAQUS® V 6.3 Standard 3-D [28] was used predict the bolt load relaxation using the quasi-elastic analysis. A global–local submodeling technique was used to model the bolted composite joint, where the displacements around the bolted joint section in the global model (a one-piece beam with no bolt) are used to drive the refined local model. This technique has the advantage that more detailed results in the vicinity of the bolted joint can be obtained with fewer elements than with the full model. All models were developed using C3D8 and C3D6 type 3-D solid elements. Since the model is symmetric about the

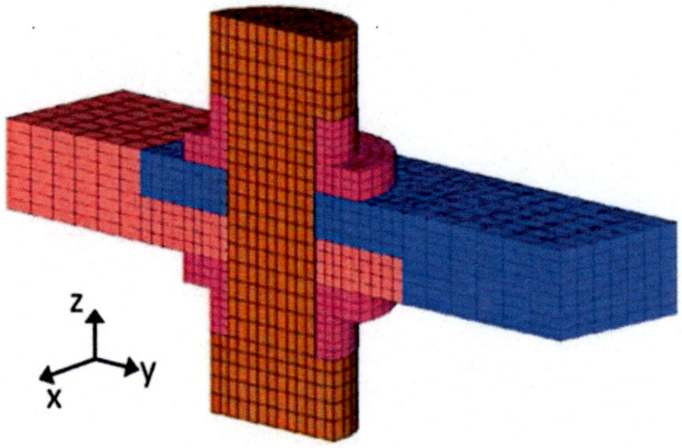

Fig. 5.8 Bolt FEA model (Local)

vertical midplane of the beam, only a half model was used, and symmetric boundary conditions were applied. The threads in the bolt were neglected in the FEA models, and the bolt was assumed to be a solid cylinder. Thus, possible bolt load relaxation due to plastic deformation and/or thread slip in the threads was not included in the models. The solid bolt simulation requires that contact surfaces be defined between all the surfaces that are in contact, and these surfaces were modeled using the contact pair approach in ABAQUS. The contact pairs are defined from free element faces. Since the sliding between the surfaces was expected to be small, the 'small sliding' option was used in all analyses. Friction coefficients were set to 0.2 for all contact surfaces, as used by Ireman [29].

5.4.1 Time Dependent Material Property Characterization

The three-dimensional (3-D) elastic lamina properties needed for the finite element analysis were calculated from composite micromechanics equations [30] (for example: the rule of mixtures to calculate the longitudinal modulus and Tsai-Hahn equations to calculate the transverse modulus and shear modulus) using the fiber and matrix properties together with the following assumptions: (1) the fibers are linearly elastic, (2) the matrix is linearly viscoelastic, with its creep compliance described by a power law, (3) the composite is specially orthotropic and transversely isotropic, and (4) the viscoelastic response depends only on the time elapsed since application of the load (i.e. the material is assumed to be non-aging). The analysis consisted of three parts: first, since the viscoelastic creep and relaxation data for the epoxy resin used in the prepreg is not available, Beckwith's [31] measured linear viscoelastic properties for Shell 58–68 epoxy at 75° F were assumed. Next, these properties were used in the FEA to predict the bolt load relaxation for an epoxy beam under different loading conditions using the quasi-elastic approach (for model validation see [26]). Since ABAQUS viscoelastic modeling capability is limited to isotropic materials, it was necessary to validate the quasi-elastic approach (explained in the next section). Lastly, Beckwith's [31] creep test results were extrapolated out to 50 h from the available data, by using the empirical power law equation for creep compliance:

$$D(t) = D_0 + D_1 t^n \tag{5.1}$$

where, from [31],

$D(t)$ = time dependent isotropic creep compliance of matrix
D_0 = initial elastic compliance of matrix = 2.726×10^{-4} (MPa)$^{-1}$
D_1 = creep coefficient of matrix = 1.0×10^{-3} (MPa)$^{-1}$ (min)$^{-n}$
t = time in minutes
n = dimensionless creep exponent = 0.19

The time-dependent viscoelastic properties of the composite joint were assumed to depend only on the time-dependent properties of the epoxy matrix material. Based on the linear viscoelastic assumption, a time-dependent matrix modulus, $E_m(t)$, was estimated from the following equation:

$$E_m(t) \approx \frac{1}{D(t)} \tag{5.2}$$

Table 5.1 Micromechanics estimate of time-dependent material properties for unidirectional carbon/epoxy[a]

Time (Hrs)	$E_m(t)$ (GPa)	$E_1(t)$ (GPa)	$E_2(t) = E_3(t)$ (GPa)	$G_{12}(t) = G_{13}(t)$ (GPa)	$\upsilon_{12}(t) = \upsilon_{13(t)}(t) = \upsilon_{23}(t)$	$G_{23}(t)$ (GPa)
0	3.662	134.5	12.07	3.612	0.30	2.535
1	3.607	134.4	11.66	3.469	0.307	2.446
2	3.515	134.4	11.55	3.449	0.308	2.434
4	3.495	134.4	11.51	3.428	0.309	2.421
6	3.482	134.4	11.48	3.414	0.309	2.412
8	3.473	134.4	11.45	3.403	0.310	2.405
10	3.465	134.4	11.40	3.395	0.310	2.40
15	3.450	134.4	11.37	3.379	0.311	2.390
20	3.439	134.4	11.34	3.366	0.312	2.382
25	3.430	134.4	11.32	3.356	0.312	2.376
30	3.422	134.4	11.55	3.348	0.313	2.371

[a] Toray Composites America T800 carbon fibers with assumed viscoelastic compliance for Shell 58–68 epoxy matrix material [31]

It was assumed that the power law constants; D_0, D_1 and n at room temperature are the same as those measured by Beckwith for Shell 58–68 epoxy [31]. By using these constants and the power law, the time-dependent creep compliance, $D(t)$, for the epoxy material was calculated. The time-dependent Young's modulus of the matrix material, $E_m(t)$, (see Table 5.1) was found from Eq. 5.2. The tensile modulus data for T800 carbon fibers was obtained from the fiber manufacturer, Toray Composites America, Inc. as $E_f = 294$ GPa and was assumed to be independent of time. Viscoelastic properties of the lamina (i.e., longitudinal modulus $E_1(t)$, transverse moduli $E_2(t)$ and $E_3(t)$, Poisson's ratios $\upsilon_{12}(t)$, $\upsilon_{13}(t)$ and $\upsilon_{23}(t)$, longitudinal shear modulus $G_{12}(t)$, and transverse shear moduli $G_{13}(t)$ and $G_{23}(t)$) were estimated using elastic fiber properties and time-dependent viscoelastic resin properties through an application of the Elastic-Viscoelastic Correspondence Principle to the micromechanics equations [30] with a fiber volume fraction, $v_f = 0.45$ (see Table 5.1). The fiber volume fraction for the composite laminate was indirectly estimated using a combined experimental/numerical technique. Measured load–displacement response from static 3 point bend tests was compared with the predicted response (FE models) initially assuming the fiber volume fraction to be 0.6. The difference between the predicted and measured response was then minimized using fiber volume fraction as curve fitting parameter. These properties were then used in the FEA to calculate the bolt load relaxation. Due to the limitations of ABAQUS [28], the quasi-elastic approach [30] was used to predict the bolt load relaxation for the orthotropic composite beams. In this approach, the viscoelastic solutions were approximated by a series of elastic solutions corresponding to different elastic properties at different times, while the stresses were assumed to be constant within each time increment.

As in the experiments, the external load was applied as a concentrated nodal force at midspan. pre-tension in the bolt was applied in a separate loading step by defining a pre-tension section in the bolts. Assembly loads were transmitted across the pre-tension section by means of a pre-tension node. Preload was applied by giving an initial displacement (in the direction parallel to the bolt axis) to this node. Bolt preload was maintained by using the fixed option under boundary conditions, and was monitored by checking the total force output on that node. The model geometry and boundary conditions were chosen to simulate the experiments, as shown in Fig. 5.2. For the composite beams, both experiments and FEA (using the global–local model and quasi-elastic approach) were conducted, but only experiments were conducted for steel joints.

Bolt load relaxation was predicted for a period of 30 h in composite bolted joints using a quasi-elastic analysis and the material properties listed in Table 5.1 for various preload under the following types of loading: (1) bolt preloading in the absence of external beam loading, (2) bolt preload in the presence of a static beam load of 250 N, and (3) bolt preload in the presence of dynamic loading of amplitude 250 N at a frequency of 1 Hz. The dynamic beam load was applied using the periodic loading option in ABAQUS for 5 cycles under each quasi-elastic step and mean bolt load data was recorded in each case.

The predicted bolt load relaxation captures the relaxation in composite joints and is well within the measured relaxation curves (Fig. 5.3). The relaxation in bolt load is predicted using the quasi-elastic approach with the material properties listed in Table 5.1. It is observed from Fig. 5.3 that the predicted relaxation is not affected by increasing preloads, but the relaxation is higher than that measured in steel joints. This contradicts the experimental observation that relaxation decreases with increasing bolt preload. There are several possible reasons for this disagreement. First, the material model does not capture the viscoelastic effects in the polymer matrix material at the micromechanical level. The micromechanical analysis referred to in the discussion following Eq. 5.2 was based on "mechanics of materials" type models which do not take into account the details of the in-situ micromechanical stress and strain distributions in the viscoelastic polymer matrix. A 3-D finite element micromechanics model which takes into account the micromechanical geometry is needed to accurately simulate the effects of such parameters as boundary conditions at the bolt-composite interface on the viscoelastic relaxation of the composite.

Figures 5.4 and 5.5 show the comparison of predicted bolt load relaxation for composite joints under external static and dynamic loads. Again, the predicted relaxation falls within the measured relaxation curves for both static and dynamic loads. Also, it is observed that the predicted bolt loads are shifted slightly with the application of the static and dynamic beam loads, but otherwise the relaxation curves are not significantly affected by the external loads. Again, the mean bolt load is plotted for the dynamic analysis.

5.5 Conclusions

- Experiments have been employed to study the effects of various bolt preloads, along with the effects of static and dynamic external loads on bolt load relaxation in composite bolted joints.
- Experiments show that for any external loading condition the bolt load relaxation decreases with increasing initial bolt preload, and these findings emphasize the importance of bolt preload selection.
- If the bolt preloads are small enough (as a percentage of bolt failure load), applied static and dynamic beam loads at 1 Hz frequency increase the magnitude of bolt load relaxation. However, for higher bolt preloads the bolt load relaxation decreases for both static and dynamic loads.
- It is observed that increasing the frequency of the external dynamic load from 1 to 5 Hz increases the rate of relaxation, and that the friction-induced heating may be at least partially responsible for this effect.
- Finite element models for bolt load relaxation in bolted composite joints based on a global/local quasi-elastic approach show reasonably good agreement with experiments except that the experimentally observed decrease in relaxation with increased bolt preload is not predicted by the models.
- The FEA predictions of bolt load relaxation agree well with the experimental observations when subjected to external static and dynamic loads, however, more detailed modeling of the polymer matrix behavior at the micromechanical level and possible time-dependent boundary conditions at the bolt-composite interface are needed to understand the experimentally observed relationship between bolt preload and bolt force relaxation.
- Comparing the bolt load relaxation in steel and composite joints for the duration of 30 h, it is clear that only about 1/3 of the total relaxation in the composite joint is due to viscoelastic behavior of the polymer matrix material, while the remaining 2/3 is apparently due to the other relaxation mechanisms such as plasticity and/or slip in the bolt threads, which probably occur in steel joints as well.

Acknowledgements Most of the work reported here was done in the Advanced Composites Research Laboratory at Wayne State University under the sponsorship of the U.S. Air Force Office of Scientific Research grant No. FA9550-04-1-0042. The program manager was Dr. Victor Guirguitui.

References

1. Griffiths B (2005) Boeing sets pace for composite usage in large civil aircraft. High-Perform Composite 13:68–71
2. Weber A (2010) Composite joining pros and cons. Assembly 53(6):38–43
3. Broughton W, Crocker LE, Gower MRL (2002) Design requirements for bonded and bolted composite structure. National Physical Laboratory Materials Center, NPL report MATC(A)65, Jan 2002
4. Thoppul SD, Finegan J, Gibson RF (2009) Mechanics of mechanically fastened joints in polymer-matrix composite structures – A review. Compos Sci Technol 69(3–4):301–329
5. Torque limits for standard threaded fasteners (1993) MSFC-STD-486, NASA Marshall Space Flight Center
6. Zhao Y, Ford D, Richardson S (2001) Torque limit for fasteners in composites, Technical Report, NASA/MSFC/ED23
7. Thomas FP, Zhao Y (2005) Torque limits for composites joined with mechanical fasteners, AIAA 2005–2351. In: Proceeding 46th AIAA/ASME/AHS/ASCE/ASC structural dynamics and materials conference AIAA, Reston, VA
8. Thomas FP (2006) Experimental observation for determining the maximum torque values to apply to composite components mechanically joined with fasteners (MSFC center Director's discretionary fund report, project No. 03–13). NASA/TM-2006-214323
9. McCarthy MA, Lawlor P, O'Donnell C, Harris K, Kelly P, Cunningham JP (2005) Measurement of bolt pre-load in torqued composite joints. Strain 41:109–112
10. Shivakumar KN, Crews JH Jr (1983) Bolt clampup relaxation in a graphite/epoxy laminate. Long-term behavior of composites, ASTM symposium, Williamsburg, ASTM STP 813, pp 5–22
11. Shivakumar KN, Crews JH Jr (1982) An equation for bolt clamp-up relaxation in transient environments. Compos Technol Rev 4(4):132–135

12. Schmitt RR, Horn WJ (1990) Viscoelastic relaxation in bolted thermoplastic composite joints. In: 35th international SAMPE symposium and exhibition, was published by SAMPE, Covina, CA pp 1336–1347
13. Horn WJ, Schmitt RR (1993) Relaxation in bolted thermoplastic composite joints. AIAA J 3(3):485–494
14. Zhao H, Gibson RF (1995) Influence of clamping force relaxation on vibration damping measurements for polymer composite cantilever beams. In: Proceedings SEM spring conference, Grand Rapids, pp 735–738
15. Chen H-S (2001) The static and fatigue strength of bolted joints in composites with hygrothermal cycling. Compos Struct 52(3–4):295–306
16. Chen HS, Kung HK (2002) A hygrothermal sensitivity evaluation on the clamp-up torque of bolted composite joint. In: Sun CT, Kim H (eds) Proceedings of the American Society for composites 17th technical conference, CRC Press LLC
17. Niu MC (1992) Composite airframe structures practical design information and data, 1st edn. Conmilit Press LTD, Hong Kong, pp 301–302
18. Monogram Aerospace Fasteners Online catalog, Access on Jan. 31, 2012. Jan. 31, 2012, http://www.monogramaerospace.com/files/active/0/Monogram%20Catalog.pdf
19. Raju KS, Kumar B, Kolachalamma A, Buhrman AJ (2005) Blind-fastener applications in composite structures: NIAR/Industry/State of Kansas (NIS). NIS 05–003
20. Kolachalama A, Raju KS (2006) Viscoelastic clamp-up relaxation of blind fastener joints in composites. In: 2nd annual symposium: graduate research and scholarly projects conference proceedings held at Wichita State University, Wichita, 28 Apr 2006
21. Caccese V, Berube KA, Fernandez J, Melo D, Kabche JP (2009) Influence of stress relaxation on clamp-up force in hybrid composite-to-metal bolted joints. Compos Struct 89:285–293
22. Pelletier KN, Caccese V, Berube KA (2005) Influence of stress relaxation in hybrid composite/metal bolted connections. Report No. UM-MACH-RPT-01-02, Prepared for the Office of Naval Research under grant No. N00014-01-1-0916
23. Fernandez M, Caccese V, Vel SS (2009) Effect of temperature and viscoelastic creep on the clamp-up load in hybrid composite/metal bolted joints. Report No. C-2004-015-RPT-03, Prepared for the Office of Naval Research under grant No. N00014-05-1-0735
24. Caccese V, Mewer R, Vel SS (2004) Detection of bolt load loss in hybrid composite/metal bolted connections. Eng Struct 26:895–906
25. Bickford JH (1995) An introduction to the design and behavior of bolted joints, 3rd edn. CRC Press, Taylor and Francis, Boca Raton
26. Thoppul SD, Gibson RF, Ibrahim RA (2008) Phenomenological modeling and numerical simulation of relaxation in bolted composite joints. J Compos Mater 42(17):1709–1729
27. Hypermesh® V 5.0, User's manual, Altair Engineering, Inc
28. ABAQUS® V 6.3, User's Manual and Theory Manual, Hibbit, Karlson and Sorensen, Inc
29. Ireman T (1998) Three-dimensional stress analysis of bolted single-lap composite joints. Compos Struct 43:195–216
30. Gibson RF (2011) Principles of composite material mechanics, 3rd edn. CRC Press, Taylor and Francis Group, Boca Raton
31. Beckwith SW (1984) Viscoelastic creep behavior of filament-wound case materials. J Spacecr Rocket 21(6):546–552

Chapter 6
New Composite Timbers, Full Field Analysis of Adhesive Behavior

Boris Clouet, Régis Pommier, and Michel Danis

Abstract Green gluing technology may have a large interest in wood industry. It allows energy saving so reducing costs. In order to fulfill requirements in relevant building codes and standards, these composite timbers have to present sufficient structural performance and shape stability. The drying step during the process of green gluing may lead to local strains and stresses development on the adhesive interface. This strains and stresses development could impact the shape stability of the product and induce joint failure. However, wood spatial microstructure may absorb these developments. It has to be noticed that, for an organic material like wood, the measurement of hygromechanical properties must be done without any interactions between the material and the measurement system. In order to process in this way, an optical measurement system such as the Digital Image Correlation can be used. Results give us the local behavior of the interface, during shrinkage and swelling of wood. The strains and stresses of the adhesive bond will be calculated from the measurement by the mean of an inverse method. Then, the structural performance of this new bonded material will be presented.

Keywords Green wood gluing • digital image correlation • Drying

6.1 Introduction

Nowadays, about 50% of wood products are processes by gluing technology. This technology is very convenient to get homogenized wood properties and standard desired dimensions. It also provides a way of using low quality wood, which otherwise would become waste or fuel. Maritime Pine is a low quality raw material, which has got a lot of knots and cracks. Only 25% of the sawn lumbers are considered "defects free" and are done for the carpentry; the rest being used for packaging. Nevertheless, the length of lumbers used for the framework cannot exceed 2.40 m and are unable to be used in construction that requires lengths above 3 m. In order to use this wood in industrial conditions, it has to be modified, using the gluing technologies.

Glued laminated beams may provide large lengths up to 30 m. In order to manufacture these Engineered Wood Products, the wood needs to be dried before gluing, leading to a waste of energy and time in the process. Nevertheless, for a few years, new adhesives formulations have been developed to allow the gluing of wood at a green state (above the fiber saturation point) [1]. The two manufacturing processes (dry gluing and green gluing) differ significantly and are presented in and Figs. 6.1 and 6.2 [2].

The major distinction between the two processes is in the drying step. In the dry gluing process, the boards needs to be dried at moisture contents between 8% and 16% before gluing whereas the green gluing process allows bonding at high

B. Clouet (✉)
University Bordeaux, I2M, UMR 5295, F-33400 Talence, France
e-mail: boris.clouet@etu.u-bordeaux1.fr

R. Pommier
CNRS, I2M, UMR 5295, F-33400 Talence, France
e-mail: regis.pommier@u-bordeaux1.fr

M. Danis
Arts et Métiers ParisTech, I2M, UMR 5295, F-33400 Talence, France
e-mail: michel.danis@iut.u-bordeaux1.fr

E. Patterson et al. (eds.), *Composite Materials and Joining Technologies for Composites, Volume 7: Proceedings of the 2012 Annual Conference on Experimental and Applied Mechanics*, Conference Proceedings of the Society for Experimental Mechanics Series 44, DOI 10.1007/978-1-4614-4553-1_6,

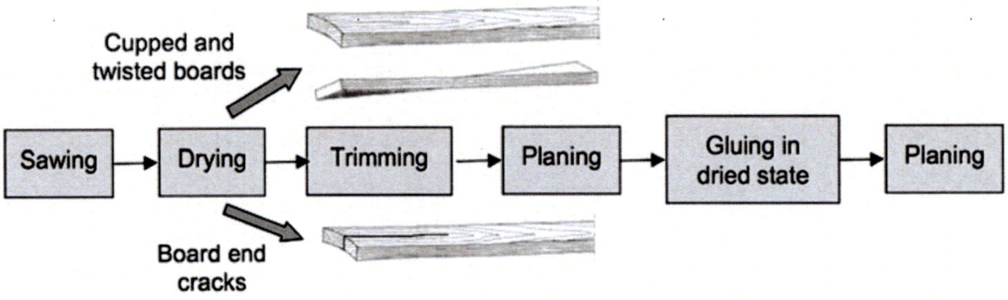

Fig. 6.1 Dry gluing manufacturing process

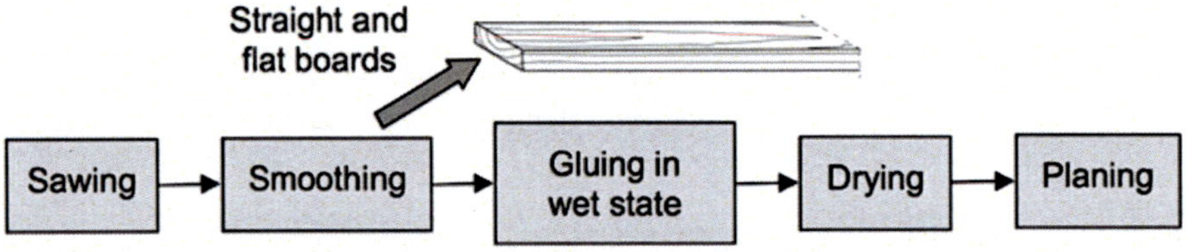

Fig. 6.2 Green gluing manufacturing process

moisture contents (above saturation point). The drying step induces different distortions and cracks of wood. An edging is necessary to remove these cracks, which induces a reduction of boards lengths and a loss of sawing yield. Thus, a planning before flat gluing is required, which reduces the material yield even further. In a distinct way, the green gluing process is carried out using straight and flat boards for which only smoothing is required, providing less material loss.

More than the technical benefits brought by the green gluing process, there are important advantages in the quality of the final product which deserve to be discussed. From previous research it is well known that several wood characteristics that influence the structural properties of sawn timber vary in a distinct way according to their position in the lumber [2–4]. In this way, it has been shown that the modulus of elasticity in softwood increases significantly from pith to outwards of logs sections. Since there is a correlation between the stiffness and strength in wood, this means that the strength at the bark of a log is larger than at the pith. Accordingly, side boards possess excellent structural properties. Added value would be achieved when using such boards as laminations in engineered wood products rather than for packaging. Källsner and Pettersson [5] showed that the performance of bond lines between wet-glued pairs of side boards was very good and that the strength of the wet-glued side boards pairs was better than for corresponding structural timber. Serrano et al. [2] found that green-glued bonds can fulfill the strength requirement for glulam.

The green gluing process also permits to get a best shape stability during drying. Serrano et al [2] found that the shape stability of green glued laminated beams after drying was satisfying. Petersson et al. showed that there was a potential for improvement as regards the shape stability of the glued products [6]. Ormarsson [7] simulated the distortions that occur in sawn timber and glued products during drying. He found optimal configurations that provide fewer distortions of the products during the step of drying. These numerous drying models were developed to predict the deformations and stresses of wood, but few were confirmed by experimental tests. Svensson and Martennsson [8] studied the fundamental hygro-mechanical behaviour of wood applied on convective drying of sawn timber. A material model describing the behaviour of wood during drying was used to simulate drying stresses in timber. The model was then confirmed by experimental destructive tests. Several destructive tests may be used to study the material behavior during drying. The best known techniques to evaluate the drying stresses are prong tests, slicing or dissecting small specimens. These tests provide a snapshot of differential shrinkage in the moisture gradients, which is related to the stress gradients in the boards at different drying steps. For an organic material like wood, the measurement of hygromechanical properties will be more efficient if there is no interactions between the material and the measurement system.

One of the most promising experimental techniques for this kind of problems is noncontact optical measurement based on Digital Image Correlation principles (DIC) [9]. This techniques provides full-field strain maps of materials under various loading regimes. Kang et al. [10, 11] show that DIC units provide valuable information about the development of drying strains and checks under simulated kiln drying conditions. It could also provide efficient results to validate theoretical simulations of wood drying.

The objective of our study is to visualize and measure the progressive deformation in a green – glued wood specimen subjected to drying conditions using the Digital Image Correlation technique. A test apparatus was designed in order to dry a green-glued engineered wood product. The specimens were partially sealed so that the deformation that normally occurs in the middle sections of the dried pieces of lumber could be measured.

6.2 Materials and Methods

The material selected is Maritime Pine (Pinus Pinaster) from the forest of Landes in the South West of France. The selected boards are taken from a common industrial sawmill. Clear wood was chosen from these boards. The thickness of the boards after surfacing is 30 mm and their width is 120 mm. The boards are all well above the fiber saturation point (30% moisture content) before gluing. The two parts of the engineered wood product are cut from the same board. The orientation of the growth rings is chosen in order to get advantageous shape stability according to the previous studies [4]. A patented one-component Polyurethane for green gluing has been used to glue these boards.

In order to measure the shape stability of the specimen in drying conditions, moisture gradients in midsections of regular beams are simulated in the fiber direction. We seal all but the top and bottom surfaces with an Emulsion Polymer Isocyanate (EPI) adhesive. In this way, the moisture transport in the samples is restricted to one direction. It also allows a more accurate measurement of the deformations in the cross section. The final dimensions of the specimen are 60 mm in thickness, 120 mm in width and 100 mm in length.

The experimental apparatus consists of a climate chamber and an optical measurement system (Fig. 6.3). A polycarbonate glass replaces the original door of the climate chamber. The measurement can be done without any interaction with the hot and humid environment of the chamber around the sample. Containers are filled with water in order to get a satisfying relative humidity in the chamber during drying. The temperature and relative humidity are recorded all along the drying test. The conditions inside the chamber are stabilised 50°C and 70% relative humidity resulting in about 11% equilibrium moisture content. The deformation of wood is associated with its moisture content. A monitor specimen is placed in the climate chamber and is daily weighed and measured. This measurement permits us to get the evolution of the moisture content of the studied sample. Indeed, the use of an optical measurement method does not allow any moves of the sample during the tests without disturbing the measurement.

All deformations are determined with a VIC3D optical measurement system. This system consists of two triggered cameras which are focused on the exposed specimen cross section in order to provide stereoscopic images of the specimen deformations. A random black speckle pattern on the surface of the specimen is necessary to enhance the accuracy of optical measurement. This pattern is obtained by spray-painting of the cross section exposed to the cameras. Analytical software based on the DIC principle allows determinations of strains on the surface of the specimens by comparing the positions of subsets of the speckle pattern in the consecutive images. The images present a resolution of 67 μm/pixel. The calibration is

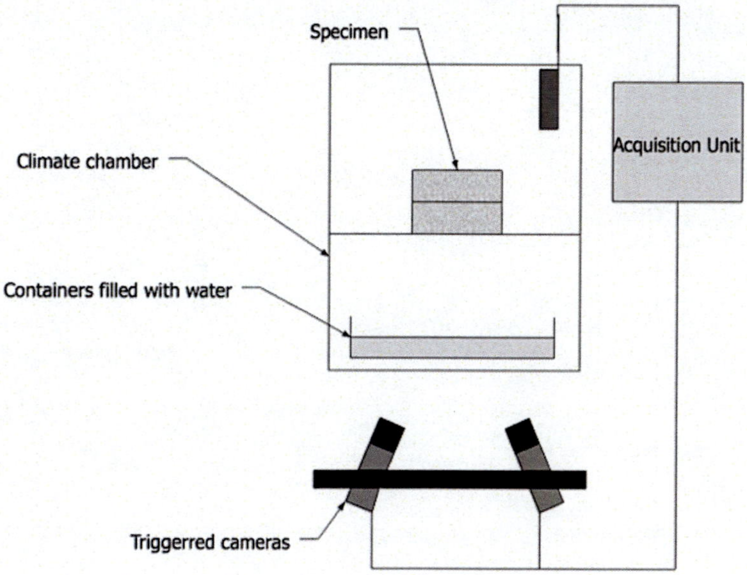

Fig. 6.3 Simplified diagram of the experimental set-up

done through the polycarbonate glass, the projection error given by the calibration results is 0.0255 pixel. The subset size is 29 pixels and the step is 7 pixels. The maximal interval confidence for the match is 0.00426 mm.

6.3 Results and Discussions

Moisture contents (MC) during drying are predicted from a monitor specimen that is sawn from the same board and has the dimensions of the studied specimen. The initial and final moisture contents are respectively 60% and 16.2% for the monitor specimen and 55% and 16,6% for the specimen that was studied. The DIC software calculates all components of local displacements and local strains vectors on the cross-section surface. It allows visual inspection and a quantitative analysis of strains development on the cross-section of the specimen. The principal strains εxx and εyy roughly correspond to the radial and tangential directions. The use of the principal strains components ε1 and ε2 presents much better correspondence with the radial and tangential strains, respectively [10]. The Fig. 6.4 shows the mean development of the principal strains occurring during drying the specimen. We observe a development of the principal strains, although the average moisture content of the sample is well above the fiber saturation point. It is explained by the formation of a moisture content gradient inside the wood during drying. Indeed, whereas the core of the specimen is still wet, its top and bottom surfaces present a moisture content that is under the fiber saturation point. That leads to the development of stresses gradients, which are revealed by strains evolution. During the early period of drying, the ratio between tangential and radial strains is much more than expected. At 30% moisture content, the ratio radial-to-tangential deformations is about 1/3, whereas the literature presents a ratio around 1/2. This ratio decreases during the test to approach the theoretical value. The slight decreases of deformations after 250 h of drying could be explained by the relaxation of stresses inside our specimen. The explanation is the stabilization of the moisture content; the specimen is at its equilibrium and no more stresses gradients occur inside the wood.

The Figs. 6.5 and 6.6 show the full field maps of major strain ε1 (radial strains) and minor strains ε2 (tangential strains) at different moisture contents.

These measurements roughly correspond to the wood drying theory. The edge surfaces are drying faster than the core of the sample. At early stage of drying, we observe compressive strains on the edge surfaces. Nevertheless, the core of the sample is still wet and prevents the surface to shrink. This induces tensile stresses in the edge layers and compressive stresses in the core layers. When the core of the sample starts to dry, this behavior is reversed.

The color-coded maps of the Fig. 6.7 represents the development of shear strains during the test. It illustrates the stresses gradients that occur in the specimen. We can observe that the shear strains around the adhesive interface increased much more than in other zones of the sample. The Fig. 6.8 shows that these strains are 10 times higher in around the adhesive

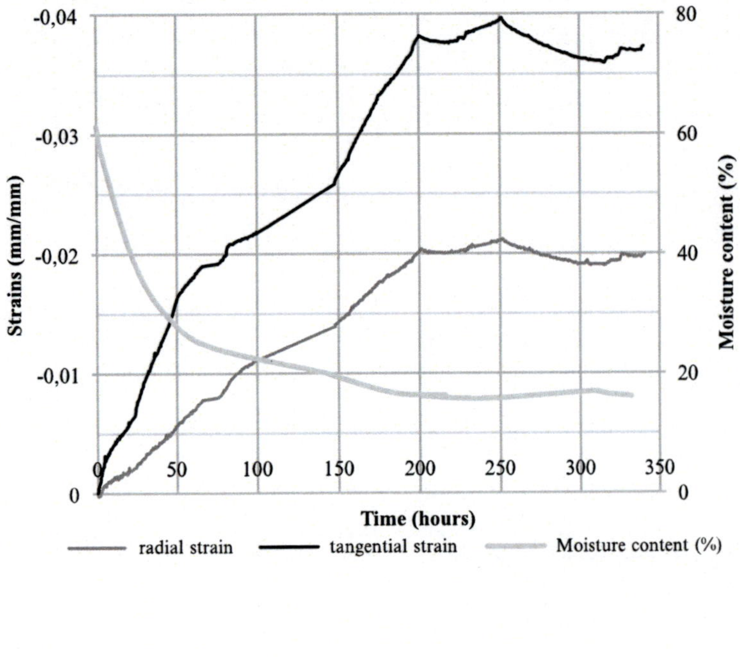

Fig. 6.4 Plots of the mean values of the principal strains and evolution of the moisture content during drying

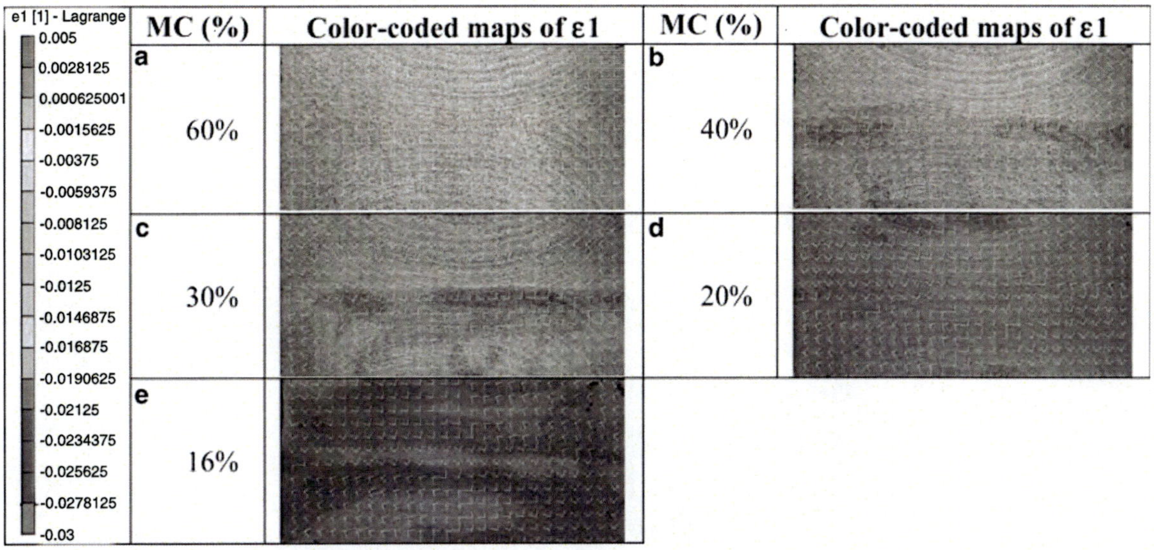

Fig. 6.5 Full field maps of major strains ε1 (radial strains) for several moisture contents

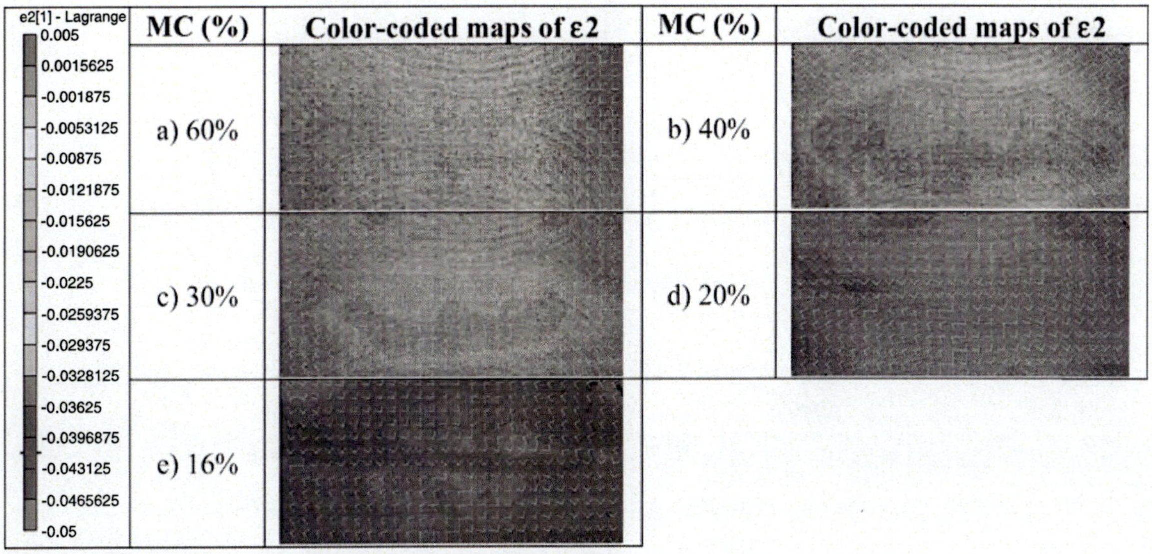

Fig. 6.6 Full field maps of minor strains ε2 (tangential strains) for several moisture contents

interface than in the wood. This behavior is due to the adhesive, which restrains the shrinkage of the wood in this zone. This induces shear stresses on the adhesive interfaces. It can be notified that these stresses did not lead to an adhesive failure.

The strains gradients in different zones of the thickness of the specimen have been studied. The plots of the strains profiles at different moisture content for lines A, B and C (Fig. 6.9) are presented Fig. 6.10.

We notice that the strains gradients increase significantly under the fiber saturation point. In the central zone (line B), the shear strains around the bond line are much more higher than in the edge zones where it almost does not exist. It may be explain by the difference in the growth rings orientation. Indeed, the growth rings in the middle of the section are parallel to the adhesive interface. It induces a larger tangential shrinkage than in the other zones, where the growth rings form an angle around 45°, thus the shrinkage in these zones is supposed to be smaller. In comparison, the strains on the edge zones grow as the distance from the adhesive interface increase.

It is important to mention that although there is a strain concentration around the adhesive interface, it does not induce any damage. The stresses that occur close to the bond line are less that the ultimate strength of the material in this area.

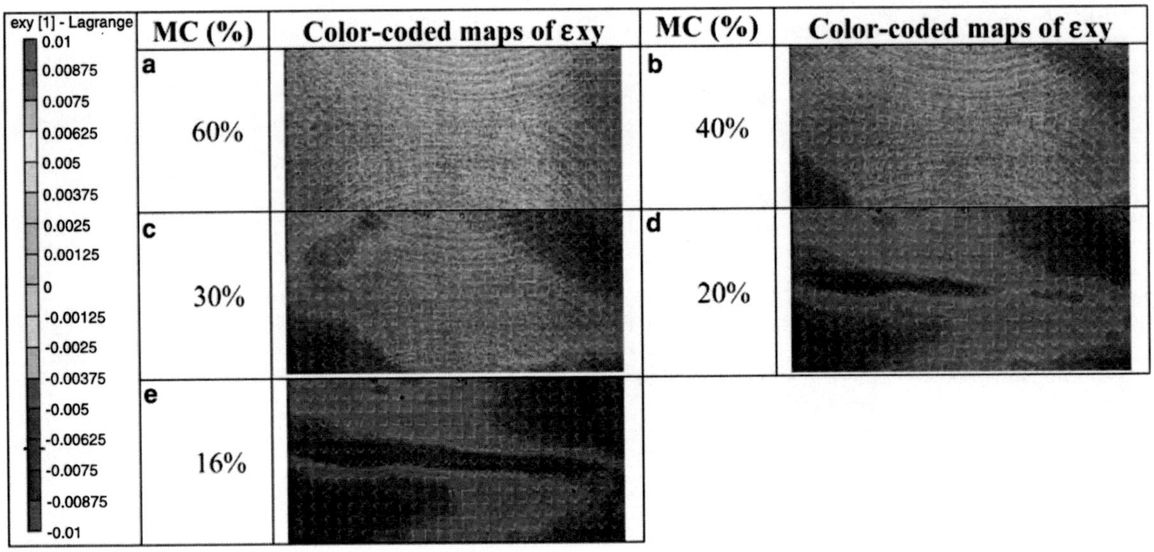

Fig. 6.7 Full field maps of shear strains εxy for several moisture contents

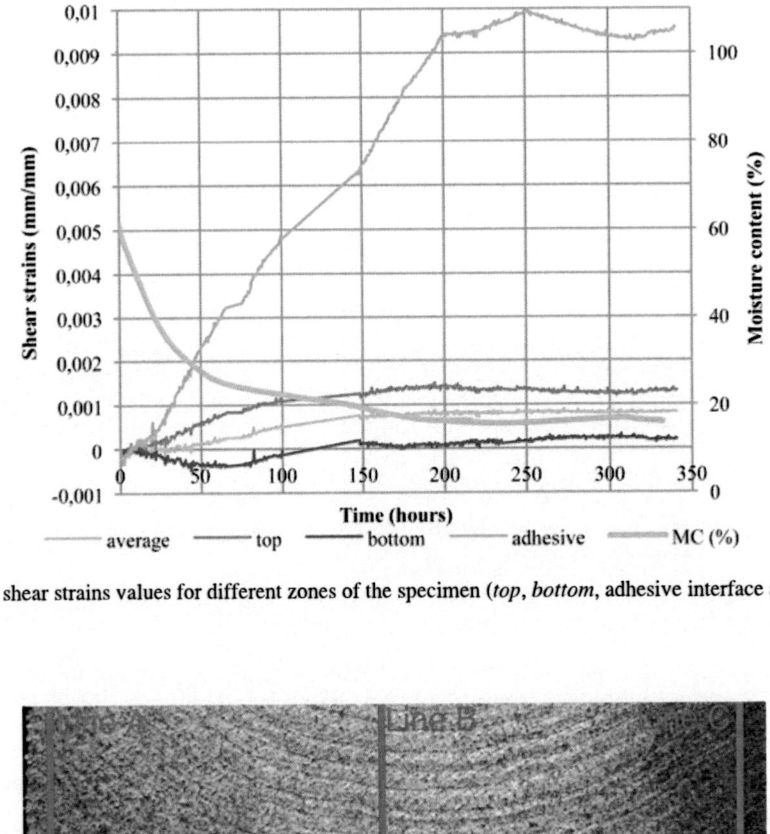

Fig. 6.8 Plots of the mean shear strains values for different zones of the specimen (*top*, *bottom*, adhesive interface and average) and evolution of the moisture content

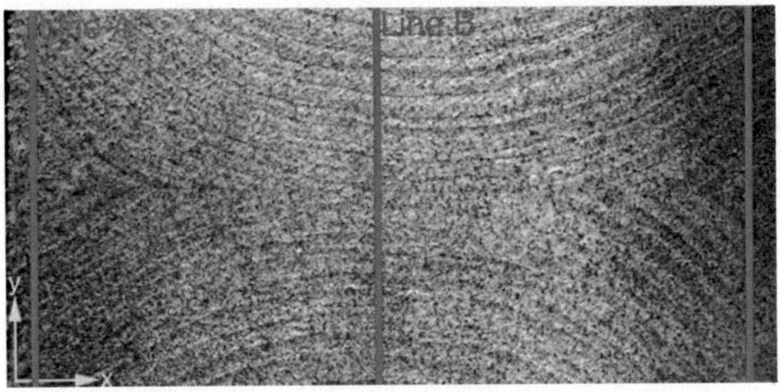

Fig. 6.9 Position of the *lines* for the study of the strains gradients in the specimen thickness

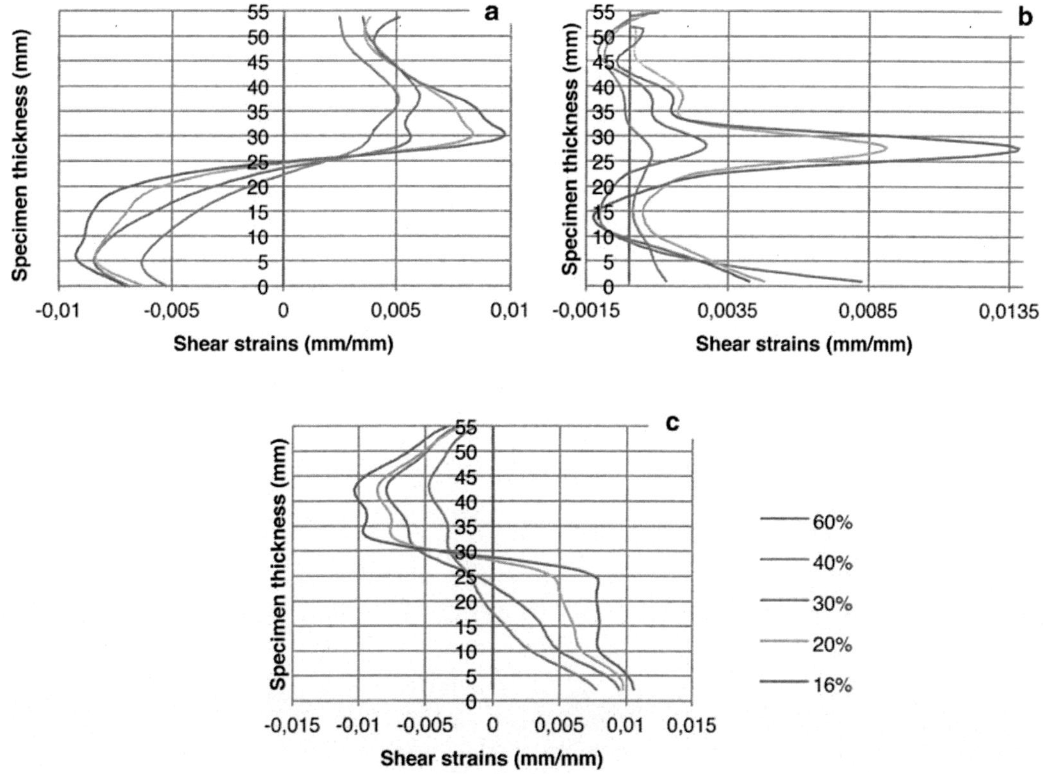

Fig. 6.10 Shear strains profiles across the specimen thickness at different moisture contents in the different zones ((**a**) line A, (**b**) line B, (**c**) line C)

The behavior of timbers frame structure is highly dependent on this point. Other experiences are necessary to confirm this adhesive behavior, like moisture content cycles and the mechanosorption effect on the joint.

It is also important to mention that the strains profiles are symmetrical in respect with the bond line, meaning that our drying is homogeneous.

6.4 Conclusion

The behavior of a green-glued wood product has been investigated using the DIC system as a measurement system. The pattern of strain curves across the thickness of the specimen reveals a symmetrical behavior on each part of the glued product. Thus the apparatus designed for this study provides a homogeneous drying of the sample. The measurement technique based on the DIC principles brings direct quantification of strains of drying green-glued wood products. The color-coded maps of principal strains clearly illustrate the progress of the deformation gradients in wood. A distinction between behavior of wood and of adhesive interface may be done. The measurement provides efficient results that can be used for a model validation. The shrinkage and swelling measurements of wood under different climate cycles were not processed, either the stress modeling.

References

1. Daudé G (2002) Synthèse d'une colle pour le collage du Pin Maritime vert, Brevet FR 2842818
2. Serrano E, Oscarsson J, Enquist B, Sterley M, Petersson H, Källsner B (2010) Green-glued laminated beams – high performance and added value. In: Proceedings (Poster session) of 11th world conference on timber engineering, Riva des Garda, 21–24 June, pp 829–830
3. Pommier R, Garbay G, Elbez G (2009) Rapport scientifique projet ABOVE

4. Steffen A, Johansson C-J, Wormuth E-W (1997) Study of the relationship between flat wise and edgewise moduli of elasticity of sawn timber as a means to improve mechanical strength grading technology. Holz als Roh- und Werkstoff 55(2–4):245–253
5. Källsner B, Pettersson H (2005) Bending strength of timber beams made of green glued pairs of side boards. In: COST Action E34 Bonding of wood, conference/workshop on green gluing of wood – process – products – market at SP Swedish National Testing And Research Institute, Boras, 7–8 Apr 2005
6. Petterson H, Källsner B, Ormarsson S (2005) Shape stability of laminated planks made of green-glued pairs of side boards. In: COST action E34 bonding of wood, conference/workshop on green gluing of wood – process – products – market at SP Swedish National Testing and research Institute, Boras, 7–8 Apr 2005
7. Ormarsson S (2000) A numerical study of the shape stability of sawn timber subjected to moisture variation: part 3: influence of annual ring orientation. Wood Sci Technol 34(3):207–219
8. Svensson S, Martennsson A (2002) Simulation of drying stresses in wood: part II. Convective air drying of sawn timber. Holz als Roh- und Werkstoff 60(1):72–80
9. Muszynski L (2006) Empirical data for modeling: methodological aspects in experimentation involving hygromechanical characteristics of wood. Dry Technol 24(9):1115–1120
10. Kang H-Y, Muszynski L, Milota M-R (2011) Optical measurement of deformations in drying lumber. Dry Technol 29(2):127–134
11. Kang H-Y, Muszynski L, Milota M-R (2006) Optical measurement of non-uniform drying strains and drying check formation in refractory species. In: Proceedings of FPS conference on quality drying for the 21st century, Bellingham, 15–17 Nov, p 240

Chapter 7
Cylindrical Bending of Bonded Layered Thick Composites

Sayed A. Nassar, Jianghui Mao, Xianjie Yang, and Douglas Templeton

Abstract In this paper, a mathematical model is developed to describe the mechanical behavior of adhesively bonded multi-layer thick composites. Based on the proposed novel deformation field, the stress field is derived for the layered composite joint. Unlike conventional shell theory, the proposed model developed takes into account the transverse shear deformation in the composite and adhesive layers; the adhesive layers are normally much more vulnerable to cracking. The validity and accuracy of the proposed model is investigated by comparing the results with finite element prediction for various loading scenarios.

Keywords Adhesive bonding • Thick composite • Layered composite • Finite element analysis

7.1 Introduction

Compared with mechanical joint, the adhesive bonded joint has lots of advantages. Basically, it could join materials of all sizes and shapes, for example, the very thin and quite delicate material, such as foils, could be joined to each other or to heavier parts. Also, the adhesive bonding can ensure smooth, unbroken surface contours without protrusions, gaps or holes. Moreover, because continuous bonding can be obtained throughout a joint, good load distribution and fatigue resistance can be obtained. In view of these merits of adhesive-bonded joint, it's worthy to carry out research on this area.

In fact, a lot of research has already been done on this area, either experimentally, or theoretically. Taking into account of adhesive plasticity, adherend stiffness imbalance and thermal mismatch between the adherends, Hart-Smith [1] developed an explicit analytical solution for the static load carrying capacity of double-lap adhesive-bonded joints. Markolefas et al. [2] extended the model to account for the cyclic loading of joint. Mortensen et al. [3] presented a unified approach for the analysis and design of adhesive bonded joints. Roman [4] carried out extensive experimental study to determine the mechanical properties of various adhesives used for joining pultruded GFRP. Gustafson et al. [5] reported the influence of adhesive parameters on the outcome of cohesive zone finite element simulations. Some researchers [1, 6] also considered damage failure of adhesive bonding. However, all of these mentioned researches were either single lap or double lap joint, whether or not these analyses would be applied to multilayer joint consists of layers greater than two is still unknown.

There were some studies considering arbitrary number of bonded layers, such as Reissner [7], Mindlin [8], and Whitney [9]. Another example is Nayfeh et al. [10], in which three materials are considered to form the multi-layer joint. Although layers larger than 2 were considered, they didn't take the adhesive layers into account, i.e. the effect of adhesive layers on joint was neglected in their models. Since the adhesive layers are much weak than adherend layers in general, neglecting of it will cause unpractical prediction for high loading condition.

To achieve certain strength requirement, it usually requires large number of stacking of same plate in a joint, for example, the bullet-proof window. In the first place, the transparency is desired, secondly, it requires that the first few lays withstand the impact of high speed bullet, it also need some layers to stop the cracking and hold the fragment in place to prevent

S.A. Nassar (✉) • J. Mao • X. Yang
Department of Mechanical Engineering, Fastening and Joining Research Institute (FAJRI), Oakland University, Rochester, MI 48309, USA
e-mail: nassar@oakland.edu

D. Templeton
US Army TARDEC, Warren, MI 48397, USA

E. Patterson et al. (eds.), *Composite Materials and Joining Technologies for Composites, Volume 7: Proceedings of the 2012 Annual Conference on Experimental and Applied Mechanics*, Conference Proceedings of the Society for Experimental Mechanics Series 44,
DOI 10.1007/978-1-4614-4553-1_7, © The Society for Experimental Mechanics, Inc. 2013

personal injury behind the window. In a view of this application, a mathematical model is derived to delineate the mechanical behavior of adhesive bonded multi-layer thick joint consists of eight layers of STARPHIRE glass in this paper. The joint is approximate 108.6 mm thick, and made of eight layers of STARPHIRE glass and they are adhesive bonded by Polyurethane via an autoclave laminating process. The model accounts for the transverse shear deformation and adhesive layers. Details of the developed model are explained in Sect. 7.2. Case study is carried out in Sect. 7.3, which also shows the validity and accuracy of the developed model when comparing to Finite element analysis results using shell theory and continuum theory. Conclusions are drawn in Sect. 7.4. The current study is applicable to static loading situation, future investigations will address on the dynamic loading and impact.

7.2 Mathematical Model

In this section, the problem of interest is depicted first. A novel deformation field is adopted, using continuum mechanics method, the closed form solution is sought for deformation variables, the other variables, such as strain, stress, force and moment, then, can be derived during loading history.

7.2.1 Problem Statement

The problem considered in this paper is an adhesive bonded multi-layer thick joint as shown in Fig. 7.1. It consists of eight layers of adherends made of STARPHIRE glass, and bonded alternately by seven layers of adhesive made of polyurethane. Each adherend layer has a thickness of 12.7 mm, and 1 mm for each adhesive layer. The joint is subjected to a load along Z direction.

7.2.2 Formulation

Considering that the joint shown in Fig. 7.1 is simple supported at both end, and subjected to a load at top surface, under cylindrical bending, the field variables are irrelevant to y component. A novel deformation field is used as shown in Eqs. 7.1 and 7.2.

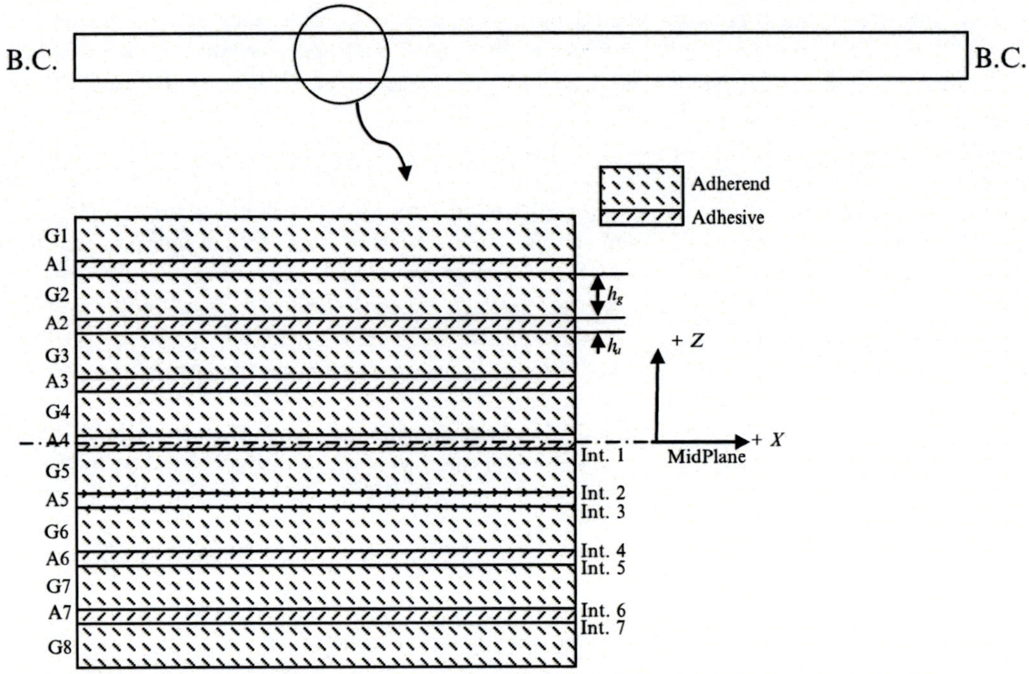

Fig. 7.1 Adhesive bonded multi-layer thick joint

$$u = (z - h_k)\phi(x) \tag{7.1}$$

$$w = w(x) \tag{7.2}$$

where h_k is the central height of each layer,

For finite deformation, the strain displacement relations are given by

$$\varepsilon_x = \frac{\partial u}{\partial x} = (z - h_k)\phi_{,x} \tag{7.3}$$

$$\varepsilon_{xz} = \frac{\partial u}{\partial z} + \frac{\partial w}{\partial x} = \phi + w_{,x} \tag{7.4}$$

Other strain components not specified are zero.

Hooke's law gives the stress strain relation as

$$\begin{bmatrix} \sigma_x \\ \sigma_y \\ \sigma_z \\ \sigma_{yz} \\ \sigma_{xz} \\ \sigma_{xy} \end{bmatrix} = \begin{bmatrix} C_{11} & C_{12} & C_{13} & 0 & 0 & 0 \\ C_{21} & C_{22} & C_{23} & 0 & 0 & 0 \\ C_{31} & C_{32} & C_{33} & 0 & 0 & 0 \\ 0 & 0 & 0 & C_{44} & 0 & 0 \\ 0 & 0 & 0 & 0 & C_{55} & 0 \\ 0 & 0 & 0 & 0 & 0 & C_{66} \end{bmatrix} \begin{bmatrix} \varepsilon_x \\ \varepsilon_y \\ \varepsilon_z \\ \varepsilon_{yz} \\ \varepsilon_{xz} \\ \varepsilon_{xy} \end{bmatrix} \tag{7.5}$$

7.2.3 Governing Equations

Neglecting the rotational inertia terms, the stress tensor for the k th. layer of the adhesive bonded joint must satisfy the following relations

$$\frac{\partial \sigma_x^{(k)}}{\partial x} + \frac{\partial \sigma_{xy}^{(k)}}{\partial y} + \frac{\partial \sigma_{xz}^{(k)}}{\partial z} = 0 \tag{7.6}$$

$$\frac{\partial \sigma_{xy}^{(k)}}{\partial x} + \frac{\partial \sigma_y^{(k)}}{\partial y} + \frac{\partial \sigma_{yz}^{(k)}}{\partial z} = 0 \tag{7.7}$$

$$\frac{\partial \sigma_{xz}^{(k)}}{\partial x} + \frac{\partial \sigma_{yz}^{(k)}}{\partial y} + \frac{\partial \sigma_z^{(k)}}{\partial z} = 0 \tag{7.8}$$

We proceed by integrating Eq. 7.6 with respect to z, considering Eq. 7.5, one obtains

$$\frac{\partial N_x}{\partial x} = 0 \tag{7.9a}$$

where

$$N_x = \int_{-h/2}^{h/2} \sigma_x^{(k)} dz = (B_1 - h_k A_1)\phi_{,x} \tag{7.9b}$$

$$B_1 = \int_{-h/2}^{h/2} C_{11} z dz \tag{7.9c}$$

$$A_1 = \int_{-h/2}^{h/2} C_{11} dz \tag{7.9d}$$

Now multiplying Eq. 7.6 by z and integrating with respect to z over the joint thickness yields

$$\frac{\partial M_x}{\partial x} - Q_x = 0 \tag{7.10}$$

where

$$M_x = \int_{-h/2}^{h/2} \sigma_x^{(k)} z dz = \phi_{,x}(C_1 - h_k B_1) \tag{7.10a}$$

$$C_1 = \int_{-h/2}^{h/2} C_{11} z^2 dz \tag{7.10b}$$

$$Q_x = \int_{-h/2}^{h/2} \sigma_{xz}^{(k)} dz \tag{7.10c}$$

Integrating Eq. 7.8 with respect to z,

$$\frac{\partial Q_x}{\partial x} + q = 0 \tag{7.11a}$$

$$q = \sigma_z\left(\frac{h}{2}\right) - \sigma_z\left(-\frac{h}{2}\right) \tag{7.11b}$$

On the other hand, from Eq. 7.10c,

$$Q_x = \int_{-h/2}^{h/2} \sigma_{xz}^{(k)} dz = (\phi + w_{,x}) A_5 \tag{7.12a}$$

$$A_5 = \int_{-h/2}^{h/2} C_{55} dz \tag{7.12b}$$

Following [7–9], we introduce a parameter k in this constitutive relation for both shear force and moment resultants.

$$\begin{bmatrix} Q_x \\ M_x \end{bmatrix} = k \begin{bmatrix} A_5 & 0 \\ 0 & C_1 - h_k B_1 \end{bmatrix} \begin{bmatrix} \phi + w_{,x} \\ \phi_{,x} \end{bmatrix} \tag{7.13}$$

Using Eqs. 7.9b and 7.13, the following equations can be derived from Eqs. 7.9a, 7.10, and 7.11a.

$$(B_1 - h_k A_1)\phi_{,xx} = 0 \tag{7.14}$$

$$(C_1 - h_k B_1)\phi_{,xx} - A_5\phi - A_5 w_{,x} = 0 \tag{7.15}$$

$$A_5 k \phi_{,x} + A_5 k w_{,xx} + q = 0 \tag{7.16}$$

q is the applied load at the adhesive bonded joint. Since any function can be expanded using Fourier sin series, we can express q as:

$$q(x) = \sum_{m=1}^{\infty} q_m \sin\left(\frac{m\pi}{L} x\right) \tag{7.17}$$

A set of solution satisfies the boundary conditions could be

$$\phi(x) = \sum_{m=1}^{\infty} a_m \cos\left(\frac{m\pi}{L}x\right) \tag{7.18}$$

$$w(x) = \sum_{m=1}^{\infty} b_m \sin\left(\frac{m\pi}{L}x\right) \tag{7.19}$$

Substituting Eqs. 7.17, 7.18, and 7.19 into Eqs. 7.15 and 7.16, the following linear equations could be obtained in matrix form.

$$\begin{bmatrix} \mathbf{H}_1 & \mathbf{0} & \cdots & \mathbf{0} \\ \mathbf{0} & \mathbf{H}_2 & \cdots & \mathbf{0} \\ \vdots & \vdots & \ddots & \vdots \\ \mathbf{0} & \mathbf{0} & \cdots & \mathbf{H}_m \end{bmatrix} \begin{bmatrix} \mathbf{X}_1 \\ \mathbf{X}_2 \\ \vdots \\ \mathbf{X}_m \end{bmatrix} = \begin{bmatrix} \mathbf{Q}_1 \\ \mathbf{Q}_2 \\ \vdots \\ \mathbf{Q}_m \end{bmatrix} \tag{7.20a}$$

where

$$\mathbf{H}_m = \begin{bmatrix} C_1\left(\frac{m\pi}{L}\right)^2 + A_5 & A_5\frac{m\pi}{L} \\ A_5k\frac{m\pi}{L} & A_5k\left(\frac{m\pi}{L}\right)^2 \end{bmatrix}, \quad m = 1, 2, \cdots, \tag{7.20b}$$

$$\mathbf{X}_m = \begin{bmatrix} a_m \\ b_m \end{bmatrix}, \quad m = 1, 2, \ldots, \quad \mathbf{Q}_m = \begin{bmatrix} 0 \\ q_m \end{bmatrix}, \quad m = 1, 2, \cdots. \tag{7.20c}$$

Solution to Eq. 7.20a is given by

$$\mathbf{X}_m = \begin{bmatrix} a_m \\ b_m \end{bmatrix} = \frac{1}{k}\begin{bmatrix} -\frac{1}{C_1}\frac{L}{m\pi} \\ \frac{1}{A_5} + \frac{1}{C_1}\left(\frac{L}{m\pi}\right)^2 \end{bmatrix}\left(\frac{L}{m\pi}\right)^2 q_m, \text{ m} = 1, 2, \cdots. \tag{7.21}$$

7.3 Numerical Illustration

In this section, 108.6 mm $\times$ 1,000 mm joint consisting of eight layers of 12.7 mm thick glass and seven layers of 1 mm thick polyurethane is subjected to two loading scenarios (Fig. 7.2) in order to illustrate the validity of the developed model. In scenario one, the adhesive bonded joint is subjected sin load, and an exponential load is applied in scenario two. k is determined as in [9], and it is found that $k = 39$. The deflection and stress distribution are compared with finite element results. The finite element calculation is carried out using commercial software ABAQUS, two theories are used, i.e. shell theory and continuum theory. For the former one, SC8R element is used, and C3D20 is used for the later one. Totally, there are 28,833 nodes and 6,400 element in the finite element model. Material properties are tabulated in Table 7.1.

7.3.1 Scenario 1: Sinusoidal Load

In this scenario, only 1 term is kept in the Fourier expansion, i.e. $m = 1$. Comparison of deflection in the bottom of joint is shown in Fig. 7.3. From Fig. 7.3, the analytical solution correlates well with shell theory prediction, and the continuum

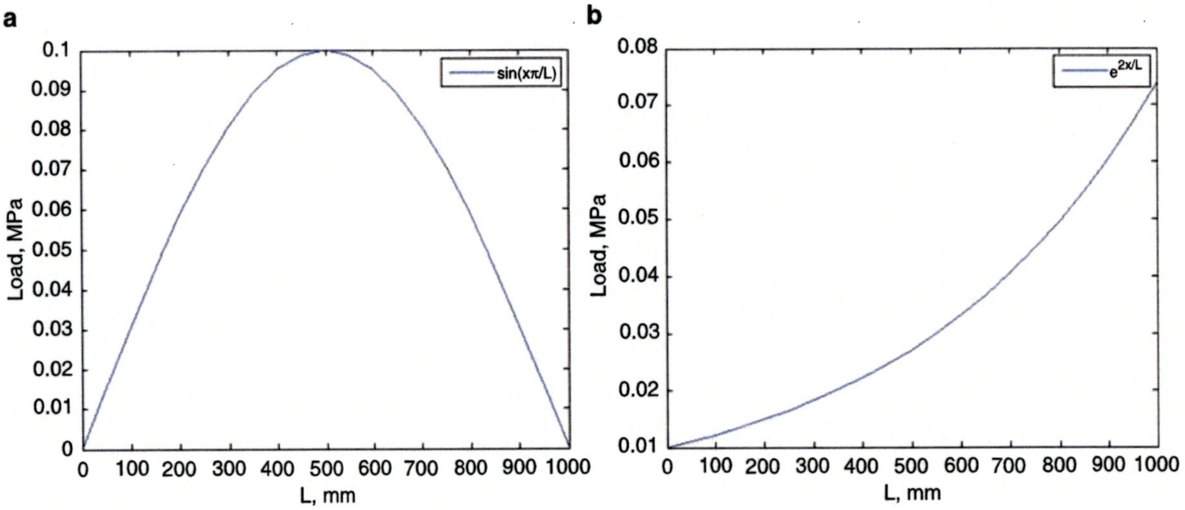

Fig. 7.2 Loading conditions (**a**) scenarios 1 sin load, (**b**) scenarios 2 exponential load

Table 7.1 Material properties

Material	Young's modulus (MPa)	Poisson's ratio
Glass	73,100	0.22
Polyurethane	3.17	0.499

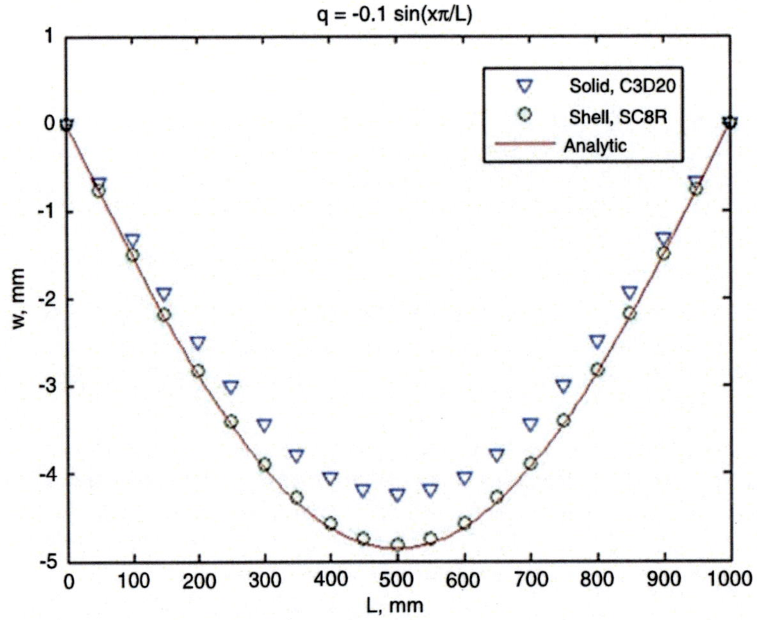

Fig. 7.3 Deflection distribution

theory underestimates the deflection about 11.8% at maximum. Comparison of u distribution in joint bottom is illustrated in Fig. 7.4. The analytical prediction is a little bit lower than the other two predictions in absolute values. Figures 7.5 and 7.6 are the stress distribution in joint bottom predicted by different theories. The analytical solution correlates well with shell theory solution for σ_{xx}, but is higher than the other two solutions for σ_{yy}.

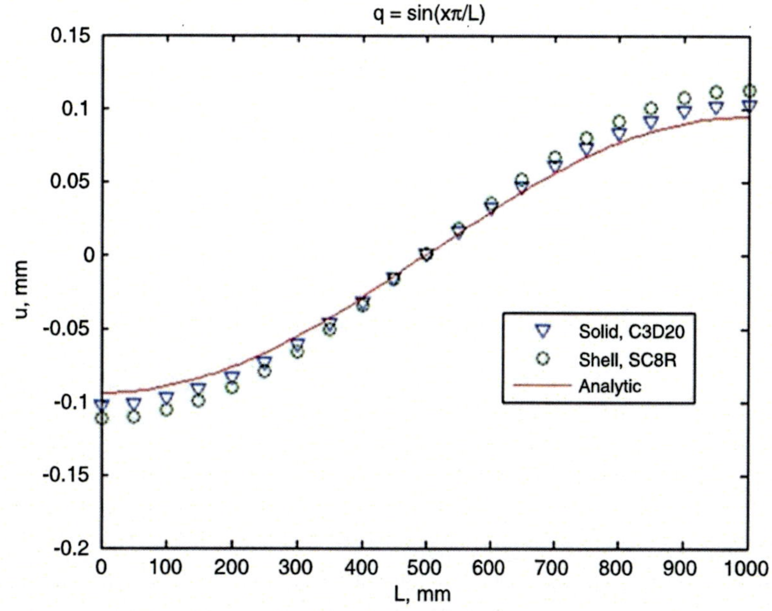

Fig. 7.4 *u* distribution

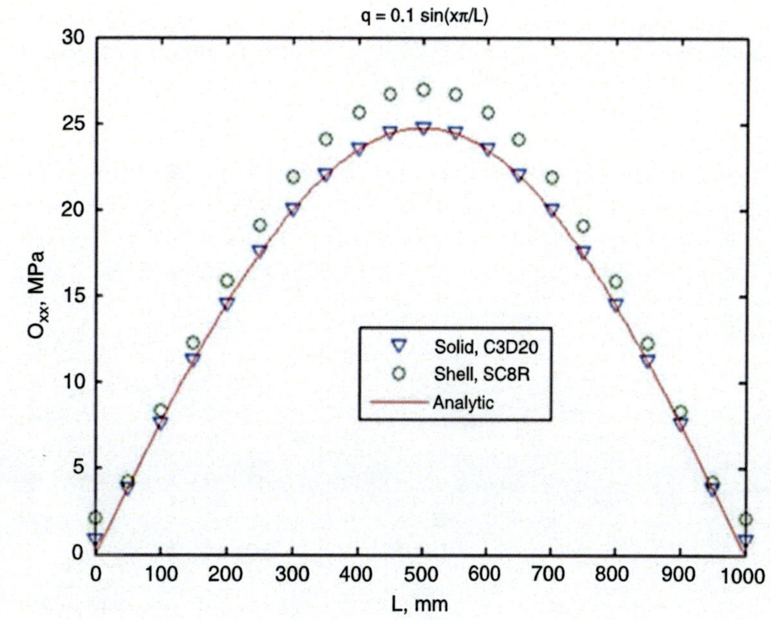

Fig. 7.5 σ_{xx} distribution

7.3.2 Scenario 2: Exponential Load

In scenario 2, the exponential load is expanded using Fourier sin series as

$$q(x) = q_0 e^{2x/L} = q_0 \sum_{m=0}^{\infty} \frac{m\pi}{2} \frac{1 - (-1)^m e^2}{1 + (m\pi/2)^2} \sin\frac{m\pi}{L}x \tag{7.22}$$

where $q_0 = -0.01$ MPa and $m = 10$ (discussed later in Sect. 7.3.3) is used in this scenario.

The comparisons of deformation field in joint bottom are depicted in Figs. 7.7 and 7.8. As in scenario 1, the analytical solution correlates well with shell theory solution for deflection as shown in Fig. 7.7 and is close to continuum theory solution for u as shown in Fig. 7.8. Besides, the maximum deflection is shifted a little bit to the right of the middle as shown in both Figs. 7.7 and 7.8. This could also been seen in Figs. 7.9 and 7.10, where the stress distribution are shown.

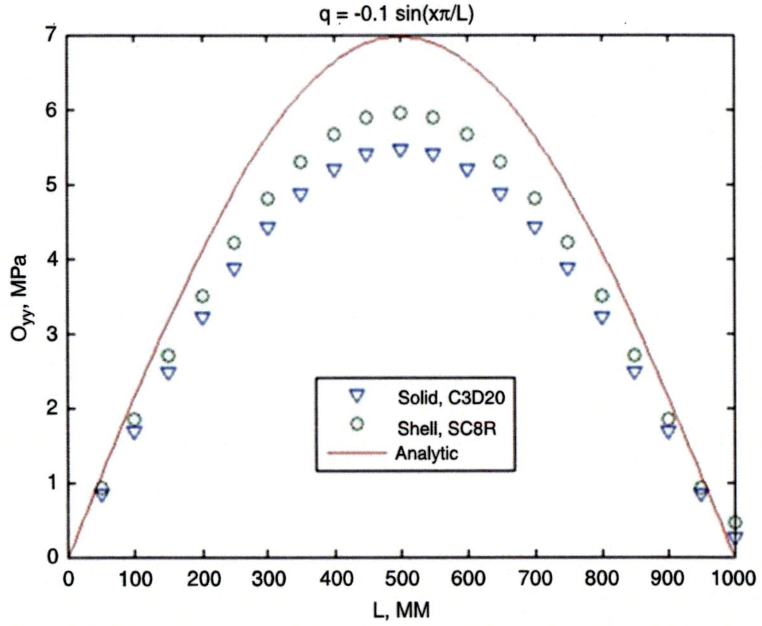

Fig. 7.6 σ_{yy} distribution

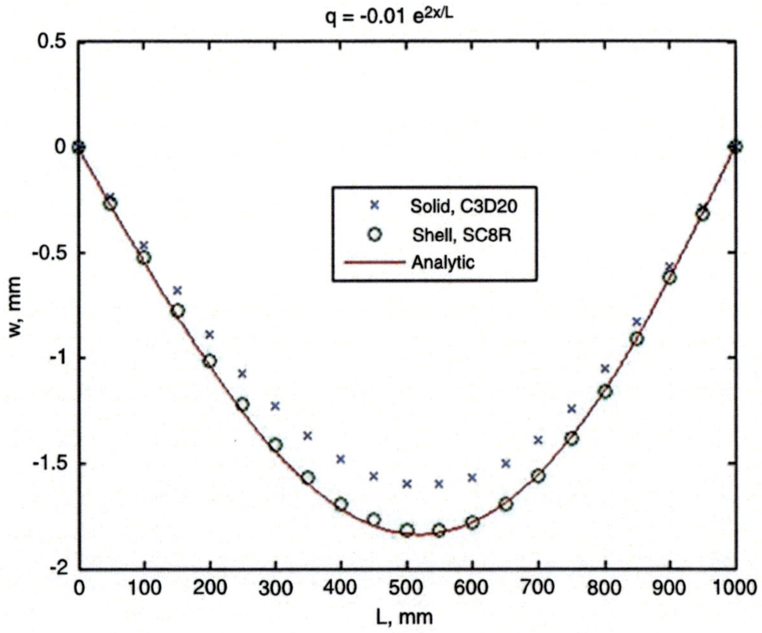

Fig. 7.7 Deflection distribution

The analytically predicted results agrees well with continuum theory for σ_{xx} as shown in Fig. 7.9 and a little higher than the tow theory predictions for σ_{yy} as shown in Fig. 7.10.

7.3.3 *Effect of Fourier Expansion Term m*

Theoretically, to make the function exact, the Fourier expansion should consist of infinite number of terms m. this subsection studies the effect of m under the exponential load condition. Figure 7.11 illustrates the effect of m for exponential load. When $m = 1$, the deflection distribution along joint length is symmetric, when $m = 10$, it shift a little bit to the right of middle, increases to 100 doesn't make the deflection distribution change a lot, so we can conclude that $m = 10$ could be used with high accuracy.

Fig. 7.8 u distribution

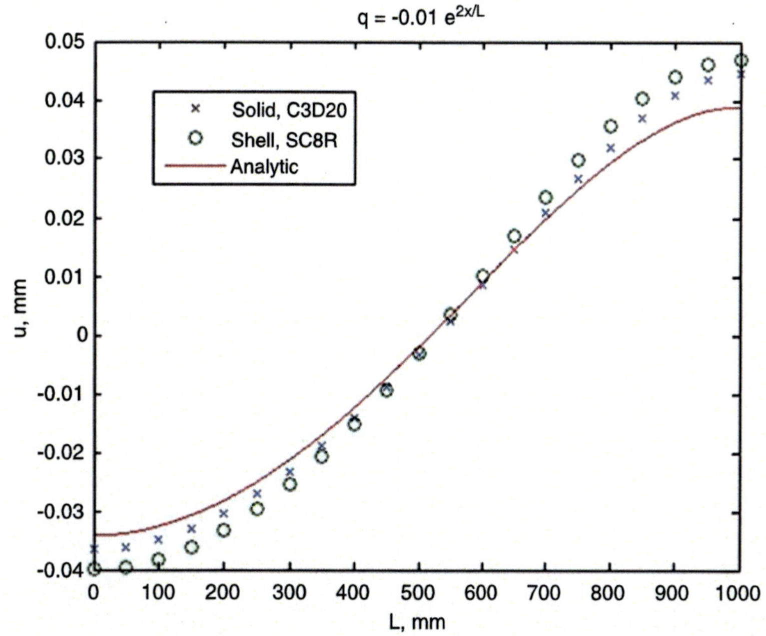

Fig. 7.9 σ_{xx} distribution

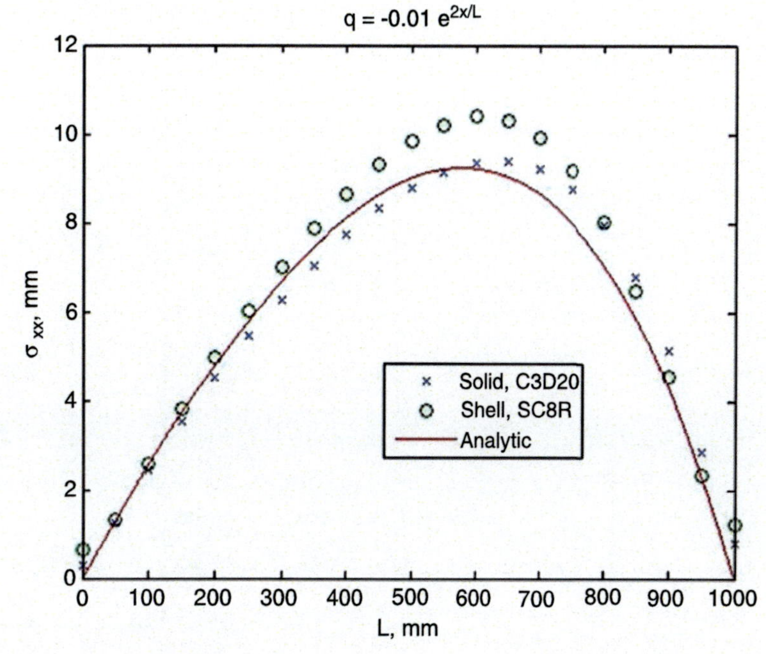

7.3.4 Interfacial Shear Stress

As already pointed out in [9], the calculation of shear stress using the constitutive relation in Eq. 7.5 will cause the inconsistence at the adherend/adhesive interface due to the distinct difference in material properties. This is clearly unrealistic, a practical way to overcome it is calculating shear stress using eqn. outlined in Eqs. 7.6, 7.7, and 7.8.

The calculated shear stresses are compared in Fig. 7.12 for seven interfaces shown in Fig. 7.1. Exponential load and m = 10 are used in the study. From Fig. 7.12, the interface shear stress near the middle plane is lower than the one away from the middle plane. Shear stresses change sign at approximately L = 590 mm. Since the Int. 7 is the furthest interface from the middle plane, it has the highest interface shear stress. The failure is expected to happen at this interface first.

Fig. 7.10 σ_{yy} distribution

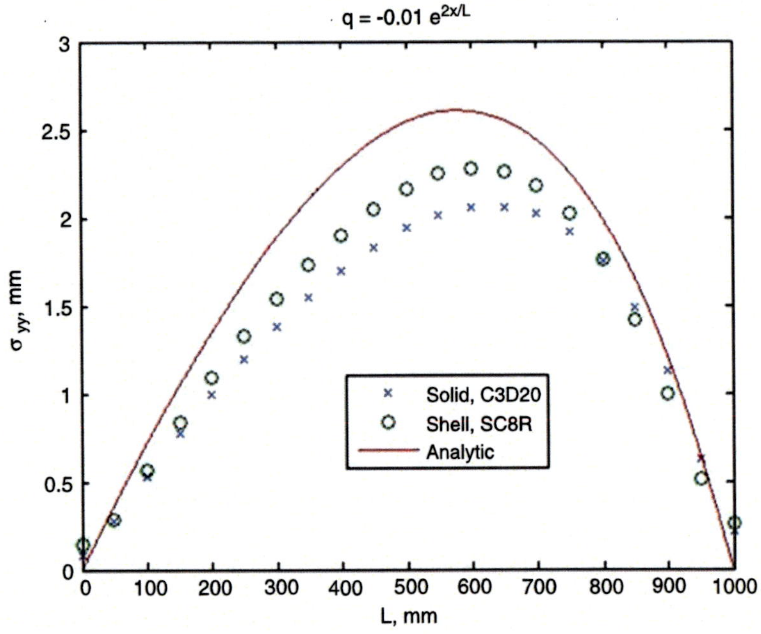

Fig. 7.11 Effect of m

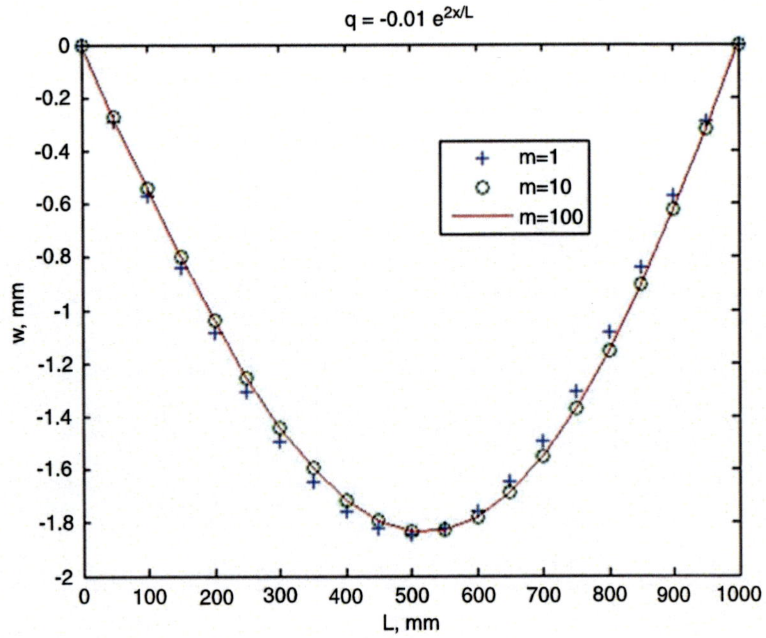

7.4 Conclusion

A high order theory is developed in this paper to delineate the mechanical behavior of adhesive bonded multi-layer thick joint. The model takes into account of transverse shear deformation and the adhesive layers. As the given load function would be expanded by Fourier series, the analytical solution of the problem is obtained using trigonometric functions. The prediction given by the proposed theory correlates well with the finite element results. Study shows that the terms used in the Fourier series for load function doesn't have significant effect on the deformation outcome in exponential loading case. In addition, the failure most likely to occur first is at the furthest interface from the middle plane.

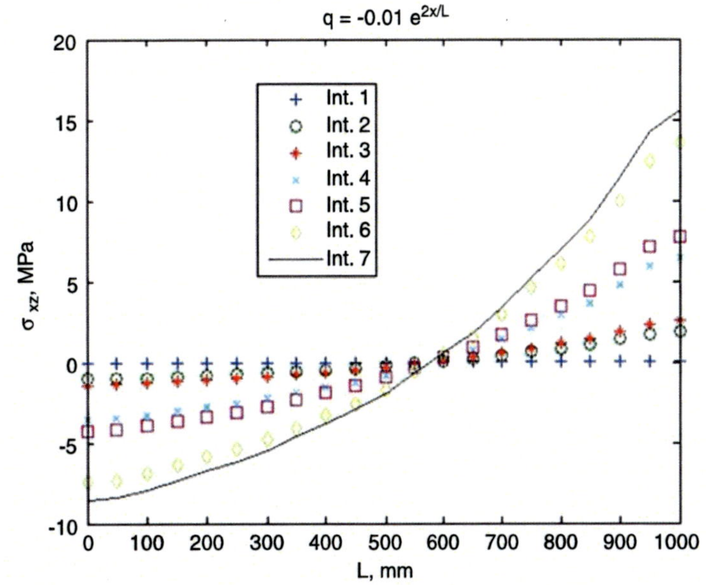

Fig. 7.12 Interface shear stress

References

1. Hart-Smith L (1973) Adhesive-bonded double-lap joints. NASA Langley report CR-112235, Douglas Aircraft Co
2. Markolefas S, Papathanassiou TK (2009) Stress redistributions in adhesively bonded double-lap joints, with elastic-perfectly plastic adhesive behavior, subject to axial lap-shear cyclic loading. Int J Adhes Adhes 29:737–744
3. Mortensen F, Thomsen O (2002) Analysis of adhesive bonded joints: a unified approach. Compos Sci Technol 62:1011–1031
4. Roman JD (2005) Experiments on double-lap joints with epoxy, polyurethane and ADP adhesives. Technical report CCLab2000-Appendix A, B
5. Gustafson PA, Wass AM (2009) The influence of adhesive constitutive parameters in cohesive zone finite element models of adhesively bonded joints. Int J Solid Struct 46:2201–2215
6. Nassar SA, Mao J, Yang X, Templeton D (2011) Effect of adhesives on the mechanical behavior of thick composite joints. In: Proceedings of the ASME 2011 pressure vessels and piping division conference, Baltimore
7. Reissner E (1945) The effect of transverse shear deformation on the bending of elastic plates. J Appl Mech 12:69–77
8. Mindlin R (1951) Influence of rotatory inertia and shear on flexural motions of isotropic, elastic plates. J Appl Mech 18:336–343
9. Whitney J (1987) Structural analysis of laminated anisotropic plates. Technomic Publishing Company, West Hemisphere
10. Nayfeh A, Nassar EAA (1982) The influence of bonding agents on the thermo-mechanically induced interfacial stresses in laminated composites. Fibre Sci Technol 16:157–174

Chapter 8
A Novel Method to Attach Membranes Uniformly on MAV Wings

Yaakov J. Abudaram, Sean Rohde, James Paul Hubner, and Peter Ifju

Abstract Highly compliant wings have been used for MAV platforms, where the wing structure is determined by some combination of carbon fiber composites and a membrane skin, adhered in between the layers of composite material. The wing topology can be tailored to obtain the desired change in aerodynamic performance through passive shape adaptation. Pre-tension of the membrane plays a major role in the static and dynamic response of membrane wings and controls the overall deflections. In the past, the methods used to apply pretension when fabricating MAV wings were rudimentary. A new technique of attaching membranes firmly on wing structures is introduced, which involves the application of a technology known as corona treatment combined with another repeatable method of tensioning silicone membranes on any given frame geometry. Corona treatment provided a means of increasing adhesion of silicone on carbon fiber through the use of a high-frequency high-voltage air plasma discharge. The silicone membrane is co-cured with carbon fiber under vacuum pressure at an elevated temperature. After cool down, the membrane is tensioned.

Keywords Micro air vehicles • MAVs • Membrane wings • Silicone rubber film • Carbon fiber • Wing manufacturing • Corona treatment • Width tapered double cantilever beam • Digital image correlation • Visual image correlation • Stress • Strain • Thermal expansion • Pre-tension

8.1 Introduction

Micro air vehicles, or "MAVs", are defined as an aircraft with a maximum size of 15 cm and are capable operating at speeds of 25 mph or less [1]. Despite their size, MAVs are expected to be able to carry a payload of 20 g for 20–60 min [2]. The idea is a design of a small, reliable, and an inexpensive vehicle that can survey a remote area that is not easily accessible with ground transportation [3]. For example, a MAV may be placed under a missile and jettisoned right before the missile reaches the targeted area to send images of the damage to the base. A variety of wings may be envisioned for these vehicles, such as fixed, rotary, and flapping while this work is concerned with the fixed type [4].

The vehicle may greatly suffer in terms of controllability and range of operation when flying outdoors due to wind gusts [5]. Similarly, operation of MAVs can be highly challenging when flying indoors because of the air ducts and vents. Two types of mechanisms are introduced to overcome this issue: utilizing a passive mechanism or an active mechanism. Latter idea can be highly complex and expensive and in most cases a passive mechanism may be sufficient enough for a given mission. The wing design can be modified depending on the desired change in aerodynamic performance through passive shape adaptation [6].

A perimeter reinforced MAV wing fabricated at University of Florida is shown in Fig. 8.1a. Perimeter of the membrane skin is restricted by the thin strip of carbon fiber. This configuration allows the membrane to bulge out at increasing angles of attack and higher speeds. Result is an increase in lifting efficiency with high C_L and low $C_{M\alpha}$ [4]. Figure 8.1b shows a batten reinforced wing where the trailing edge of the membrane is not constrained while the membrane is supported by three

Y.J. Abudaram (✉) • S. Rohde • P. Ifju
University of Florida, Gainesville, FL 32607, USA
e-mail: abudaram@ufl.edu

J.P. Hubner
University of Alabama, Tuscaloosa, AL 35487, USA

E. Patterson et al. (eds.), *Composite Materials and Joining Technologies for Composites, Volume 7: Proceedings of the 2012 Annual Conference on Experimental and Applied Mechanics*, Conference Proceedings of the Society for Experimental Mechanics Series 44,
DOI 10.1007/978-1-4614-4553-1_8,

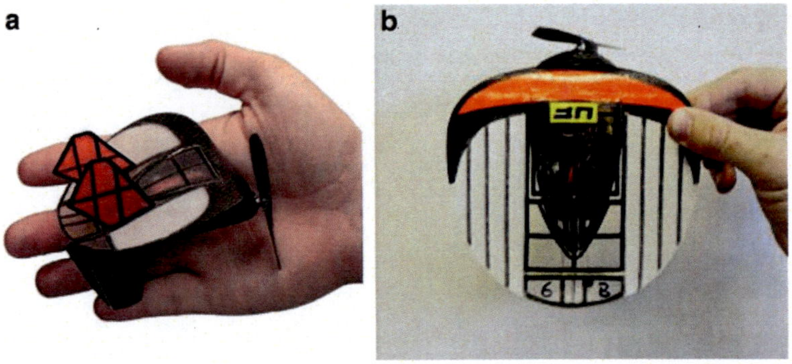

Fig. 8.1 Two different types of airplanes that are fabricated in micro air vehicle laboratory at the University of Florida (**a**) Perimeter Reinforced (**b**) Batten Reinforced

two-layer unidirectional carbon fiber battens on each side of the wing. In this case when the MAV flies into a head-on gust, the trailing edge flexes upwards resulting in adaptive washout. The effect is a near-constant lift history with exceptional smoothness even in gusty conditions [7].

Numerous materials have been found applicable to be used as membranes on the aforementioned wing geometries [3]. Most widely used material has been latex due to its benefits (elongation properties, ease of fabrication, and passive shape adaptation [8]). Nevertheless, latex material does not have appreciable shelf life and degrades at a rapid rate under hot and humid conditions. Further disadvantages include the degradation of the material when touched or contacted with various fluids, such as alcohol, oil, or water [9]. Rudimentary methods used to be applied when applying the latex membrane on wing skeleton: the latex rubber was stretched to be pinned on a particle board and adhered with glue on the carbon fiber wing; thereby, the pre-tensions achieved on various models were neither repeatable nor measurable.

Static and dynamic response of membrane wings are significantly affected by the pretension. Overall deflections (amount of camber on a perimeter reinforced wing, trailing edge vibration and reflex on a perimeter reinforced wing, etc.) are controlled by the tightness of the membrane [9].

The current work focuses on altering the material from latex to silicone to vastly improve durability. A method to quantify the pretension is discussed in detail by Abudaram et al. [9], in which this paper will use a similar technique to measure the strains developed on a silicone membrane that is co-cured with carbon fiber composite in an oven. One drawback is poor adhesion of most adhesives to silicone rubber. The solution to this manufacturing problem is corona treatment. The goal of surface treatment is to alter the mechanical properties at the surface of the silicone material to achieve a particular desired characteristic. In the cases to be considered, the primary characteristic of the materials is its surface energy. Surface energy affects the ability of a material to bond to an outer layer, which can be increased by roughening the surface area.

Despite being an integral part of the manufacture of polymers, corona treatment is rarely discussed. The name corona comes from the halo like appearance it gives off. Early tests confirmed its existence, but the potential for commercial application was not fully realized until the 1950s. Although this unique technique has been used for over 50 years, the principle behind how corona treatment improves adhesion is not entirely understood. Four generally accepted theories remain behind how the air plasma affects the surface of the substrate. The formation of electrets, the removal of weak boundary layers, an increase in roughness caused by small pits, and the deposition of chemically charged hydroxide ions on the surface are the most common explanations [10]. Electrets are non-conducting substances, like polymers, which become partially charged when placed under an electric field. The air plasma is generated by creating a steep electrical gradient. The surface to be treated is placed near the air plasma so the electrets hypothesis is a possible explanation. The weak boundary layer theory asserts that failures of adhesives are primarily the cause of weak boundary layers, which can be the result of air pockets between the surfaces or impurities in the adhesive [11]. The third explanation for the improved adhesion of surfaces treated with air plasma is attributed to an increase in roughness. Common practice is to sand a surface down before an adhesive is to be applied in order increase the strength of the bond. The theory behind this is that the surface energy increases as the surface area increases. A change from a flat surface to a rough one increases the surface area of the material on the microscopic level while on a macroscopic level the material appears as the same size. The increased surface energy means that a greater energy must be applied to the system to break the bonds and cause fracture. The fourth theory of how corona treatment improves adhesion to materials is that the air plasma results in a separation of molecules from their electrons, which creates an abundance of charged free radicals to be deposited on the nearest surface. In the case of these applications this surface is the substrate being altered. The deposited charged particles now exhibit a strong static attraction to foreign bodies. Since these theories are not mutually exclusive, the actual mechanism may in fact be a combination of all four of the aforementioned causes.

To analyze how effective the corona treatment is between silicone and prepreg carbon fiber layers a width tapered double cantilever beam test will be applied. Jyoti et al. [12] has pointed out the difficulty of measuring the crack length during dynamic fracture toughness testing on double cantilever (DCB) specimens. A practical solution was found by Daniel et al. [13] with the study of the quasi-static Mode I energy release rate of width tapered specimens, which allowed the energy release rate to be independent from the crack length. There also exists a similar technique of thickness tapered DCB; however, this type is cumbersome to manufacture.

8.2 Experimental Techniques

8.2.1 Fabrication of Wings and Pre-tensioning Method

Step 1. A preexisting wing mold is used to construct the wing. A light coat of spray glue is applied.

Step 2. Teflon film is laid on the mold for easy separation of the composite wing after the curing process is completed.

Step 3. Two layers of 45° oriented bi-directional prepreg carbon fiber layers are cut out utilizing a template. Another layer of carbon fiber is cut out only for the leading edge for added strength.

Step 4. A random speckling pattern is applied on one side of the silicone membrane while both sides are corona treated.

Step 5. The perimeter of the silicone membrane is sandwiched between the layers of the composite. A picture is acquired by the VIC system as a reference image.

Step 6. The entire assembly is vacuumed to be cured in the oven at 130°C for 4 h.

Step 7. A similar procedure is followed for a batten reinforced wing, except that unidirectional carbon fiber is used for the construction of the battens. On Step 7 the silicone is tensioned and the final picture is obtained utilizing the VIC cameras for analysis. Five perimeter reinforced and two batten reinforced wings are manufactured for analysis and comparison; however, future work will involve a greater number of batten reinforced wings (Fig. 8.2).

Fig. 8.2 Experimental steps involved in the construction and analysis of the MAV wings

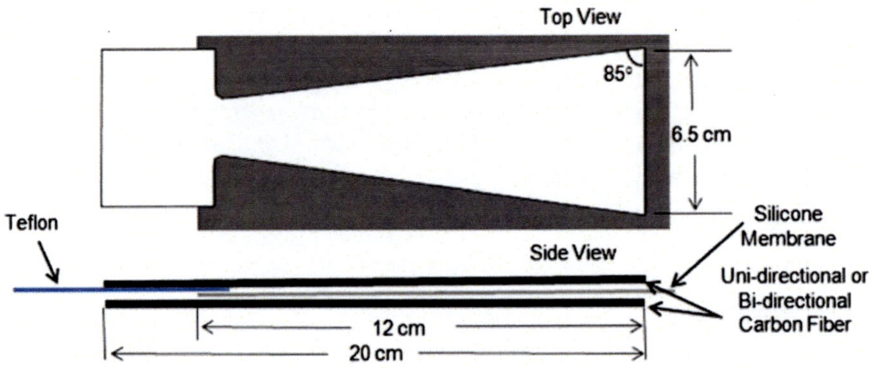

Fig. 8.3 Preparation of tapered double cantilever beam

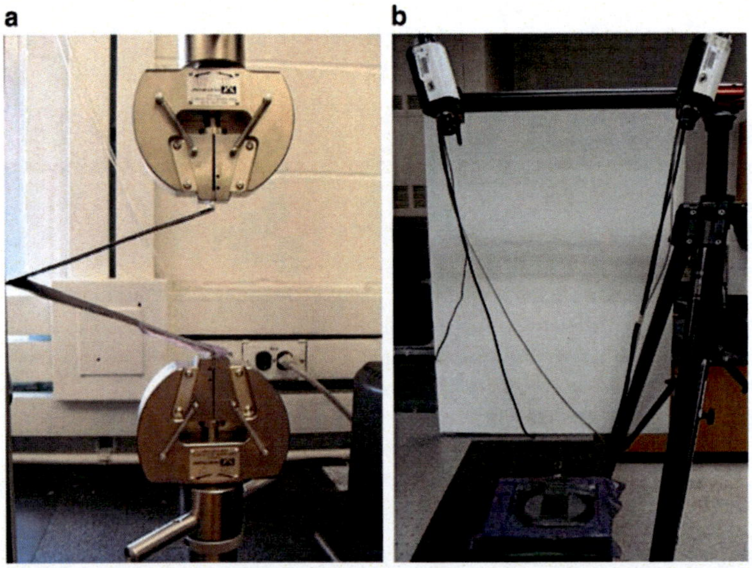

Fig. 8.4 Experimental setup (**a**) analysis of the DCBT (**b**) analysis of the deformed silicone membrane

8.2.2 Fabrication of Width Tapered Double Cantilever Beam

The reasons for the choice of the width tapered double cantilever beams WTDCB specimen are discussed in the introduction. Four groups of WTDCB are manufactured according to the illustration shown in Fig. 8.3, each group containing six specimens: plain silicone with bi-directional and unidirectional carbon fiber and corona treated silicone with bi-directional and unidirectional carbon fiber. Silicone membrane is sandwiched between two six-layer carbon fiber composites. A thin film of Teflon is applied to initiate the crack at the starting point of the beam. Finally, the entire assembly is cured in the oven under vacuum pressure to be analyzed in the Instron machine.

8.2.3 Complete Experimental Arrangements for Analysis

The experimental setup for the testing of corona treatment consists of two hinges adhered on both sides of the double cantilever beam and hinges clamped on the two jaws of the Instron 5567 as seen on Fig. 8.4a. The Instron machine is capable of measuring loads up to 30 kN with maximum and minimum speeds of 500 mm/min and 0.001 mm/min respectively. The V-clamp design provides an increasing compression force on the hinges with an increasing vertical tensile force applied on the jaws to avoid slipping. Crack is initiated with a Teflon film while the crack length (mm) and the required force (N) are recorded as the jaws pull the two sides of the beam.

Table 8.1 Properties of silicone membrane

Property	Value (Unit)
Elastic modulus (E)	385 kPa
Poisson's ratio	0.43
Thermal expansion coefficient(CTE)	$2.8*10^{-4}$/C
Thickness	0.28 mm
Operating temperature range	−62°C–218°C

Figure 8.4b shows the arrangement to measure the strains developed on the speckled specimen with two VIC cameras connected to the computer through a DAC device. The cameras are calibrated to acquire pictures of the membrane before and after the curing of the carbon fiber.

8.2.4 Properties of Silicone Membrane

Abudaram et al. [9] calculated the elastic modulus and Poisson's ratio values experimentally utilizing the VIC system by hanging weights at the end of a long strip of silicone membrane (12:1 ratio) and recorded the strain values in the x and y directions. The values provided for the elastic modulus is calculated by curve fitting the data points according to the Eulerian approach. The thermal coefficient of expansion was computed in a similar manner by recording the temperatures with the infrared thermometer and strain values in x and y directions utilizing the VIC system. The thickness of the silicone membrane was measured with digital calipers while the operating temperature range was provided by the manufacturer (Table 8.1).

8.2.5 Visual Image Correlation System

All deformation will be measured with visual image correlation (VIC), a non-contacting full-field measurement technique originally developed by researchers at the University of South Carolina. The underlying principle is to calculate the displacement field of a test specimen by tracking the deformation of a random speckling pattern applied to the surface. Two pre-calibrated cameras digitally acquire this pattern before and after loading, using stereo-triangulation techniques. The VIC system then tries to find a region (in the image of the deformed specimen that maximizes a normalized cross-correlation function corresponding to a small subset of the reference image. The reference image is taken when no load is applied to the structure). The image space is iteratively swept by the parameters of the cross-correlation function to transform the coordinates of the original reference frame to coordinates within the deformed image. As it is unlikely that the deformed coordinates will directly fall onto the sampling grid of the reference image, accurate grey-value interpolation schemes are implemented to achieve optimal sub-pixel accuracy without bias.

The twin cameras are connected with a PC via an IEEE 1394 firewire cable, and a specialized unit is used to synchronize the camera triggers for instantaneous shots. A standard acquisition board installed in the computer carries out digitalization of the images, and the image processing is carried out by custom software, provided by Correlated Solutions, Inc. Typical data results obtained from the VIC system consist of geometry of the surface in discrete coordinates (x, y, z) and the corresponding displacements (u, v, w). The VIC system places a grid point every N pixels, where N is user defined. A final post-processing option involves calculating the in-plane strains (ε_{xx}, ε_{yy}, and ε_{xy}). This is done by mapping the displacement field onto an unstructured triangular mesh, and conducting the appropriate numerical differentiation, in which the complete definition of finite strains is used.

8.3 Results and Discussion

8.3.1 WTDCB Test for Corona Treatment

Separation speed on the Instron machine was set to 10 mm/s and average forces from crack initiation until the complete crack propagation are reported for all specimens on Table 8.2. Standard deviations on treated samples may have been caused by the application of the corona treatment: the amount of time, plasma charge distance and speed from the silicone surface,

Table 8.2 Average forces (n) recorded on all specimens

	Untreated unidirectional (N)	Treated unidirectional (N)	Untreated bidirectional (N)	Treated unidirectional (N)
Specimen #1	29.7	30.3	19.8	32.9
Specimen #2	21.1	33.9	18.8	22.0
Specimen #3	21.9	30.6	20.6	24.4
Specimen #4	30.5	29.7	20.0	NA[a]
Specimen #5	21.2	37.2	21.2	20.0
Specimen #6	32.2	30.7	18.9	21.9
Average	26.1	32.1	19.9	24.2
Standard Dev.	5.22	2.91	0.94	5.09

[a]Results for treated unidirectional specimen #4 are not reported on this table due to a manufacturing error

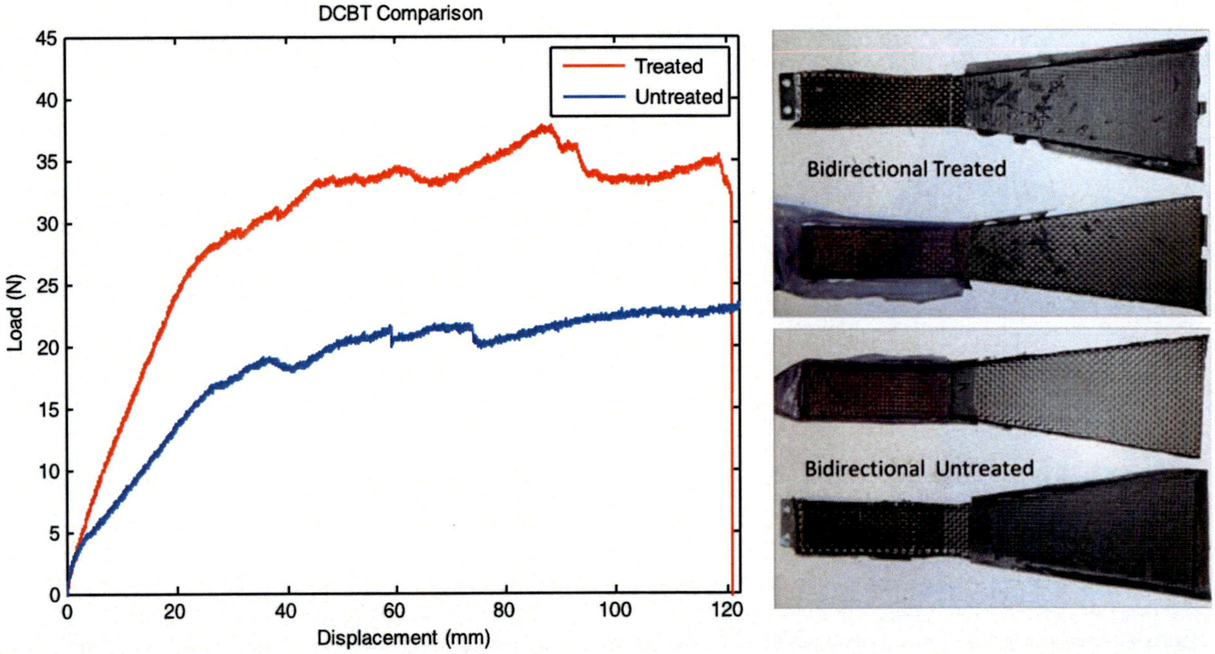

Fig. 8.5 WTDCB tests for corona treated and untreated silicone on bidirectional carbon fiber

uniform application on each specimen during the fabrication plays crucial role in the adhering process. The high deviation on untreated unidirectional specimens is probably due to the uneven thicknesses composite layers.

The data clearly shows that more energy is required to separate all corona treated specimens. Corona treated and untreated specimens with maximum averages are reported on Figs. 8.5 and 8.6 along with pictures after separation for further discussion. Untreated samples are separated in a much smoother fashion in comparison to corona treated samples. The dynamics involved in the detachment of corona treated samples are extremely complex due to adhesive and cohesive effects.

8.3.2 Membrane Pre-tensioning on Wing

Before acquiring pictures, the wing samples were waited to cool down to room temperature. Normal strains in x and y directions along with shearing strains are reported in Fig. 8.7 for a typical wing among five experiments. The strain resolution of the VIC system is estimated to be 500 με, a relatively high value (compared to strain gages, for example) due to the fact that the data is obtained by appropriately differentiating the displacement fields. The VIC system is not expected to have the resolution to capture the accurate strain information in the carbon fiber frame; therefore, only the membrane deformation was taken into account for calculations.

The membrane expanded due to the vacuum pressure and the high temperature in the oven. When cooled down to room temperature, the wing was naturally pretensioned. The anisotropy of strain is clearly evident from Fig. 8.7 as the wing has a

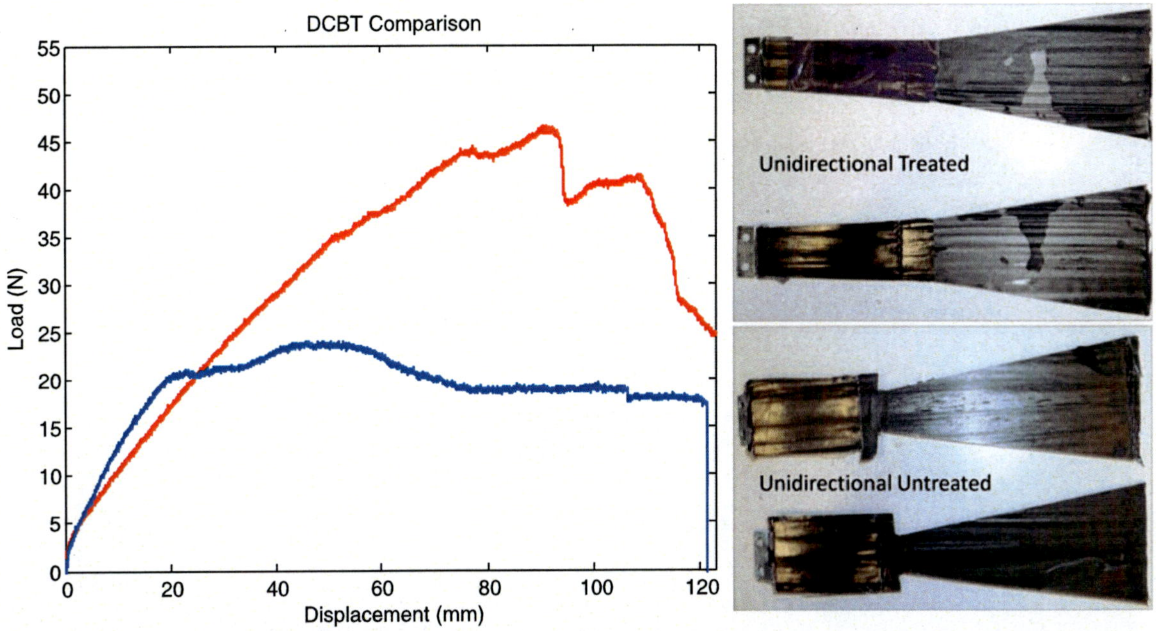

Fig. 8.6 WTDCB tests for corona treated and untreated silicone on unidirectional carbon fiber

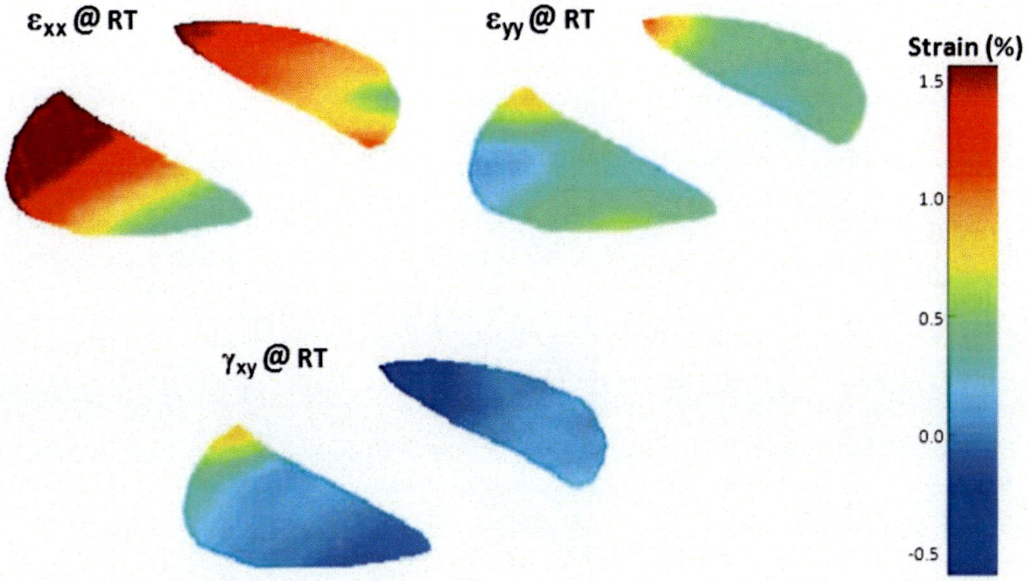

Fig. 8.7 Acquired strain distribution on silicone membrane at the *bottom* of the perimeter reinforced wing

camber and the frame geometry is not symmetrical. Note that images on Fig. 8.7 show the bottom of the wing. Higher strains are computed toward the leading edge while this effect is muted toward the trailing edge for normal strains. Insignificant amounts of shear strains are recorded for all samples.

All strain values are then converted to stresses using the values on Table 8.1 as seen in Fig. 8.8. Stresses in normal directions range from approximately 3 kPa to over 9 kPa while much higher averages are recorded for stresses in x direction. This effect could be due to a shorter span in the chordwise direction in comparison to the longitudinal direction.

Average strains on the right and left sides of the five wings were calculated in x and y directions along with standard deviations as seen in Fig. 8.9. For each wing slightly higher strain values in the chordwise direction were measured and the highest standard deviation among all samples was calculated in this direction as well. The rest of the variations were within the same percentage, which shows the method is reasonably repeatable.

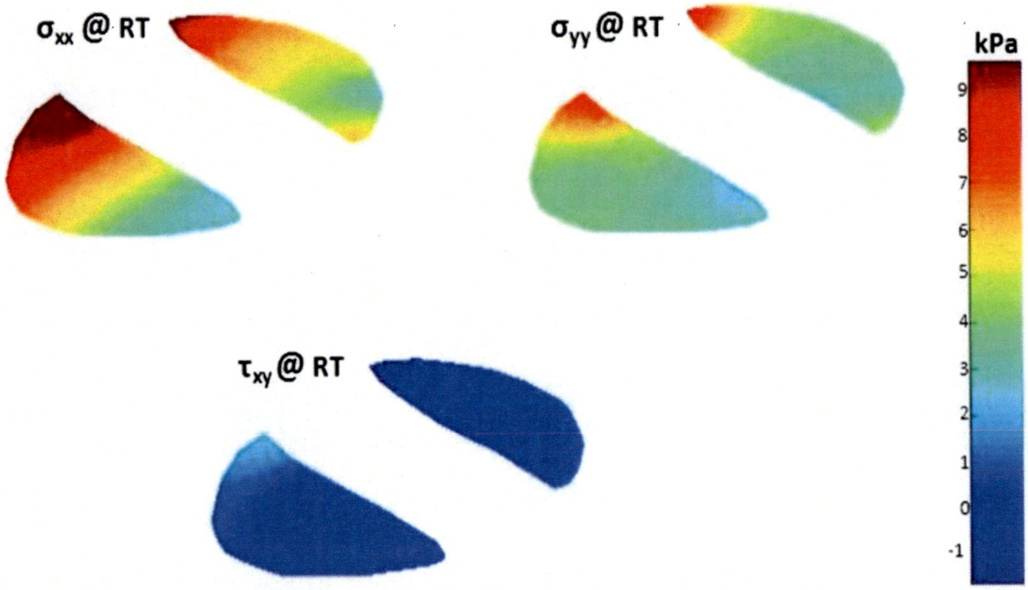

Fig. 8.8 Calculated pre-tension distribution on silicone membrane at the *bottom* of the perimeter reinforced wing

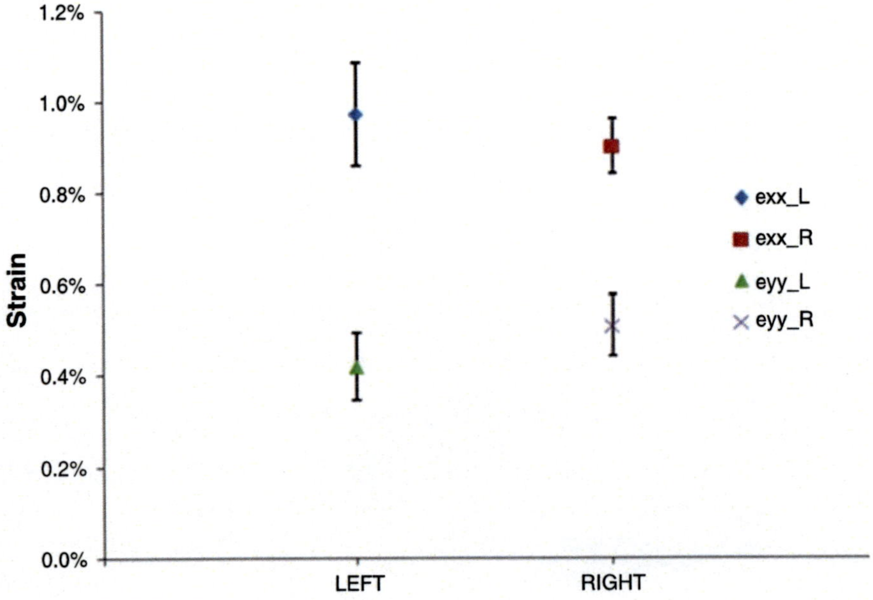

Fig. 8.9 Average strains developed on membrane in x and y directions on *left* and *right* hand side of perimeter reinforced wing

8.4 Conclusion

Various samples for corona treated and untreated silicone/carbon fiber specimens were manufactured. The samples were analyzed via Instron machine. The bonding technique is then applied to a series of MAV wings to be co-cured in the oven under vacuum pressure and the results are processed utilizing visual image correlation system:

1. When corona treated specimens were compared to their non-treated counterparts, the method showed a significant effectiveness in bonding.
2. Unidirectional carbon fiber specimens required more tensile force to separate whether they are treated or not.
3. The fabrication technique to build a wing with corona treated silicone membrane is explained.

4. VIC system showed the membrane is tensioned on both sides of the wing. The average strains were measured for all samples to show the repeatability.
5. The pretension was quantified via analysis.

Future work will involve results for batten reinforced MAV wings in a similar fashion. The wings may be cured at various temperatures to see the effect of temperature increase during the curing process on the strained developed at the room temperature.

Acknowledgements This work is supported, in part, by the Air Force Office of Scientific Research under Grant FA9550-10-1-0152.

References

1. Mueller TJ (2000) Proceedings of the conference on fixed flapping and rotary wing vehicles at very low Reynolds numbers, Notre Dame University, Indiana, 5–7 June 2000
2. McMichael JM, Col. Francis MS, USAF (1997) Micro air vehicles – toward a new dimension in flight from http://www.fas.org/irp/program/collect/docs/mav_auvsi.htm
3. Ifju PG, Ettinger S, Jenkins D, Martinez L (2001) Composite materials for micro air vehicles (MAV's). SAMPE J 37(4):7–12
4. Stanford B, Ifju P, Albertani R, Shyy W (2008) Fixed membrane wings for micro air vehicles: experimental characterization, numerical modeling, and tailoring. Prog Aerospace Sci 44:258–294
5. Pisano W, Lawrence D (2008) Autonomous gust insensitive aircraft. In: AIAA guidance, navigation, and control conference, Honolulu, 18–21 Aug 2008
6. Stanford B, Ifju P (2009) Aeroelastic topology optimization of membrane structures for micro air vehicles, structural and multidisciplinary optimization. doi: 10.1007/soo158-008-0292-x
7. Abudaram Y, Stanford B, Ifju P (2009) Wind tunnel testing of load-alleviating membrane wings at low reynolds numbers. In: Proceedings of 47th AIAA sciences meeting including the New Horizons Forum and aerospace exposition conference, Orlando, 5–11 Jan 2009
8. Stanford B, Sytsma M, Albertani R, Viieru D, Shyy W, Ifju P (2007) Static aeroelastic model validation of membrane micro air vehicle wings. AIAA J 45(12):2828–2837
9. Abudaram Y, Ifju PG, Hubner JP, Ukeiley L (2012) Controlling pre-tension of silicone membranes on micro air vehicle flexible wings. In: 50th AIAA sciences meeting, Nashville, 9–12 Jan 2012
10. Zhang D, Sun Q, Wadsworth LC (1998) Mechanism of corona treatment on polyolefin films. Polymer Eng Sci 38(6):965–970
11. Bikerman J (1967) Causes of poor adhesion: weak boundary layers. Ind Eng Chem 59(9):40–44
12. Joyoti A, Gibson RF, Newaz GM (2005) Experimental studies of mode I energy release rate in adhesively bonded width tapered composite DCB specimens. Compos Sci Technol 65:9–18
13. Daniel IM, Shareef I, Aliyu AA (1985) Rate effect on delamination fracture toughness of a toughened graphite/epoxy. American Soc Test, Houston, Texas,13-15 March 1985 937:260–274

Chapter 9
A Model for Fracture Characterization of Adhesively-Bonded Joints

Jianghui Mao, Sayed A. Nassar, and Xianjie Yang

Abstract In this paper, an analytical model is proposed to characterize the fracture behavior of an adhesively bonded joint loaded in peel. Unlike previous theories, the current model includes some new parameters, such as young's modulus of adhesive material, thickness of adhesive layer, and parameter *c* in the stress function, which will be determined by solving a nonlinear equation. The new developed model also takes into account the crack tip rotation. To verify the current work, comprehensive comparison is made between previous theories, experimental data, and current model. Result shows that the current model is better than previous theories in regards to the coincidence with experimental data. The effect of each key parameter on the prediction is also discussed in this paper.

Keywords Adhesive bonding • Composite joint • COD • Strain energy release rate (SERR)

9.1 Introduction

Adhesively bonded joints, especially composite laminated joints, have been widely used in automotive and aerospace industry nowadays. Various kinds of failure modes have been identified in everyday use, among which delamination is the most life-limiting mechanism in bonding joint. The analysis of adhesive joint is usually based on through-thickness averages of stress and strain, then, the basic variables are the peel stress, σ, the shear stress, τ, the opening distance, w, and the shear deformation, v, of the adhesive layer. The shear loading case for adhesive joint has been studied in Nassar et al. [1], current study will focus on the peel loading case using Double Cantilever Beam (DCB) model.

The adhesively bonded joint loaded in peel has been intensively studied. Whitney [2] carried out experiments on composite joint and discussed a number of approaches for data reduction schemes, El-Zein [3] tested a composite joint in elevated temperature and proposed a theory for determining critical load, the developed model only took adherend into account, and crack tip opening was neglect. Hashemi et al. [4] analyzed some DCB composite joints using corrected beam theory, Johnson et al. [5] investigated the environmental durability of a series of bonded systems, Shindo et al. [6] experimentally and analytically investigated the cryogenic Mode I interlaminar fracture behavior and toughness of SL-E woven glass-epoxy laminate, Jyoti et al. [7] experimentally explored the effect of adhesive layer to the critical energy release rate of specimen, Xie et al. [8] discussed methodologies and techniques used for characterizing metal/epoxy interfaces, latter, Pradeep et al. [9] used the evaluation methodologies for assessing the interfacial integrity using FEA, the methodology considered the effect of adhesive properties such as Young's modulus and thickness, however the crack tip rotation wasn't taken into account. In contrary to the above mentioned researches, increasing number of researches (Barenblatt [10], Yang et al. [11], Andersson et al. [12] and Andersson et al. [13], Ouyang et al. [14]) used cohesive zone model (CZM) to solve the adhesively bonded joint. The CZM was critically dependent on accurate constitutive models, i.e. stress-elongation (σ-w) relation of adhesive layer, it had been proven very hard to correlate the bulk properties of adhesives with their behavior in a thin and constrained layer [15], also, a stress-elongation relation works fine for this adherend material may not work for other adherend material, hence repeated works were need to determine stress-elongation relation for other adherend material,

J. Mao (✉) • S.A. Nassar • X. Yang
Department of Mechanical Engineering, Fastening and Joining Research Institute (FAJRI), Oakland University, Rochester, MI 48309, USA
e-mail: nassar@oakland.edu

E. Patterson et al. (eds.), *Composite Materials and Joining Technologies for Composites, Volume 7: Proceedings of the 2012 Annual Conference on Experimental and Applied Mechanics,* Conference Proceedings of the Society for Experimental Mechanics Series 44,
DOI 10.1007/978-1-4614-4553-1_9,

It's well known that the fracture energy and morphology varies with the thickness of the adhesive layer [16], previous studies either neglect the existence of adhesive layer [2, 5, 17], or model the adhesive layer as spring [18–20], or neglect the crack tip rotation [9, 17], others using stress-elongation relation to represent the property of adhesive layer [12–14], they were suitable for modeling very thin adhesive or interfacial behavior of adhesively bonded joint, however, due to the lack of consideration about the adhesive material properties and geometry properties, the results obtained by previous were very limited to specific cases. In this study, a model is developed which takes into account of the adhesive material properties and adhesive thickness and their geometry configuration, also a novel parameter c is introduced, and it's determined by solving a nonlinear equation, in addition, crack tip rotation is considered. Various cases were compared and discussed to verify the novel developed model. The effective of each key parameter in the model is discussed.

9.2 Analytical Model

The geometric model for the analysis is presented in Fig. 9.1. Two adherends with same size, length l, height h, and width b are bonded by an adhesive layer with one end leaving unbounded and this end is subjected to peel load P as in Fig. 9.1a. The adhesive layer has a thickness of h_a, and length l-a. the coordinate is originated at the center of left end in Fig.9.1a. The free body diagram of adherend is shown in Fig. 9.1b.

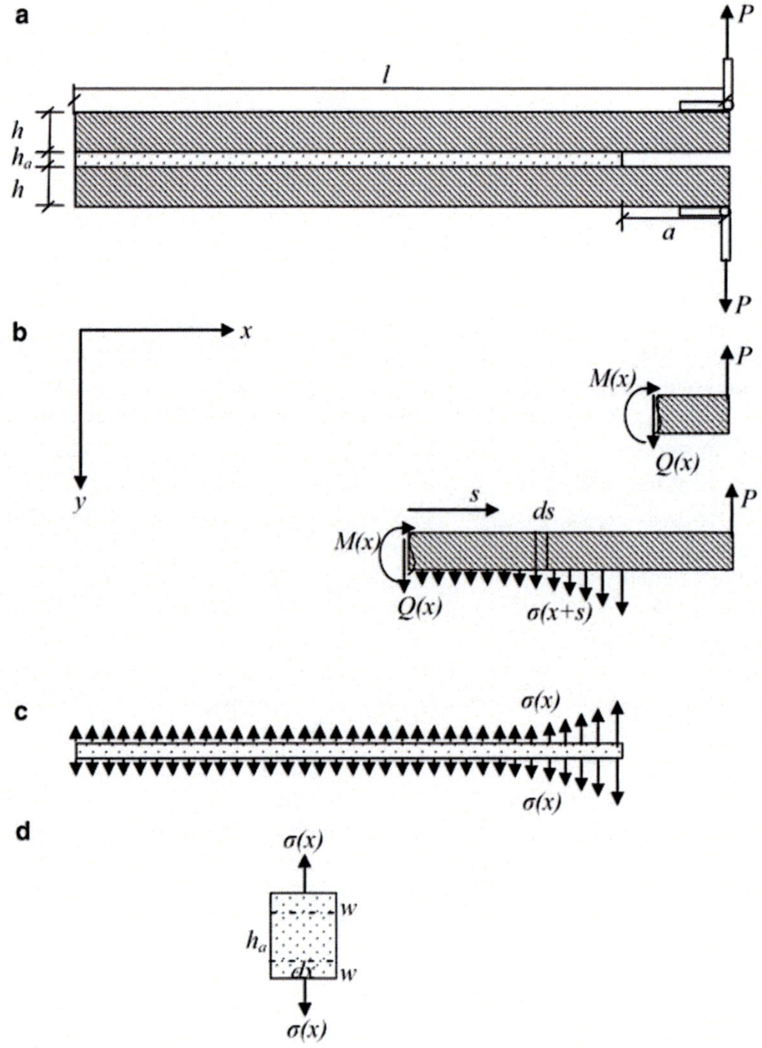

Fig. 9.1 Schematic of analysis model

For the unbounded adherend portion, the moment is expressed as

$$M(x) = P(l - x) \qquad l - a \leq x \leq l \tag{9.1}$$

For the bonded ahderend portion, the moment can be given as

$$M(x) = P(l - x) - b\int_0^{l-a-x} s\sigma(x+s)ds \qquad 0 \leq x \leq l - a \tag{9.2}$$

as can be seen in Fig. 9.1b, s is a new coordinate parallels with x-axis, and originated at x.

Force equilibrium of upper adherend gives

$$b\int_0^{l-a} \sigma(x)dx = P \tag{9.3}$$

where $\sigma(x)$is the stress distribution along the bonding interface, and it is assumed to be

$$\sigma(x) = \sigma_0 e^{c(x-l+a)} \tag{9.4}$$

where σ_0 is the stress at the crack tip, and c is the material parameter.

From Eqs.9.3 and 9.4,

$$\frac{b\sigma_0}{c}\left[1 - e^{c(a-l)}\right] = P \tag{9.5}$$

The moment and deflection relationship of adherend can be give as

$$w''(x) = -\frac{M(x)}{E_x^{(1)} I} \tag{9.6}$$

where $E_x^{(1)}$ is the adherend Young's modulus in x direction. I is the Moment of Inertia.

9.2.1 *Modeling of Bonded Region* ($0 \leq x \leq l - a$)

For the upper half adhesive, the elongation in y direction is obtained as

$$w(x) = -\frac{h_a}{2}\frac{\sigma(x)}{E_y^{(2)}} \qquad 0 \leq x \leq l - a \tag{9.7}$$

where $E_y^{(2)}$ is the adhesive Young's modulus in y direction.

First derivative of Eq. 9.7 gives

$$w'(x) = -\frac{h_a c}{2E_y^{(2)}}\sigma(x) \qquad 0 \leq x \leq l - a \tag{9.8}$$

Substitute Eq. 9.2 into Eq. 9.6, and using Eqs. 9.4 and 9.5, one obtains,

$$\frac{\left(\frac{1}{ce^{c(l-a)}} + \frac{E_x^{(1)} I h_a}{2E_y^{(2)} e^{c(l-a)}}\frac{c^3}{b}\right)e^{cx} - x + l - a - \frac{1}{c}}{1 - e^{c(a-l)}} = l - x \tag{9.9}$$

From Eq. 9.9, c is depended on x. Following Nayfeh et al. [21], a continuum mixture format is carried out to eliminating the x-dependence by integrating both side of Eq. 9.9 from 0 to l-a, and divided by l-a, then, a nonlinear equation can be obtained from Eq. 9.9 for determining parameter c as follows

$$\frac{1}{c^2(l-a)}+\frac{E_x^{(1)}Ih_a}{2E_y^2}\frac{c^2}{b(l-a)}+\frac{\frac{1}{2}(l-a)-\frac{1}{c}}{1-e^{c(a-l)}}=\frac{1}{2}(l+a) \tag{9.10}$$

Equation 9.10 can be solved by Newton–Raphson method. Theoretically, there are four solutions for this equation, but the realistic one should be real number, and not far away from 0, otherwise it will cause numerical problem.

9.2.2 *Modeling of Unbounded Region* ($l - a \le x \le l$)

From Eqs. 9.1 and 9.6

$$E_x^{(1)}Iw' = \frac{1}{2}Px^2 - Plx + B_1 \tag{9.11}$$

$$E_x^{(1)}Iw = \frac{1}{6}Px^3 - \frac{1}{2}Plx^2 + B_1x + B_2 \tag{9.12}$$

B_1 and B_2 are constant, to be determined by boundary condition. The continuity condition at $x = l - a$ gives

$$\frac{1}{2}P(l-a)^2 - Pl(l-a) + B_1 = -\frac{E_x^{(1)}Ih_a}{2E_y^{(2)}}c\sigma_0 \tag{9.13}$$

$$\frac{1}{6}P(l-a)^3 - \frac{1}{2}Pl(l-a)^2 + B_1(l-a) + B_2 = -\frac{E_x^{(1)}Ih_a}{2E_y^{(2)}}\sigma_0 \tag{9.14}$$

Equations 9.13 and 9.14 give the value of B_1 and B_2 as

$$B_1 = -\frac{E_x^{(1)}Ih_a}{2E_y^{(2)}}c\sigma_0 - \frac{1}{2}P(l-a)^2 + Pl(l-a) \tag{9.15}$$

$$B_2 = -\frac{E_x^{(1)}Ih_a}{2E_y^{(2)}}\sigma_0[1-c(l-a)] + \frac{1}{3}P(l-a)^3 - \frac{1}{2}Pl(l-a)^2 \tag{9.16}$$

Then, the deflection of unbounded region of adherend can be determined as

$$\begin{aligned} w = \frac{1}{E_x^{(1)}I}\Bigg\{&\frac{1}{6}Px^3 - \frac{1}{2}Plx^2 + \left[-\frac{E_x^{(1)}Ih_a}{2E_y^2}c\sigma_0 - \frac{1}{2}P(l-a)^2 + Pl(l-a)\right]x \\ &- \frac{E_x^{(1)}Ih_a}{2E_y^{(2)}}\sigma_0[1-c(l-a)] + \frac{1}{3}P(l-a)^3 - \frac{1}{2}Pl(l-a)^2\Bigg\} \qquad l-a \le x \le l \end{aligned} \tag{9.17}$$

9.2.3 Joint Fracture Properties

The loadline deflection will be given by

$$w_P = -\frac{1}{E_x^{(1)}I}\left[\frac{E_x^{(1)}Ih_a}{2E_y^{(2)}}\sigma_0(ca+1)+\frac{1}{3}Pl^3+P(l-a)^2l-Pl^2(l-a)-\frac{1}{3}P(l-a)^3\right] \tag{9.18}$$

The Crack Open Displacement *COD* of loadline for the joint is obtained as

$$COD = \frac{2}{E_x^{(1)}I}\left[\frac{E_x^{(1)}Ih_a}{2E_y^{(2)}}\sigma_0(ca+1)+\frac{1}{3}Pl^3+P(l-a)^2l-Pl^2(l-a)-\frac{1}{3}P(l-a)^3\right] \tag{9.19}$$

and Crack Tip Opening Displacement *CTOD* is

$$CTOD = \frac{h_a\sigma_0}{E_y^{(2)}} \tag{9.20}$$

From Eqs. 9.5 and 9.9, the compliance of the joint can be obtained. In the determination of c, the continuum mixture theory is used, which eliminates the x-dependence of c. A closed examination finds that using the continuum mixture theory, the prediction tends to deviate from experiment, the deviation is related to crack length and joint length as *(l-a)/l/k*, which is called the correction factor in this paper, and k is a constant, which can be determined experimentally or using FEA method. With the correction factor in consideration for bonded joint region, the compliance of the joint can be expressed as

$$C = \frac{2}{E_x^{(1)}I}\left[\frac{E_x^{(1)}Ih_ac(ca+1)(l-a)}{2E_y^{(2)}b[1-e^{c(a-l)}]lk}+\frac{1}{3}l^3+(l-a)^2l-l^2(l-a)-\frac{1}{3}(l-a)^3\right] \tag{9.21}$$

The Strain Energy Release Rate (SERR) of the joint can be given as

$$G_I = \frac{P^2}{2b}\frac{dC}{da} = \frac{P^2}{E_x^{(1)}Ib}\left\{\frac{E_x^{(1)}Ih_ac^2(l-a)}{2E_y^{(2)}blk}\frac{1+cae^{c(a-l)}}{[1-e^{c(a-l)}]^2}+a^2\right\} \tag{9.22}$$

In Eqs. 9.21 and 9.22, the correction factor is incorporated directly after derivation.

Compared to previous studies (shown in Table 9.1, Summary of different theories), such as beam theory (crack tip rotation is not considered), El-Zein [3], more variables are taken in to account, such as Young's modulus of adhesive material and thickness of adhesive layer, although Penado [18] considered these parameters, he modeled the adhesive layer as spring, while our model treated the adhesive layer more naturally as continuum material, and the parameter c is introduced in the model, in addition, the crack tip rotation is taken into account. The advantages of introducing these variables will be illustrated subsequently.

9.3 Results and Discussion

One distinct difference of current theory is the introduction of c, which will be determined by Eq. 9.10 using Newton–Raphson method. Another one is k in the correction factor, FEA method is used, and it's determined as $k = 22$.

To compare various theories listed in Table 9.1 Summary of different theories with current theory, three joints are considered. The experimental results are obtained from various sources, the material properties of each components are listed in Table 9.2 Material properties , and specimens configuration are shown in Table 9.3 specimen geometry, unit: mm In regards to the APC-2/AS4-CFRP composite joint, the adhesive properties was unspecified [22], $E = 12.5415$ MPa, and $v = 0.4$ is assumed.

Table 9.1 Summary of different theories

Theories	Compliance	Strain energy release rate
Beam theory	$C = \frac{8a^3}{Ebh^3}$	$G_I = \frac{3P\delta}{2ba}$
Area method	-	$G_I = \frac{1\Delta A}{b\Delta a} = \frac{1}{2b\Delta a}(P_1\delta_2 - P_2\delta_1)$ where ΔA is the area under the $P-\delta$ curve
El-Zein, 1988 [17]	$C = \frac{12}{bh^3}\left(\frac{a_{11}a^3}{3} - \frac{a_{26}h^3}{24}\right)$ where a_{ij} is the component of adherend compliance matrix	$G_I = \frac{6P^2a_{11}a^2}{b^2h^3}$
Penado, 1993 [18]	$C = \frac{8}{E_x^{(1)}b}\left(\frac{a}{h}\right)^3\left[1 + \frac{3}{Bh^{0.25}}\frac{h}{a} + 3\left(\frac{1}{B^2h^{0.5}} + \frac{E_x^{(1)}}{8G_x^{(1)}}\right)\left(\frac{h}{a}\right)^2 + \frac{3}{2B^3h^{0.75}}\left(\frac{h}{a}\right)^3\right]$ For small h/a, $C \approx \frac{8}{E_x^{(1)}b}\left(\frac{a}{h}\right)^3$ Superscript 1 and 2 denotes adherend and adhesive respectively, G is shear modulus of adherend, $B = \sqrt[4]{\frac{3k}{Eb}}$, $k = \frac{1}{\frac{1}{k_{adherend}} + \frac{1}{k_{adhesive}}}$, $k_{adherend} = 4\frac{E_y^{(1)}b}{h}$, $k_{adhesive} = \frac{b}{h_a}\frac{E_y^{(2)}}{1-\upsilon_{xy}\upsilon_{yx}}$	$G_I = \frac{12P^2}{E_x^{(1)}b^2h}\left(\frac{a}{h}\right)^2\left[1 + \frac{2}{Bh^{0.25}}\frac{h}{a} + \left(\frac{1}{B^2h^{0.5}} + \frac{E_x^{(1)}}{8G_x^{(1)}}\right)\left(\frac{h}{a}\right)^2\right]$ For small h/a, $G_I \approx \frac{12P^2}{E_x^{(1)}b^2h}\left(\frac{a}{h}\right)^2$
Pradeep, 2010 [9]	$C = \frac{a^3}{3}\left[\frac{1}{(EI)_1} + \frac{1}{(EI)_2}\right]$, where $(EI)_1 = D_{11}^{(1)}\left[1 - \left(\frac{D_{12}^{(1)}}{D_{11}^{(1)}}\right)^2\right]$, $(EI)_2 = \bar{D}_{11}\left[1 - \left(\frac{\bar{D}_{12}}{\bar{D}_{11}}\right)^2\right]$, $\bar{D}_{11} = D_{11}^{(2)} + D_{11}^{(3)}$; $\bar{D}_{12} = D_{12}^{(2)} + D_{12}^{(3)}$; $D_{11}^{(k)} = E_kI_k/(1-v_k^2)$, $D_{12}^{(k)} = v_kD_{11}^{(k)}$, $I_k = bh_k^3/12$, subscript $k(k = 1,2,3)$ represent upper adherend, adhesive, lower adherend, respectively	$G_I = \frac{P^2a^2}{2b}\left[\frac{1}{(EI)_1} + \frac{1}{(EI)_2}\right]$

Table 9.2 Material properties

Material		Properties		
		E_x, E_y, E_z GPa	G_{12}, G_{13}, G_{23} GPa	v_{12}, v_{13}, v_{23}
Adherend	APC-2/AS4-CFRP [22]	121, 9, 5.2	5.2, 5.2, 1.9	0.34, 0.34, 0.46
	7075-T6 Aluminum [5]	70.8	26.62	0.33
	Titanium [5]	110	41.98	0.31
Adhesive	FM®73 M [5]	2.07	0.77	0.34
	FM-X5 [5]	5	1.92	0.3

Table 9.3 Specimen geometry, unit: mm

	APC-2/AS4-CFRP joint [22]	Al/FM®73 M/Al [5]	Titanium/FM-X5/Titanium [5]
l	185	305	305
a	55	vary	vary
h	2.5	9.53	6.604
h_a	2.5	0.25	0.338
b	10	27.127	27.127

9.3.1 Critical Load

Critical load of APC-2/AS4-CFRP joint predicted by different theories is shown in Table 9.4. Note that the adhesive thickness is relatively thick in this case. From Table 9.4, all the theories give closed prediction with experiment result [22] except Penado's theory [18], and the current theory is the most closed one.

For Al/FM®73 M/Al joint, the critical load predicted is shown in Fig. 9.2. From Fig. 9.2, Beam theory and Pradeep's theory give prediction almost double the experimental result at small crack length, however, with the increase of crack length, the results predicted by these two theories improve a lot. On the other hand, El-Zein and Penado's theory overestimate the result at small crack length, but underestimated the result at larger crack length. Compared to the other theories, current theory improves a lot for the whole examined crack length region.

In regards to Titanium/FM-X5/Titanium joint, the results are shown in Fig. 9.3, From Fig. 9.3, Beam theory, El-Zein and Pradeep's theories overestimated the critical load about 70% at small crack length, with the increase of crack length, prediction improves greatly. Current theory gives the best agreement with experiment result.

Table 9.4 Comparison of different theories in calculating of critical load and Critical SERR for APC-2/AS4-CFRP joint

	Critical load, *N*	G_{IC}, 10^3 J/m^2
EXP.	65	–
Current theory	65.02813	0.813683716
Penado93 [18]	27.39576	0.342422089
Pradeep2010 [9]	65.10282	0.813726064
El-Zein88 [17]	65.09943	0.967903038
Beam theory	65.09943	0.812440909

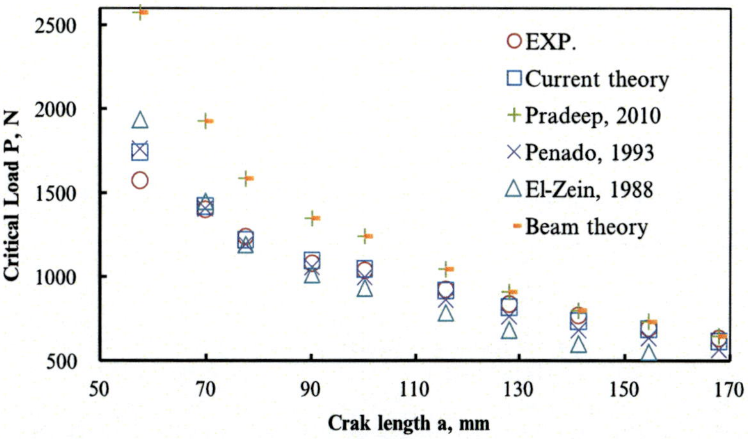

Fig. 9.2 Critical load determined by various theories for Al/FM-73 M/Al joint, EXP. is the experimental result from [5]

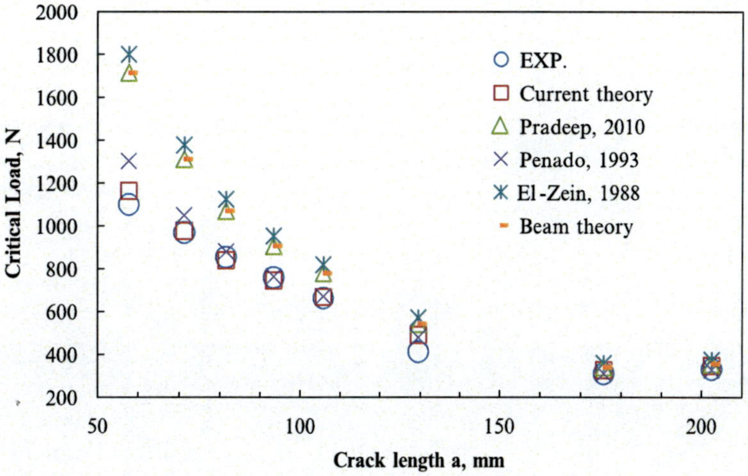

Fig. 9.3 Critical load determined by various theories for Titanium/FM-X5/Titanium joint, EXP. is the experimental result from [5]

9.3.2 Strain Energy Release Rate

Critical SERR of APC-2/AS4-CFRP joint is listed in Table 9.4. From Table 9.4, current theory gives the critical SERR very close to Pradeep, El-Zein, and beam theory. For Al/FM-73 M/Al joint, the results are listed in Table 9.5, the critical SERR varies a little with crack length, the average critical SERR is about 2.88 × 10^3 J/m^2 from experiment, current theory and Penado's theory are closer to experimental one. For Titanium/FM-X5/Titanium joint (see Table 9.6), current theory is even better than Penado's theory.

From the above comparison, current theory has wider application than others. El-Zein, and beam theory doesn't consider the adhesive Young's modulus and thickness, the predictions given by these theories are poor when adhesive Young's modulus is relatively high, for example, the Titanium/FM-X5/Titanium joint, Pradeep's theory doesn't take the crack tip rotation into account, the application into studied case is not good as shown above, Penado modeled the adhesive as spring, while the adhesive is relatively thick, the result is very bad as for APC-2/AS4-CFRP joint.

In order to investigate the effect of various variables on the model prediction, parametric study is conducted, and Al/FM-73 M/Al joint mentioned above is used in the study.

Table 9.5 Strain energy release rate calculated by various theories for Al/FM-73 M/Al joint, original experiment data from [5], Unit: 10^3 J/m^2

a, mm	EXP.[a]	Current theory	Beam theory	Pradeep (2010)	Penado (1993)	El-Zein (1988)
57.6	2.488905	3.107943	3.580419	5.855193	3.514229	2.20118
69.81	2.566863	2.935684	3.510627	4.824342	3.163628	1.813645
77.48	3.00945	2.619628	3.145701	4.023975	2.750794	1.512758
90.16	2.87214	2.808973	3.157013	3.945006	2.844677	1.483071
100.39	3.003159	3.124819	3.462503	4.144193	3.089528	1.557953
115.8	2.950404	3.150254	3.457107	3.903236	3.025984	1.467368
128.12	2.808355	3.046448	3.336363	3.61755	2.87414	1.359968
141.18	3.49567	2.969125	3.277991	3.399757	2.759343	1.278092
154.51	2.723483	3.090733	3.25183	3.434655	2.83846	1.291211
167.9	–	2.9156	3.103577	3.162394	2.653622	1.188859
Average	2.879825	2.976921	3.328313	4.03103	2.951441	1.51541

[a]EXP. Area method is used in data reduction

Table 9.6 Energy release rate calculated by various theories for Titanium/FM-X5/Titanium joint, original experiment data from [5], Unit: 10^3 J/m^2

a, mm	EXP.[a]	Theory	Beam theory	Pradeep (2010)	Penado (1993)	El-Zein (1988)
57.79	2.5759641	2.695957	3.245316916	5.059703	3.496242	5.314588
71.34	2.6274945	2.791347	3.337686341	4.517814	3.347762	4.745402
81.63	2.7653341	2.647826	3.142969859	3.936969	3.02941	4.135296
93.48	2.9477465	2.694341	3.107263735	3.708148	2.949602	3.894948
105.88	2.9099332	2.726367	2.991920052	3.520433	2.87634	3.697777
129.58	1.486955	2.162289	1.943941124	2.555824	2.166875	2.684575
175.76	1.9748004	1.738753	1.695481897	1.863362	1.649902	1.95723
202.57	–	2.597676	2.50584707	2.698627	2.428324	2.834572
Average	2.4697468	2.506819	2.746303374	3.48261	2.743057	3.658048

[a]EXP. Area method is used in data reduction

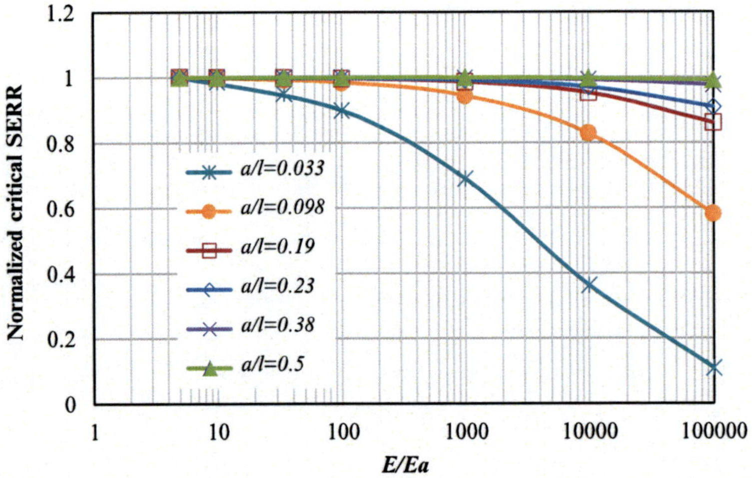

Fig. 9.4 Effect of Young's modulus on critical SERR

9.3.3 *Effect of Adhesive Young's Modulus*

Figure 9.4 shows the effect of adhesive Young's modulus on critical SERR, with *x*-axis showing the Young's modulus ratio of adherend to adhesive, and *y*-axis showing the critical SERR normalized by its initial value. The Young's modulus ratio of adherend to adhesive is accomplished by changing the adhesive Young's modulus, and keep adherend's constant. From Fig. 9.4, the adhesive Young's modulus affects the critical SERR differently at different crack length. When the crack length is relatively small compared to joint length, the change of adhesive Young's modulus affects the joint's critical SERR significantly, increase of E_a decreases the critical SERR nonlinearly. However, the influence of E_a tends to disappear with the increase of crack length. At $a/l = 0.38$, the change of E_a almost have no impact on the critical SERR, and further increase of crack length, the curves seems to be overlapping as in Fig. 9.4. The result indicates that adhesive Young's

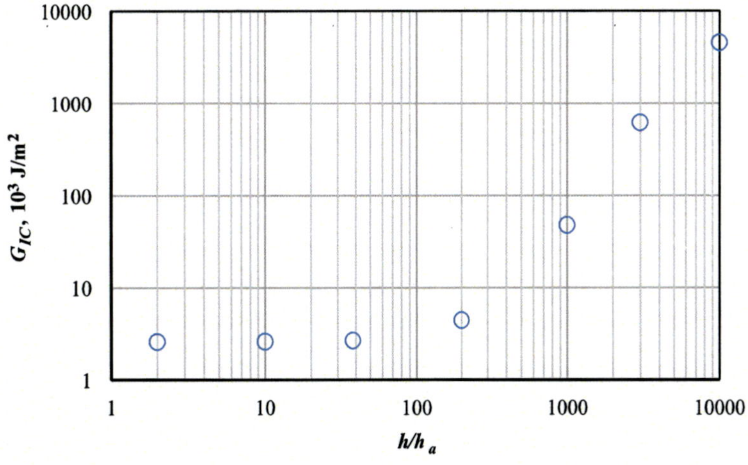

Fig. 9.5 Effect of adhesive thickness on critical SERR

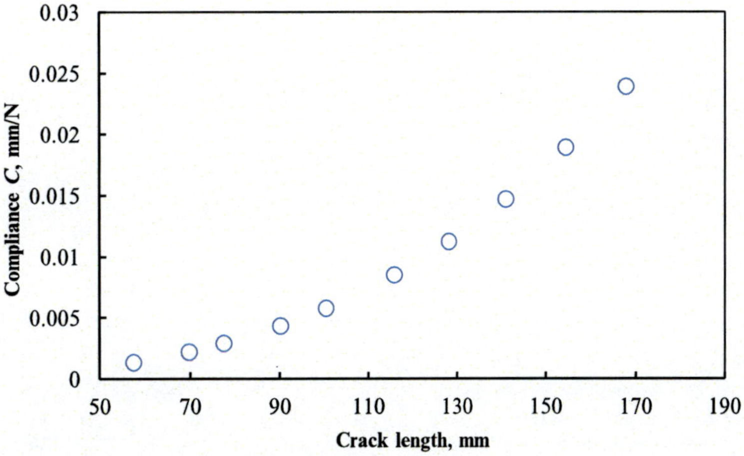

Fig. 9.6 Effect of crack length on joint compliance

modulus is necessary to be considered when the crack length is relatively small, however, in a large crack length joint, the adhesive Young's modulus is almost irrelevant.

9.3.4 Effect of Adhesive Thickness

Effect of adhesive thickness is investigated by changing the adhesive thickness at fixed adherend thickness. The result shown in Fig. 9.5 is obtained at crack length $a = 69.81$ mm. From Fig. 9.5, the change of adhesive thickness h_a has almost no influence on critical SERR when the thickness ratio of adherend to adhesive is below about 100, beyond which, however, the influence becomes significant, decrease of h_a greatly increases critical SERR. In real application, the thickness ratio of adherend to adhesive is rare beyond 100, current study provides a theoretical support for other theories which neglect the effect of adhesive thickness.

9.3.5 Effect of Crack Length

Joint compliance C versus crack length a is shown in Fig. 9.6. It is illustrated in the figure that joint compliance increases nonlinearly with the extension of crack length, the increase rate becomes larger as crack length increase. This indicates that the load is going to decrease as crack propagates as shown in Fig. 9.2, meanwhile, the critical SERR almost keeps unchanged as can been seen in Table 9.5, this is in agreement with the statement in [23].

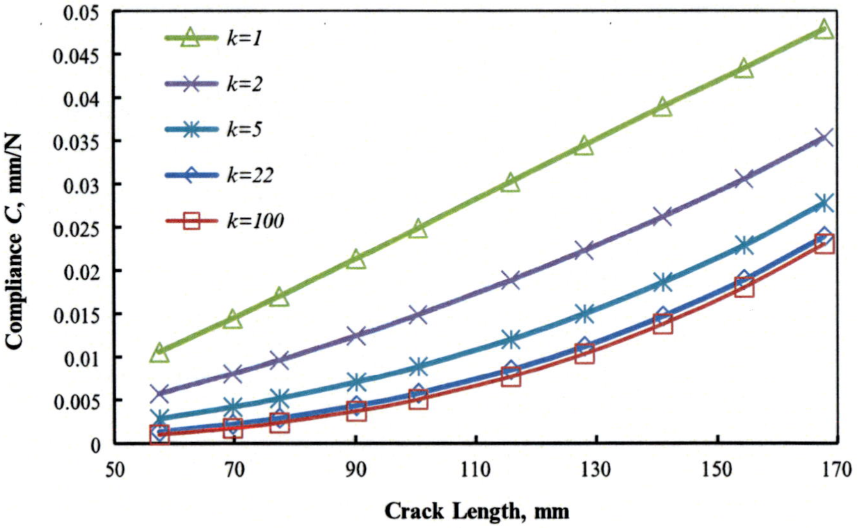

Fig. 9.7 Effect of k in correction factor on the joint compliance

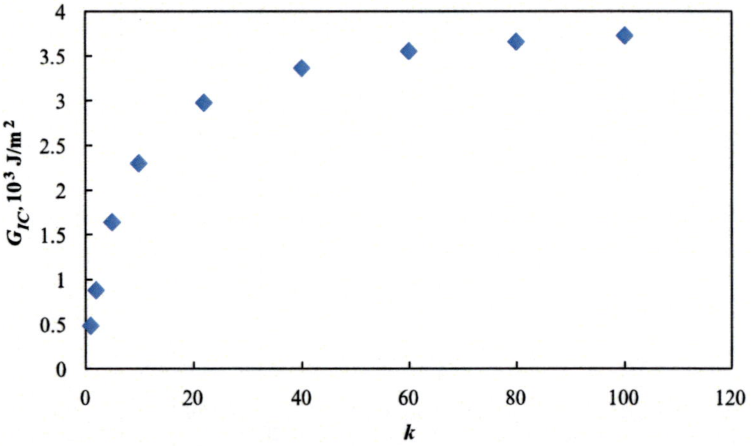

Fig. 9.8 Effect of k in correction factor on the critical SERR

9.3.6 *Effect of k in the Correction Factor*

The introduction of correction factor is mainly due to the using of continuum mixture theory for reduction of x-dependence of c. The effect of k in the correction factor is investigated. Figures 9.7 and 9.8 show, respectively, the effect of k in correction factor on the joint compliance and critical SERR. From these figures, the joint compliance is in inverse proportion to k, increase of k decreases joint compliance at a given crack length, and results in an increased nonlinearity compliance versus crack length curve, meanwhile, the curves becomes much closer. On the contrary, the critical SERR is in direct proportion to k, increase of k increases critical SERR, while the rate of increase is decreased as the tangent of the curve decrease.

9.4 Conclusion

Analytical model is developed to characterize the fracture behavior of an adhesively bonded joint loaded in peel. The model takes various variables into account, and continuum mixture theory is used for determining new introduced parameter c, also, the crack tip rotation is considered. The model has been applied to 3 different joints, current prediction has a better agreement with experiment compared with other theories. The parametric study leads to the following conclusion.

Adhesive Young's modulus is necessary to be considered when the crack length is relatively small (less than 20% of joint length), however, in a large crack length joint (above 38% of joint length), the adhesive Young's modulus is almost irrelevant.

The adhesive thickness has little influence on the critical SERR when h/h_a is approximately less than 100, but significant effect is observed beyond this value. Although it rarely happens in real application that thickness ratio of adherend to adhesive is bigger than 100, current study does provide a justification for other theories which neglect the effect of adhesive thickness.

Crack length affect joint compliance in direct proportion. With the increase of crack length, joint compliance goes up.

Constant k in the correction factor affects the joint compliance in inverse proportion at a given crack length, however, it affects the SERR in direct proportion.

References

1. Nassar S, Mao J, Yang X, Templeton D (2011) Effect of adhesives on the mechanical behavior of thick composite joints. In: Proceeding of the ASME 2011 pressure vessels & piping division conference, Baltimore, pp PVP2011-57692-1 ~ 9
2. Whitney J, Browning C, Hoogsteden W (1982) A double cantilever beam test for characterizing mode I delamination of composite materials. J Reinf Plast Compos 1:297–313
3. El-Zein M (1986) Characterization of mode I delamination of composite laminates at elevated temperatures. Southern Illinois University, Carbondale
4. Hashemi S, Kinloch AJ, Williams JG (1990) The analysis of interlaminar fracture in uniaxial fibre-polymer composites. In: Proceedings of the royal society of London. Series A, mathematical and physical sciences. The Royal Society, London, pp 173–199
5. Johnson W, Butkus L, Velentin R (1998) Applications of fracture mechanics to the durability of bonded composite joints. Springfield, Virginia: DOT/FAA/AR-97/56, National Technical Information Service (NTIS)
6. Shindo Y, Horiguchi K, Wang R, Kudo H (2001) Double cantilever beam measurement and finite element analysis of cryogenic mode I interlaminar fracture toughness of glass-cloth/epoxy laminates. J Eng Mater Technol 123:191–197
7. Jyoti A, Gibson R, Newaz G (2005) Experimental studies of mode I energy release rate in adhesively bonded width tapered composite DCB specimens. Compos Sci Technol 65:9–18
8. Xie W, Sitaraman SK (2003) Investigation of interfacial delamination of a copper-epoxy interface under monotonic and cyclic loading: experimental characterization. IEEE Trans Adv Packag 26(4):447–452
9. Pradeep K, Rao B, Sivakumar S, Balasubramanium K (2010) Interface fracture assessment on sandwich DCB specimens. J Reinf Plast Compos 29(13):1963–1977
10. Barenblatt G (1959) The formation of equilibrium cracks during brittle fracture. General ideas and hypothesis. Axisymmetrical cracks. J Appl Math Mech (PMM) 23:622–636
11. Yang Q, Thouless M, Ward S (1999) Numerical simulations of adhesively-bonded beams failing with extensive plastic deformation. J Mech Phys Solid 47:1337–1353
12. Andersson T, Stigh U (2004) The stress–elongation relation for an adhesive layer loaded in peel using equilibrium of energetic forces. Int J Solid Struct 41:413–434
13. Andersson T, Biel A (2006) On the effective constitutive properties of a thin adhesive layer loaded in peel. Int J Fract 141:227–246
14. Ouyang Z, Li G (2009) Local damage evolution of double cantilever beam specimens during crack initiation process: a natural boundary condition based method. J Appl Mech 76(5):051003-1–051003-8
15. Adams R, Coppendale J (1979) The stress–strain behaviour of axially-loaded butt joints. J adhes 10:49–62
16. Chai H (1988) Shear fracture. Int J Fract 37:137–159
17. El-Zein M, Reifsnider K (1988) Evaluation of GIC of a DCB specimen using an anisotropic solution. J Compos Technol Res 10:151–155
18. Penado F (1993) A closed form solution for the energy release rate of the double cantilever beam specimen with an adhesive layer. J Compos Mater 27:383–407
19. Jacobsen T, Sorensen B (2001) Mode I intra-laminar crack growth in composites- modelling of R-curves from measured bridging laws. Compos Part A: Appl Sci Manufact 32:1–11
20. Tamuzs V, Tarasovs S, Vilks U (2001) Progressive delamination and fiber bridging modeling in double cantilever beam composite specimens. Eng Fract Mech 68:513–525
21. Nayfeh AH, Nassar SA (1982) The influence of bonding agents on the thermomechanically induced interfacial stresses in laminated composites. Fibre Sci Technol 16:157–174
22. Meo M, Thieulot E (2005) Delamination modelling in a double cantilever beam. Compos Struct 71:429–434
23. Broek D (1982) Elementary engineering fracture mechanics. Martinus Nijhoff, Boston

Chapter 10
Probabilistic Characterization of Elastic Properties of Composites Using Digital Image Correlation Technique

Mark R. Gurvich and Patrick L. Clavette

Abstract Reliability assessment of composite structures requires probabilistic description of material properties including high-fidelity evaluation of their main statistical parameters. With respect to composite materials, the probabilistic description should capture anisotropy of random elastic properties and their possible statistical correlations. A general approach of probabilistic characterization of anisotropic stiffness based on digital image correlation (DIC) technique is developed. Since composites are anisotropic, their "traditional" probabilistic characterization usually requires expensive and time-consuming programs of multiple tests. The developed approach, however, uses only a single test per coupon to estimate several random properties in different directions since the entire 2D distributions of surface strains are measured instead of one value in conventional experiments. Major benefits of the developed approach are significant (a) simplification of characterization; (b) cost and time reduction; (c) improvement in accuracy; and (d) generation of all necessary statistical parameters. Demonstration of the developed probabilistic approach is shown on example of representative carbon-fiber reinforced composite.

Keywords Composites • Reliability • Probabilistic characterization • Digital image correlation • Elasticity • Statistical correlation

10.1 Introduction

Experimental characterization of the entire elastic behavior of composites is usually a time-consuming problem due to anisotropy of material properties and necessity to quantify numerous parameters versus only a few characteristics (e.g., modulus and Poisson's ratio) in testing of isotropic materials. More traditional testing approaches are primarily based on "one test, one parameter" philosophy. Thus, application of digital image correlation (DIC) techniques for elastic characterization of composites seems to be especially promising due to opportunity to measure all strains at considered surfaces, i.e., potentially generate *three* independent pieces of data (two normal and one shear strains) from each test. Another huge advantage of DIC is its capability to characterize the entire field of strains rather than one (or a limited number) of points in more traditional testing. Along with an opportunity to capture and understand expected *deterministic* non-uniformity of strain distributions (e.g., under bending or in the vicinity of holes), DIC can provide a potential way to quantify *statistical* non-uniformity due to inevitable randomness of material properties. Based on this motivation, the objective of this study is to (a) develop a general concept of DIC-based experimental characterization of random elastic properties of composites; (b) provide a specific solution for a case of uni-axial load; and (c) demonstrate its robust experimental implementation on example of representative composite.

M.R. Gurvich (✉) • P.L. Clavette
United Technologies Research Center, 411 Silver Lane, MS 129-73, East Hartford, CT 06108, USA
e-mail: GurvicMR@utrc.utc.com; ClavetPL@utrc.utc.com

E. Patterson et al. (eds.), *Composite Materials and Joining Technologies for Composites, Volume 7: Proceedings of the 2012 Annual Conference on Experimental and Applied Mechanics,* Conference Proceedings of the Society for Experimental Mechanics Series 44, DOI 10.1007/978-1-4614-4553-1_10,

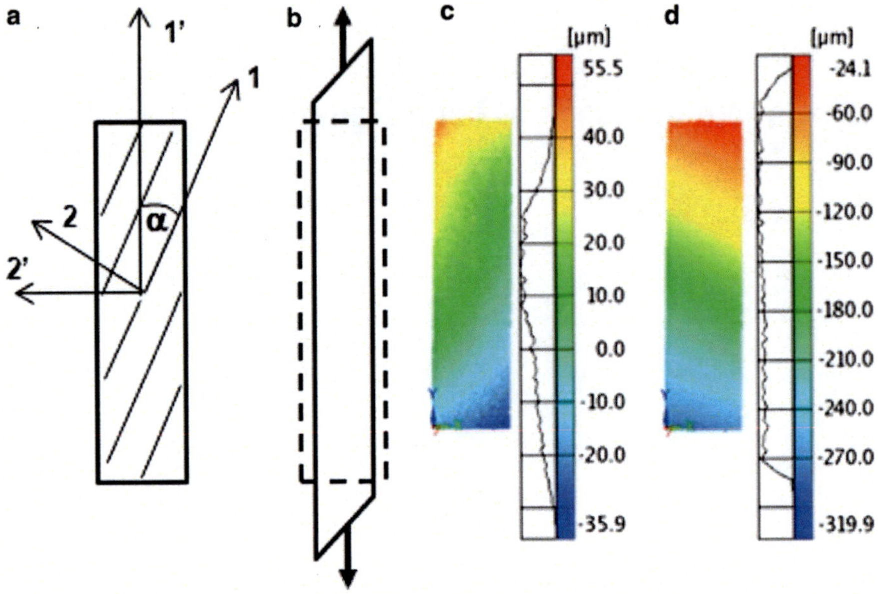

Fig. 10.1 Schemes of (**a**) composite coupon, (**b**) its deformation under uni-axial load, and representative DIC measurements of displacements (**c**) U_2' and (**d**) U_1' ($\alpha = 45°$)

10.2 Concept

Developed approach is based on consideration of orthotropic materials under conditions of in-plane loading. Probabilistic deformation $\tilde{\varepsilon}_{ij}$ of the material under random load $\tilde{\sigma}_{ij}$ can be written in a general matrix form as

$$\begin{bmatrix} \tilde{\varepsilon}_{11} \\ \tilde{\varepsilon}_{11} \\ \tilde{\varepsilon}_{12} \end{bmatrix} = \begin{bmatrix} \tilde{S}_{11} & \tilde{S}_{12} & 0 \\ & \tilde{S}_{22} & 0 \\ sym & & \tilde{S}_{66} \end{bmatrix} \begin{bmatrix} \tilde{\sigma}_{11} \\ \tilde{\sigma}_{22} \\ \tilde{\tau}_{12} \end{bmatrix}; \quad \begin{bmatrix} \tilde{\varepsilon}'_{11} \\ \tilde{\varepsilon}'_{11} \\ \tilde{\varepsilon}'_{12} \end{bmatrix} = \begin{bmatrix} \tilde{S}'_{11} & \tilde{S}'_{12} & \tilde{S}'_{16} \\ & \tilde{S}'_{22} & \tilde{S}'_{26} \\ sym & & \tilde{S}'_{66} \end{bmatrix} \begin{bmatrix} \tilde{\sigma}'_{11} \\ \tilde{\sigma}'_{22} \\ \tilde{\tau}'_{12} \end{bmatrix} \tag{10.1}$$

where $\tilde{S}_{ij}$, $\tilde{S}'_{ij}$ are random characteristics of the compliance matrix defined in the axes of symmetry and arbitrary axes, respectively. (In this study, sign "~"above a variable indicates its random nature, and sign "-"shows its average value). Deterministic relationships between S_{ij} and S'_{ij} are well-known (see, for example, [1]) and are defined by angle of transformation α (Fig. 10.1a). Thus, assuming that random compliance characteristics are distributed by the normal law, the entire problem of probabilistic elastic characterization can be formulated as evaluation of corresponding statistical parameters in the axes of symmetry, namely, average values $[\bar{S}]$ and correlation matrix $[K_S]$:

$$[\bar{S}] = \begin{bmatrix} \bar{S}_{11} \\ \bar{S}_{12} \\ \bar{S}_{22} \\ \bar{S}_{66} \end{bmatrix}; \quad [K_S] = \begin{bmatrix} K_{S11,11} & K_{S11,12} & K_{S11,22} & K_{S11,66} \\ & K_{S12,12} & K_{S12,22} & K_{S12,66} \\ & & K_{S22,22} & K_{S22,66} \\ sym & & & K_{S66,66} \end{bmatrix} \tag{10.2}$$

In case of known deterministic load, random deformation can be experimentally quantified in form of average values $\bar{\varepsilon}'_{ij}$ and corresponding correlations $K'_{\varepsilon ij,kl}$. Therefore, Eqs. 10.1 and 10.2) provide fundamental relationships of a classical linear regression problem in a generalized form as

$$y = \sum_{m=1}^{M} a_m x_m \tag{10.3}$$

where y are *known* measurements of strains (both average values $\bar{\varepsilon}'_{ij}$ and correlations $K'_{\varepsilon ij,kl}$); a_m are *unknown* statistical parameters of compliance matrix in the axes of symmetry (also both average values $\bar{S}_{ij}$ and correlations $\bar{K}_{Sij,kl}$); x_m are *known* parameters defined by load orientation, strain orientation, load parameters; and $M=14$ is the total number of unknown parameters (four for average values and ten for correlations). Classical method of statistical treatment (e.g., the least square method) can be easily applied to solve regression problem Eq. 10.3. The following section describes practical implementation of this concept in detail.

10.3 Specific Solution

Consider a case of uni-axial deterministic load σ'_{11} (Fig. 10.1b) with corresponding known statistical characteristics of deformation $\bar{\varepsilon}'_{ij}$ and $K'_{\varepsilon ij,kl}$ as a result of testing. Therefore, according to Eq. 10.1, three random characteristics of the compliance matrix can be easily defined as

$$\tilde{S}'_{11} = \tilde{\varepsilon}'_{11}/\sigma'_{11};\ \tilde{S}'_{12} = \tilde{\varepsilon}'_{22}/\sigma'_{11};\ \tilde{S}'_{16} = \tilde{\varepsilon}'_{12}/\sigma'_{11} \tag{10.4}$$

and summarized as the following partial matrices of corresponding statistical parameters:

$$[\bar{S}'] = \begin{bmatrix} \bar{S}'_{11} \\ \bar{S}'_{12} \\ \bar{S}'_{16} \end{bmatrix};\ [K'_S] = \begin{bmatrix} K'_{S11,11} & K'_{S11,12} & K'_{S11,16} \\ & K'_{S12,12} & K'_{S12,16} \\ sym & & K'_{S16,16} \end{bmatrix} \tag{10.5}$$

where

$$\bar{S}'_{11} = \bar{\varepsilon}'_{11}/\sigma'_{11};\ \bar{S}'_{12} = \bar{\varepsilon}'_{22}/\sigma'_{11};\ \bar{S}'_{16} = \bar{\varepsilon}'_{12}/\sigma'_{11} \tag{10.6}$$

$$K'_{S11,11} = K'_{\varepsilon 11,11}/\left(\sigma'_{11}\right)^2;\ K'_{S12,12} = K'_{\varepsilon 22,22}/\left(\sigma'_{11}\right)^2;\ K'_{S16,16} = K'_{\varepsilon 12,12}/\left(\sigma'_{11}\right)^2 \tag{10.7}$$

$$K'_{S11,12} = K'_{\varepsilon 11,22}/\left(\sigma'_{11}\right)^2;\ K'_{S11,16} = K'_{\varepsilon 11,12}/\left(\sigma'_{11}\right)^2;\ K'_{S12,16} = K'_{\varepsilon 22,12}/\left(\sigma'_{11}\right)^2 \tag{10.8}$$

Assuming that the randomness of material orientation can be ignored, relationships between matrices Eqs. 10.5 and 10.2 can be written in any arbitrary coordinate system as

$$[\bar{S}'] = [F][\bar{S}];\ \ [K'_S] = [F][K_S][F]_t \tag{10.9}$$

where

$$[F] = \begin{bmatrix} m^4 & 2m^2n^2 & n^4 & m^2n^2 \\ m^2n^2 & m^4+n^4 & m^2n^2 & -m^2n^2 \\ -2m^3n & 2mn(m^2-n^2) & 2mn^3 & mn(m^2-n^2) \end{bmatrix};\ \begin{matrix} m=\cos\alpha \\ n=\sin\alpha \end{matrix}; \tag{10.10}$$

Equation 10.9 are final statements of linear regression analysis to quantify *unknown* statistical parameters $[\bar{S}]$ and $[K_S]$ through *known* (measured) statistical parameters $\bar{\varepsilon}'_{ij}$ and $K'_{\varepsilon ij,kl}$. One may note that *four* unknown parameters of $[\bar{S}]$ are functions of only *three* known average parameters of deformation Eq. 10.6. To overcome this discrepancy, at least, two different angles of transformation should be considered. Similarly, *ten* unknown parameters of $[K_S]$ are defined by only *six* parameters of correlations Eqs. 10.7 and 10.8. It also indicates that, at least, two different orientations should be sufficient for the entire characterization. Consideration of more than two orientations ($\alpha \neq 0$; 90°) would only increase fidelity of statistical estimations.

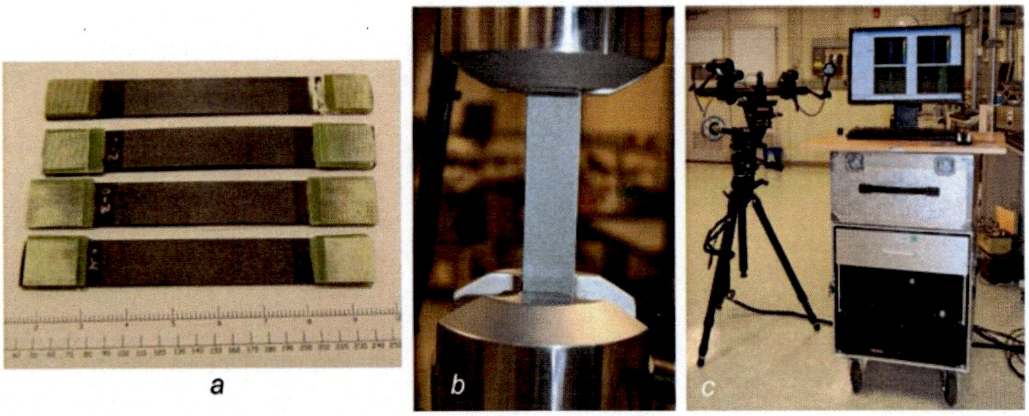

Fig. 10.2 Representative coupons (**a**), coupon under testing (**b**); and DIC implementation (**c**)

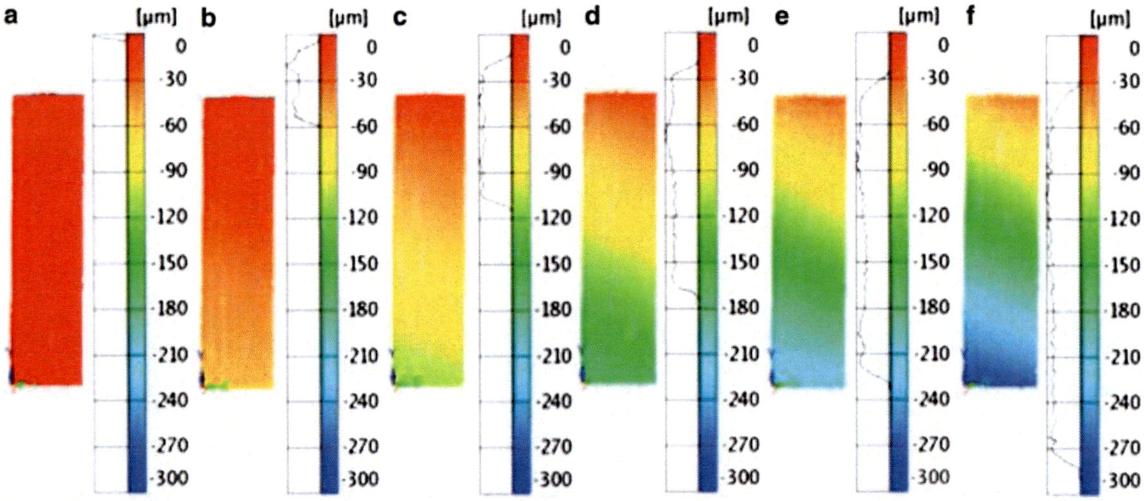

Fig. 10.3 Distributions of DIC-measured displacements U_1' as function of applied load: (**a**) 0; (**b**) 450; (**c**) 900; (**d**) 1,350; (**e**) 1,800; and (**f**) 2,250 N ($\alpha = 45°$)

10.4 Experimental Implementation

For the experimental demonstration of the developed solution, laminated carbon-fiber composite coupons with lay-up $[\alpha]_{20}$ were considered. The coupons were prepared using IM7 12000 filament count tow uni-directional carbon tape from Hexcel impregnated with Cycom 977–3 resin (ply thickness = 0.005″; the volume fraction = ~63%). The rectangular 25×200 mm coupons were considered with material orientation α = 0; 15; 30; 45; 60; 75; and 90° (Fig. 10.2a). Tensile tests were conducted on a 90 kN servo-hydraulic MTS load frame (Fig. 10.2b). Hydraulic MTS grips were used. A displacement rate of 1.25 mm/min was applied in a tensile mode to the biased specimens. The displacement was increased at a constant rate until a load of 2.25 kN was achieved. Linear-elastic behavior was observed for all considered orientations and load ranges.

During the test, full field DIC-based measurement was conducted to observe the center of the specimen gage section. The DIC equipment was a GOM Aramis 5 M (5 megapixel) stereo system (Fig. 10.2c). The DIC imaging frame rate was varied in order to achieve roughly 100–200 images for each test, being estimated from the anticipated modulus change as a function of bias angle. The imaged region was an area of ~100 mm length in the center of the approximately 150 mm long gage section. For the purposes of the analysis, a coordinate transformation was imposed to align (a) the axial direction with the left edge of the specimen; (b) the long transverse direction with the bottom of the imaged area, and (c) the short transverse direction normal to the plane surface of the specimen. Representative measured displacements at $\alpha = 45°$ are shown in Fig. 10.1c, d.

Representative DIC results for one specimen at $\alpha = 45°$ demonstrate the deformation behavior in detail in Fig. 10.3 as function of applied load. Expected non-symmetry of deformation is clearly observed with opportunity to quantify non-zero

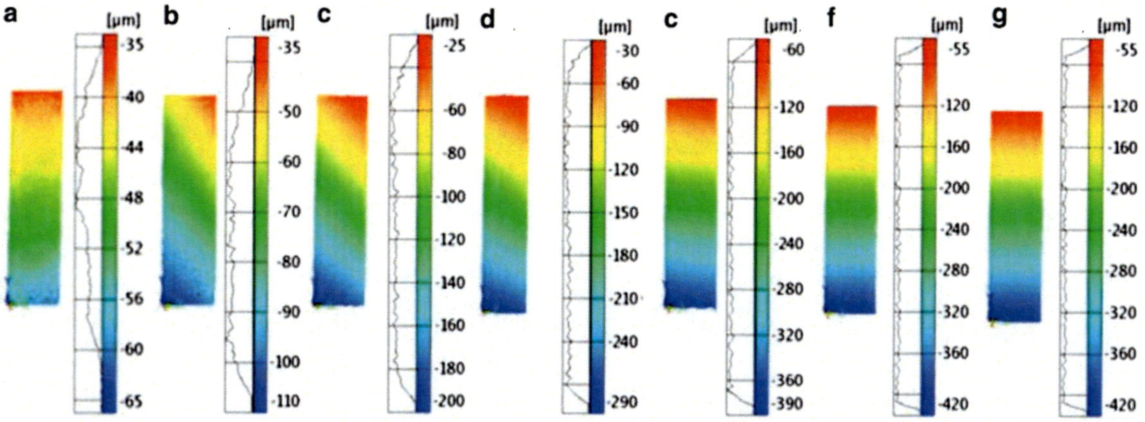

Fig. 10.4 Representative distributions of DIC-measured displacements U_1' as function of material orientation: (**a**) 0; (**b**) 15; (**c**) 30; (**d**) 45; (**e**) 60; (**f**) 75; and (**g**) 90° (tension of 2,250 N)

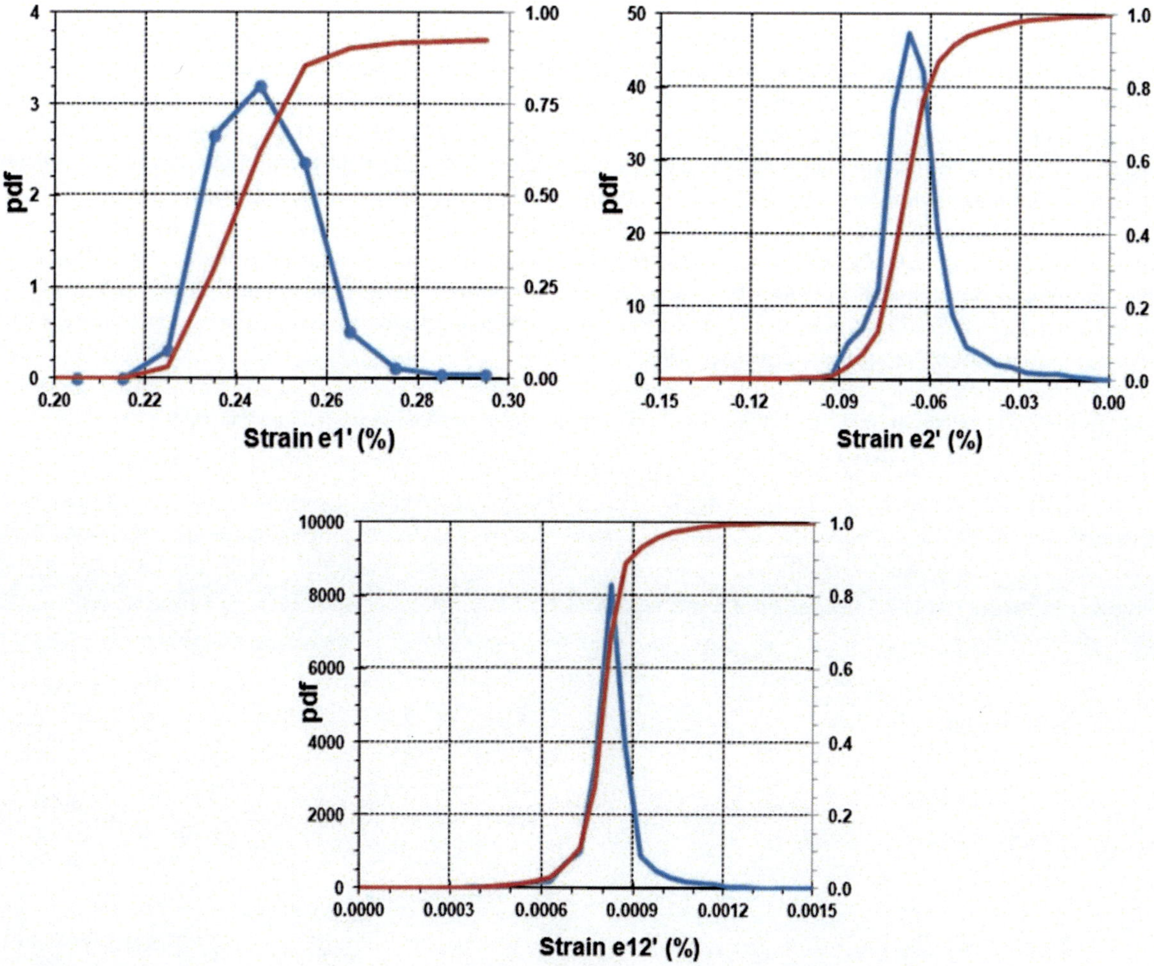

Fig. 10.5 Statistical distributions (pdf and cumulative probability) of strains for a representative test ($\alpha = 45°$)

shear strains. Variation of the deformation field as a dependence of material orientation is illustrated in Fig. 10.4. As expected, deformation at $\alpha = 0$ and 90° is practically symmetric since compliance characteristics $\bar{S}_{16}=\bar{S}_{26} = 0$ for such angles. Minor deviation from the symmetry is observed for $\alpha = 60$ and 75°. Significant shear is observed for angles between 15° and 45°. It is important to note that the iso-displacement contours do not match exactly the bias angle in considered examples. This is due to the effect of the boundary conditions at the grip section restricting the shear deformation. To mitigate this effect, much longer coupons should be considered. In other words, criteria to define coupon sizes for DIC characterization could be, in general, different than similar definitions applied in more traditional characterization of composite stiffness and strength.

Finally, representative statistical distributions of random strains are illustrated in Fig. 10.5. The distributions are calculated for one coupon only ($\alpha = 90°$) but for relatively large statistical populations of measured strains (more than 12,000 points per each distribution). In the considered example, all DIC-based measurements of strains within one coupon are taken into consideration. Other more advanced statistical approaches (e.g., sampling) can be applied as well to achieve more efficient statistical treatment, especially, in case of many coupons and too big statistical populations. This example shows that statistical parameters of deformation ($\bar{\varepsilon}'_{ij}$ and $K'_{\varepsilon ij,kl}$) can be easily quantified. Let us note that, in general, there is difference between statistical variation within one coupon and within a population of coupons. Therefore, for probabilistic characterization of *material* properties (rather than an *individual* coupon), several samples of the same material orientation should be considered in a corresponding test program with follow-up statistical treatment of *all* results together.

10.5 Conclusions

A relatively simple approach of full statistical characterization of random elastic properties of composite materials is proposed. The approach is based on unique opportunities of the DIC technique, using only one test, to (a) quantify deformation of multiple points; and (b) capture deformation in different directions. Corresponding convenient relationships between statistical characteristics of random strains and random characteristics of anisotropic compliance matrix are given as functions of material orientation. It is shown that only two considered material orientations could be sufficient for characterization of all average values as well as all correlations of random elastic properties. Experimental illustration of the approach, shown on example of representative uni-directional composite, indicates applicability and simplicity of the proposed analysis. Although the approach is illustrated for uni-axial tension, similar implementation can be suggested for more complex load cases, e.g., in-plane stress state. Also, a simplified variant of the approach can be equally applied for deterministic characterization of anisotropic elastic properties, where only average values of measured strains are used to generated average values of compliance characteristics.

Acknowledgements The authors thank the United Technologies Research Center for permission to publish this paper.

Reference

1. Bogdanovich A, Pastore CM (1996) Mechanics of textile and laminated composites: with Applications to Structural Analysis, Chapman and Hall, London, 570p

Chapter 11
Mechanics of Fiber-Reinforced Porous Polymer Composites

Sandip Haldar and Hugh A. Bruck

Abstract Fiber reinforcement has been found in natural porous structures, such as Palmetto Wood. These reinforcements appear to improve the mechanical behavior of these porous structures to enhance strength and energy absorbing capability. Similar use of fiber reinforcement has yet to be pursued in synthetic porous materials to form composites. Using multi-scale DIC, we previously investigated the macroscale and microscale mechanisms responsible for the mechanical behavior of Palmetto wood, which we used to formulate new mechanical models of the effects of fiber reinforcement on the evolution of damage and plastic strain. We have now formulated synthetic fiber-reinforced porous materials with fiber reinforcement to determine if these effects translate under quasi-static flexure using multi-scale DIC and the new mechanical models.

Keywords Fiber-reinforced sandwich composite • Flexural behavior • Damage • Digital Image Correlation (DIC)

11.1 Introduction

The multiscale mechanical characterization of Palmetto wood using Digital Image Correlation (DIC) previously revealed the role of macrofiber reinforcement in porous cellulose matrix on load transfer and failure mechanisms under quasi-static three-point bending [1, 2]. It was determined that the global bending deformations are transferred through localized bending of the fibers, while failure occurs due to pore collapse preceding shear-dominated fiber-matrix debonding. A damage model was subsequently developed to decouple the effect of the pore collapse and plastic deformation in the palmetto wood. The evolution of damage as well as plastic deformation has been investigated in detail using the model developed for both quasi-static and dynamic three-point bending [3]. Motivated by the unique mechanical behavior of Palmetto wood, foam-core sandwich composite structures have been developed to achieve enhanced mechanical behavior through the introduction of pultruded carbon fibers into the core similar to macrofibers in Palmetto wood. Therefore, in this paper the effect of the fiber-reinforcement on the mechanical performance and damage evolution of the porous polymer composite has been studied by comparing porous polymer cores with and without fiber-reinforcement in sandwich composite structures under quasi-static three point bending.

11.2 Fabrication of Sandwich Composite Structures with Fiber-Reinforced Porous Polymer Composite Cores

Sandwich composite structures were prepared from soft closed-cell foam as core and epoxy-carbon composites from The Composite Store. Carbon rods of diameter of 0.027″ from The Composite Store have been used as reinforcement in the soft foam. Epoxy resin 105 and slow hardener 206 from West System have been used as the adhesive in the interfaces between facesheet – core and carbon rod reinforcement – foam core. Two types of adhesives were prepared, namely, pure-epoxy based and nano-enhanced epoxy. West system epoxy resin 105 with 206 slow hardener has been mixed by 5:1 ratio by weight.

S. Haldar (✉) • H.A. Bruck
Department of Mechanical Engineering, University of Maryland, College Park, MD 20742, USA
e-mail: shaldar@umd.edu; bruck@umd.edu

E. Patterson et al. (eds.), *Composite Materials and Joining Technologies for Composites, Volume 7: Proceedings of the 2012 Annual Conference on Experimental and Applied Mechanics*, Conference Proceedings of the Society for Experimental Mechanics Series 44,
DOI 10.1007/978-1-4614-4553-1_11,

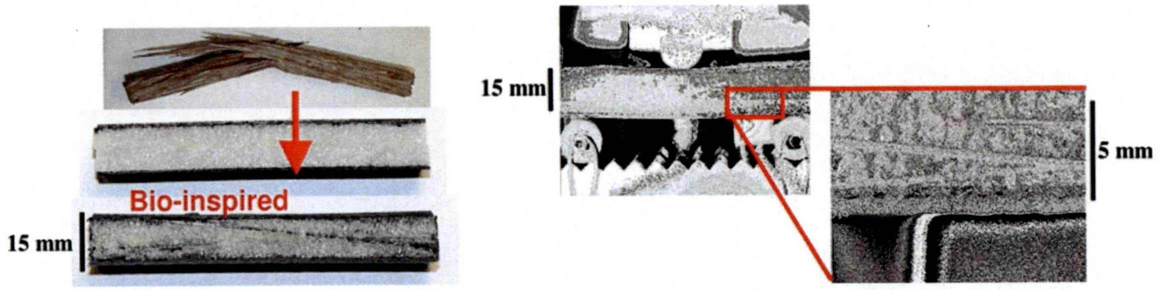

Fig. 11.1 (*Left*) Development of composite sandwich structure with reinforcement in the core and (*right*) multiscale experimental methodology

Nano-enhancement was achieved by mixing 1 wt% PR-19 grade low density carbon nanofiber (LDCNF) that was obtained from Applied Sciences. The pure epoxy adhesive was prepared by hand-mixing of the epoxy resin and hardener for 3–5 min, with a subsequent degassing for 5 min at a vacuum chamber with 22 mmHg vacuum pressure. The nano-enhanced adhesive was prepared by a hand-mixing of epoxy resin and LDCNF for 3–5 min followed by 90 min of sonication and a degassing of 3–5 min in the vacuum chamber with 22 mmHg vacuum pressure. These pure epoxy and nano-enhanced epoxy based adhesives have been used as adhesives to glue the soft foams, and facesheets of carbon strips. The foam and facesheets were manually attached by the adhesives detailed above. For another class of specimen, reinforcement of Carbon rods have been used by penetrating the carbon roads in the soft foam core after dipping them in the adhesive to achieve better adhesion with the foam. The glued and reinforced specimens were cured in a closed chamber at 80°C for 30 h under a dead weight. The bio-inspired sandwich structures fabricated for this investigation are depicted in Fig. 11.1.

11.3 Effect of Macrofiber Reinforcement on the Damage Evolution in Palmetto Wood

The damage behavior of palmetto wood under quasi-static and dynamic loading is investigated through the flexural response of the specimens. The Damage parameter is determined by $D = 1 - (E/E_0)^2$. The following expression for the evolution of damage with total flexural strain based on strain energy conservation was then fit to the experimental measurements [4],

$$D = D_{\lim}[1 - \exp(-A(\varepsilon - \varepsilon_d))] \tag{11.1}$$

In this expression, $D_{\lim}$ represents the limiting amount of damage that can occur, ε_p is the strain at which damage initiates, and A is an acceleration factor that describes the evolution of damage with strain. Results from fitting the curves to the experimentally determined damage can be seen in Fig. 11.2. Using Eq. 11.1, it was possible to study the effect of the macrofiber volume fraction and loading rate on the damage accumulation. The acceleration factor and limiting amount of damage did not appear to be affected substantially by loading rate or macrofiber concentration. However, the strain at which damage begins to accumulate did decrease by 15–30%.

In addition to the damage that accumulates with flexural strain, the mechanism of plasticity due to pore collapse must also be accounted for. Therefore, the plastic flexural strain, ε_p was related to the total flexural strain through a form of the conventional power law hardening relationship as follows,

$$\varepsilon_p = \varepsilon\left[1 - \exp\left(-B\left(\varepsilon - \varepsilon_y\right)^p\right)\right] \tag{11.2}$$

where ε_y is the yield strain and p and $B$are related to the power law hardening exponent and coefficient, respectively. The plastic strain could be determined experimentally by measuring the permanently deformed shape of the specimen. The evolution of macroscopic plastic strain with respect to the total strain could then be determined by fitting Eq. 11.2 to the data. The resulting fits can be seen in Fig. 11.3.

From the constants that were determined, it can be seen that the hardening exponent does not appear to vary with macrofiber concentration and loading rate. Instead, the hardening coefficient appears to increase by approximately 50% as the loading rate increases. The yield strain also substantially decreases by 40–50% as the loading rate increase, but does not vary as substantially with macrofiber concentration. This would tend to be consistent with the previous comment on the time required for the local strain energy level to reach a critical level that results in a strain rate dependency for the pore collapse

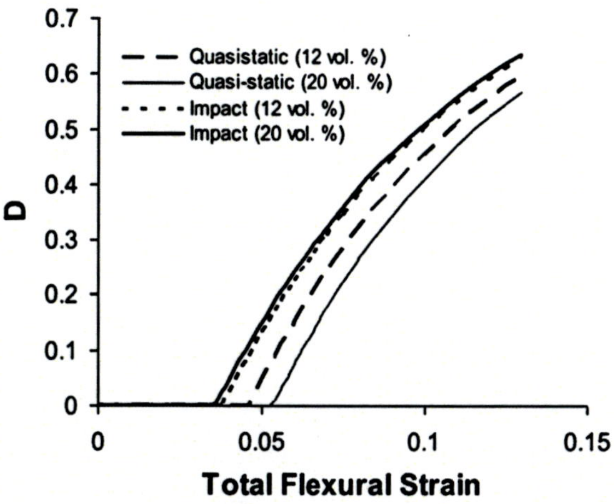

Fig. 11.2 Damage evolution with total global flexural strain in Palmetto wood of 12 and 20 vol% macrofiber [3]

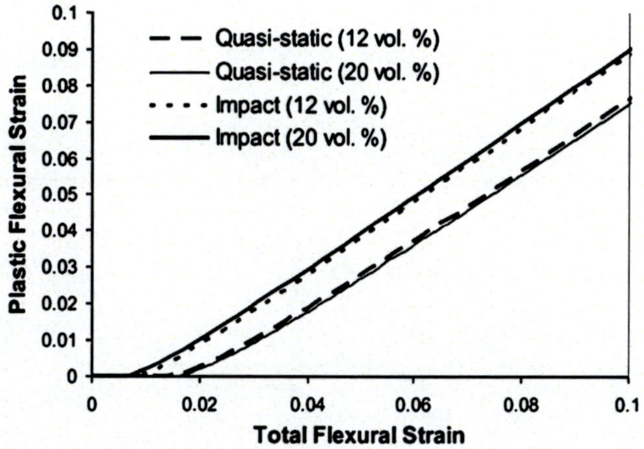

Fig. 11.3 Evolution of plastic flexural strain with global total flexural strain in Palmetto wood of 12 and 20 vol% macrofiber under quasi-static and dynamic load [3]

mechanism responsible for plasticity. Equations 11.1 and 11.2 were combined to determine the constitutive response of the Palmetto wood as follows,

$$\sigma = E_0(1 - D)^2(\varepsilon - \varepsilon_p) \tag{11.3}$$

The resulting fits to the experimental data can be seen in Fig. 11.4. It was previously determined that for quasi-static loading these values increased with macrofiber concentration according to the rule-of-mixtures. As the loading rate increases, these values increase substantially by approximately 300%, a higher increase than was observed with the Weibull model fit. The sensitivity of these values to loading rate is similar to that observed in polymeric materials. Given the fibrous nature of Palmetto wood, it is natural to assume that the dependency is related to the viscoelastic behavior of the material. Thus, the moduli could be associated with the rubbery and glassy response of the Palmetto wood at low and high loading rates, respectively.

11.4 Characterization of Sandwich Material with Reinforced Core

Sandwich composite structures have been developed using reinforcement of pultruded carbon rod in the porous polymeric foam core to replicate the structure of Palmetto and investigate the strength and energy absorbing capabilities. The sandwich composite specimens have been characterized under quasi-static three-point bend test. The quasi-static response of sandwich specimens with varying reinforcement in the foam core has been depicted in Fig. 11.5. The increase of reinforcement in the foam core

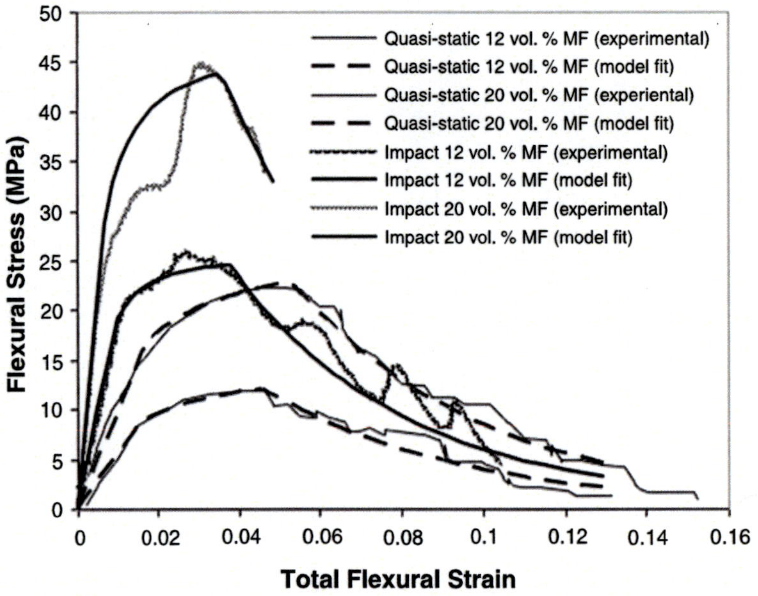

Fig. 11.4 Model fit to the experimental flexural response of Palmetto wood under quasi-static and dynamic load with 12 and 20 vol% macrofiber [3]

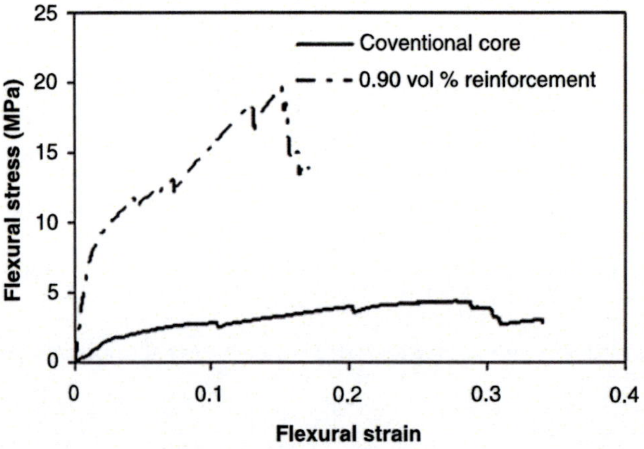

Fig. 11.5 Flexural response of the conventional sandwich and sandwich with bio-inspired core under quasi-static three point bend test

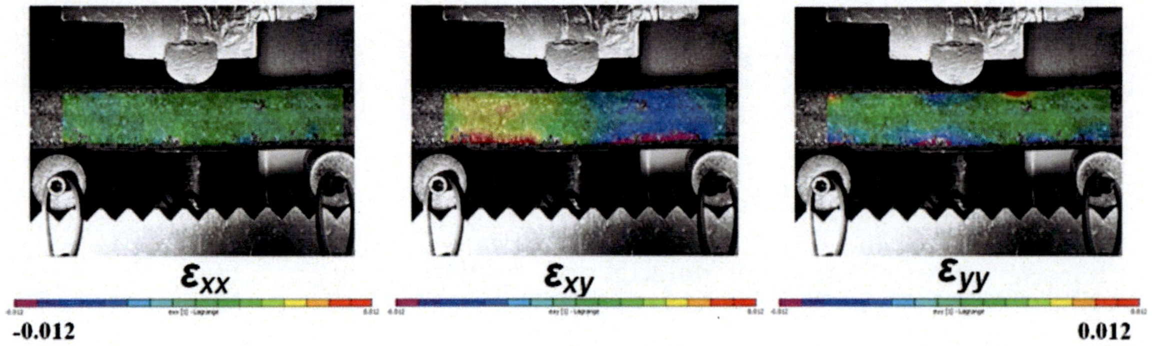

Fig. 11.6 Macroscale strain fields in sandwich composite structure under quasi-static flexure associated with global bending deformations transmitted through localized bending of the fibers

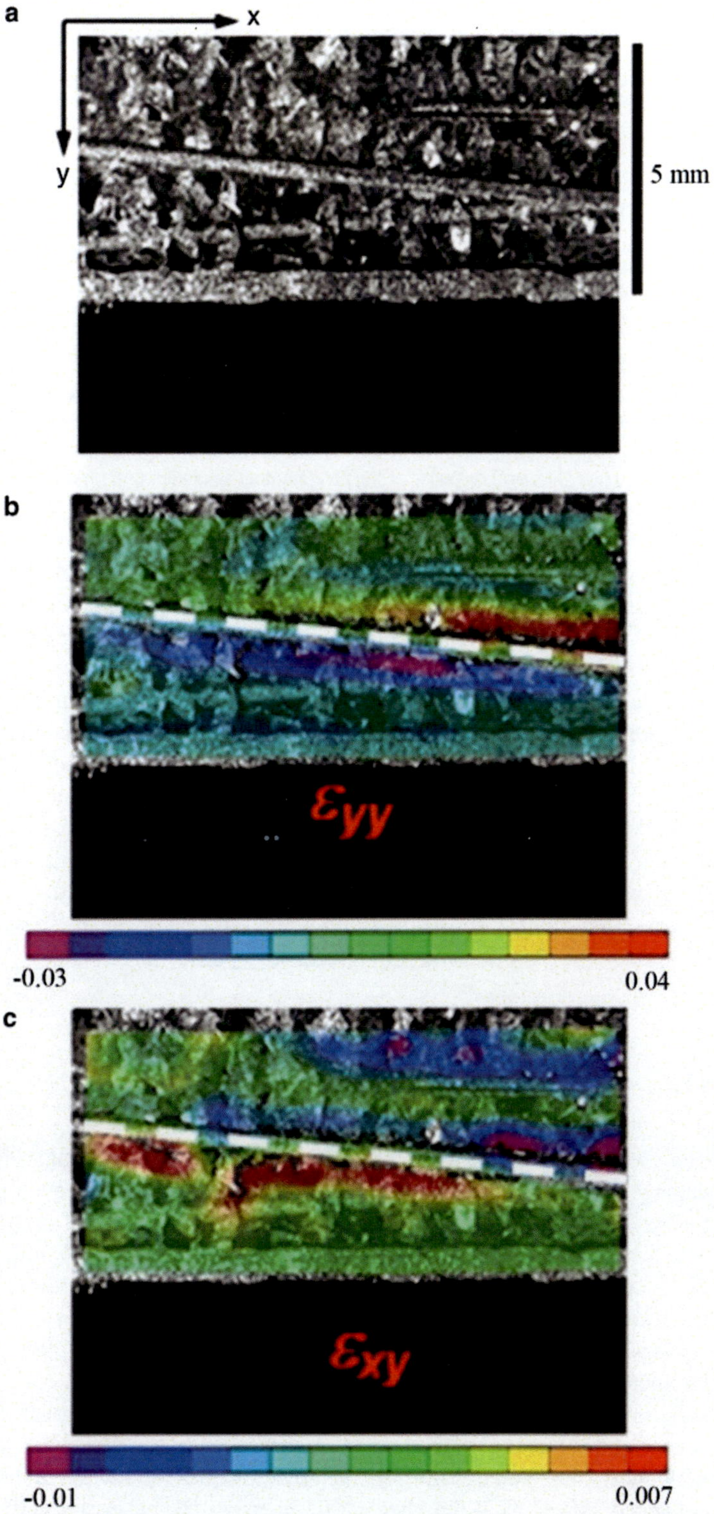

Fig. 11.7 (a) Optical image used for DIC, Local (b) Compressive and (c) Shear strain concentration around the fiber-reinforcement obtained by DIC. Inhomogeneous strain fields in sandwich composite structure at higher magnification obtained from DIC of the images captured at higher magnification associated with failure due to pore collapse and fiber-matrix debonding mechanisms

increases its stiffness as well as energy absorbance. It was determined that a reinforcement 0.90 vol% could result the increase in stiffness by around 100%. The strain fields obtained by the DIC analysis of the images captured at macroscale in Fig. 11.6 shows the deformation similar to that of homogenous material under flexure. However, the strain fields obtained from DIC of the images captured at higher magnification (~10×) in Fig. 11.7 reveals the inhomogeneous strain field in the sandwich. In Palmetto wood, similar strain fields were observed at the same magnification levels and attributed to the load transfer of the global bending

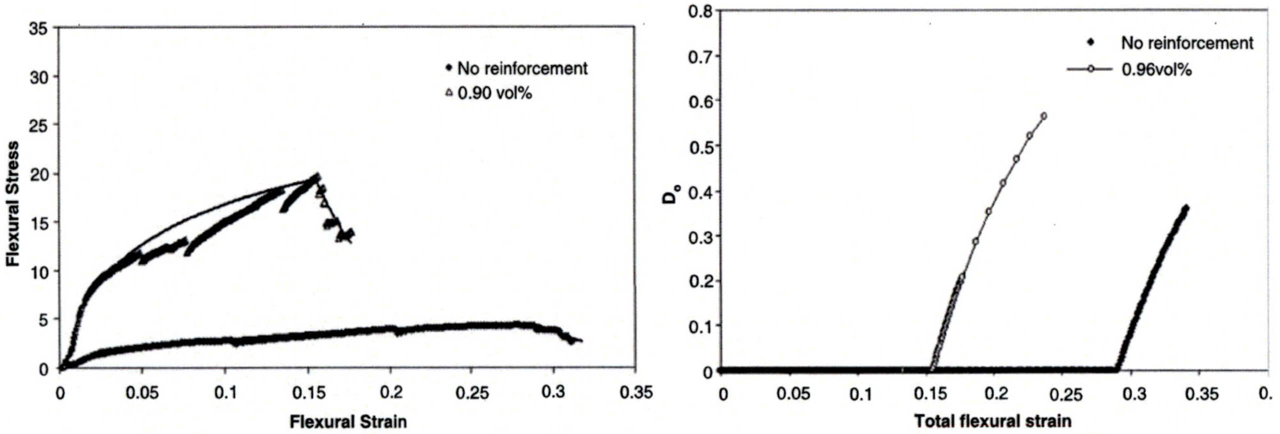

Fig. 11.8 (*Left*) Flexural response of the conventional sandwich and sandwich with bio-inspired core under quasi-static three point bend test and (*right*) damage evolution with total flexural strain in sandwich cores with different reinforcement

deformations through the localized bending of the fibers, and failure due to pore collapse and fiber-matrix debonding mechanisms [2]. Therefore, it appears that fiber-reinforcement of the sandwich cores are exhibiting similar mechanisms.

The damage model developed to investigate the damage evolution in Palmetto wood has been employed to the flexural response of the composite sandwich structures with reinforced core. The constitutive relation developed to relate the damage and plastic strain is used fit the flexural response of the composite sandwich structure. The resulting fit and experimental data is shown in Fig. 11.8. The evolution of damage with total global flexural strain is also depicted in Fig. 11.8. It is seen that the reinforcement of pultruded carbon rod in the core of the sandwich structure enhances the damage resistance by increasing the stiffness and stress at which damage and plastic deformation initiates, although it is limited by a lower damage initiation strain (~15%) than is experienced without fiber reinforcement (~30%). The lower damage initiation strain is closer to what was observed in the Palmetto wood (~4–5%), albeit it the concentration of fibers in the synthetic core is approximately 10% than in Palmetto wood. So, it is anticipated that higher loading levels will further decrease the damage initiation strain while increasing the stiffness and stress at which damage and plastic deformation initiate.

11.5 Conclusions

The damage evolution in Palmetto wood and the effect of macrofiber reinforcement in porous cellulose matrix on the damage behavior have been investigated. A model has been developed to decouple the effect of the damage due to fiber-matrix debonding and plasticity due to pore collapse on flexural strain. The model was able to reveal both faster plastic strain and damage evolution under dynamic loading in Palmetto wood, but the evolution was not substantially affected by the concentration of macrofibers. However, the stiffness and subsequent strength are affected by both the loading rate and the subsequent concentration of macrofibers.

Sandwich specimens have been developed based on using closed-cell soft-foam as a core material and carbon-epoxy laminates as facesheets. Reinforcement of pultruded carbon rods has been used in the soft foam core similar to the structure observed in Palmetto wood. Using the hierarchical structure of Palmetto wood as a template, nano-enhancement was employed in the epoxy based adhesive used in the facesheet-foam and reinforcement-foam interfaces. The preliminary quasi-static and dynamic characterization of the sandwich materials shows that the pultruded carbon fibers with nano-enhanced adhesives at the fiber-matrix interface increases the stiffness and energy absorption of the sandwich structure. The damage modeling of the behavior of sandwich composite structures with reinforcement in the core showed that the use of fiber reinforcement affects the evolution of damage in a manner that is similar to what has been observed in Palmetto wood.

Acknowledgements This work was supported by ONR award number N000140910640.

References

1. Gershon AL, Bruck HA, Xu S, Sutton MA, Tiwari V (2010) Multiscale mechanical and structural characterizations of Palmetto wood for bio-inspired hierarchically structured polymer composites. Mater Sci Eng C 30:235–244
2. Haldar S, Gheewala N, Grande-Allen J, Sutton MA, Bruck HA (2011) Multi-scale mechanical characterization of Palmetto wood as a template for biologically-inspired polymer composites. Exp Mech 51(4):575–589
3. Haldar S, Bruck HA, (2012) Characterization of dynamic damage mechanisms in Palmetto wood as biological inspiration for impact resistant polymer composites, Submitted to Mechanics of Materials
4. Chow CL, Lu TJ (1989) On evolution law of anisotropic damage. Eng Fract Mech 34(3):679–701

Chapter 12
Mechanics of Multifunctional Skin Structures

Hugh A. Bruck, Kelsey Cellon, Satyandra K. Gupta, Mark Kujawski, Ariel Perez-Rosado, Elisabeth Smela, and Miao Yu

Abstract Robotic structures are typically fabricated using discrete components that serve only a single function, such as carbon fibers or polymer composites for structural reinforcement. The ability to integrate these discrete components into a single structure provide enormous opportunities to enhance the performance of robots by reducing weight and increasing the energy storage/harvesting. We have recently been pursuing the integration of sensors and solar cells into multifunctional "skin" structures. One issue that limits the integration of these components is the effect that that the interaction between them will have on the mechanical behavior, and its subsequent impact on multifunctional performance. To minimize these effects and better understand the mechanics of multifunctional skin structures, we first pursued compliant strain sensors integrated onto the compliant wings of a flapping wing MAV to sense deformations at the wing in real time. These measurements were correlated to the thrust force of the wing. The effects of these compliant strain sensors were also determined by comparing the new thrust measurements with the original wings and with 3D Digital Image Correlation (DIC) measurements. It was determined by SRS analysis that strain sensors tended to be more sensitive to lower integral modes of the flapping frequency than the thrust measurement data, while the DIC measurements correlated more directly with thrust measurements in the time domain. Thus, it would appear that the strain sensor and DIC measurements are sensitive to different aspects of the mechanics of the multifunctional skin structures that can be taken into account when using the strain sensors for real-time sensing and control. In addition to compliant strain sensor integration, flexible Solar Cells (SCs) were also integrated onto the compliant wings. Their ability to harvest energy during flight has the potential to prolong the time of flight for more autonomous operation. The effects of the integrated SCs on the 3D shapes of the compliant wings were also characterized during flapping. The solar cells had a significant effect on the wing shape, increasing the stiffness of the wings and reducing the volume of air that the wings need to capture to generate thrust, particularly at the apex and nadir of the flapping cycle where thrust is generated by blowback. Thus, a multifunctional analysis was developed to account for the tradeoff between the extra power require to generate the necessary thrust for the MAV to stay airborne versus what is gained from integrating the SCs in order to determine how to design the multifunctional skin structures to optimize performance.

12.1 Introduction

Over the past several years, there has been an increasing interest in Miniature Air Vehicles (MAVs). The development of these vehicles have been for commercial, research, government, and military purposes. The size of these aircrafts is similar to the size of large insects and small birds. Thus, engineers have turned to nature for better designs. Biological systems are adaptive and stable, making them an ideal goal for engineers to achieve. Many biological systems are multifunctional and serve more than one purpose. For example, birds use their wings to fly; however, besides simply creating lift, these wings are self-sensing, self-healing, and have thermoregulation. If engineers can incorporate these multifunctional structures into their designs, a simple structure can serve more than one purpose and improve the overall performance of the vehicle.

Many MAVs use flapping wings to create the lift necessary to achieve flight. These designers are trying to achieve the same flapping motion as birds, bats, or insects. However, these creatures use their wings for more than simple flight. They are able to sense the pressure created against their wings, and adjust their flight accordingly. By achieving the same

H.A. Bruck (✉) • K. Cellon • S.K. Gupta • M. Kujawski • A. Perez-Rosado • E. Smela • M. Yu
Department of Mechanical Engineering, University of Maryland, College Park, MD 20742, USA
e-mail: bruck@umd.edu

E. Patterson et al. (eds.), *Composite Materials and Joining Technologies for Composites, Volume 7: Proceedings of the 2012 Annual Conference on Experimental and Applied Mechanics*, Conference Proceedings of the Society for Experimental Mechanics Series 44,
DOI 10.1007/978-1-4614-4553-1_12, © The Society for Experimental Mechanics, Inc. 2013

sensing capabilities on the wings of an MAV, more information of the vehicle's flight is known at the wing level. This information allows for a better control system so that the vehicle can autonomously achieve the required lift force to maintain height.

The best way to sense the deformation of the wing during flight is to measure the strain on the surface of the wing. A new way to measure strain must be applied that minimizes the effects of the sensors on wing deformation during flight. Conventional strain gauges add weight and affect the overall compliance of the wing. They also only measure strain at a specific points on the wing. To get an overall measurement of strain, several strain gauges must be applied across the wing. This adds even more weight and affects the overall performance of the wings.

To overcome these challenges, compliant latex sensors can be applied to the surface of wings. These sensors have been previously developed and used for other platforms. The output from these sensors were measured at different frequencies and correlated to the thrust force generated at those frequencies.

Other processes in nature also help to inspire new ways of integrating multifunctional elements into an existing structure. Many researchers are looking at ways to integrate solar cells into their designs for energy harvesting. By integrating solar cells into the wings of an MAV, the wings would also act as an energy source for the MAV. It would allow for an increased flight time, while decreasing the payload contribution of a large power source, thus potentially allowing for either a size reduction with the same performance capability, or an increase in overall payload capacity. The integration of solar cells is also a stepping-stone on the path of completely removing the need for an external power source, such as a heavy battery, which creates huge improvements in payload requirements. However, the effects of integrating these solar cells and strain sensors on MAV performance must be characterized and modeled.

In this investigation, we integrate flexible SCs and strain sensors into the compliant wings of a flapping wing MAV. Their effects on thrust generation are characterized, as are their effect on wing shape. Digital Image Correlation (DIC) was used to observe the effects of the solar cells and compliant strain sensors on the overall wing shape during flapping.

12.2 Integration of Multifunctional Elements into Skin Structures

There is a general approach to integrating multifunctional elements to existing structures.

1. *Integrate Multifunctional Elements into Structures*: In this step the materials and processes to integrate the new elements to the existing structure are determined and performed. The aim is that these materials and processes have a minimal effect on the wing's performance.
2. *Characterize Impact of the Multifunctional Elements*: The effects of the multifunctional elements are actually measured and compared to the original measurements of wings lift, thrust, and wing shape. Three different measurements must be taken to have a complete description performance.
3. *Develop Multifunctional Models*: From the observed physical characteristic of the elements tested, new models must be developed. These models must quantify the benefits of integrating the elements into the existing structure.

For this case, the sensing elements being integrated into the wings are compliant strain sensors. These strain sensors were developed by mixing electrically conductive exfoliated graphite (EG) into a latex polymer host. The conductivity of the overall material is controlled by the amount of EG in the host [1]. These elastomeric strain gages are painted (airbrushed) onto the surface of interest. They can be readily applied on different parts of the surface, allowing the strain at different parts of the wing to be measured. These strain sensors have a high gauge factor, making them ideal for this application since they will have high sensitivity. These elastomeric strains sensors also have a low stiffness making them compliant (stretchable) and thus compatible with compliant skin structures. ES removed Fig. 12.1.

By using these elastomeric sensors, we can measure the strains on a flapping wing without compromising the wing's compliant movement. In Fig. 12.1, the location of these strain sensors on the wing is shown. A grid was patterned where the largest deformation was seen visually. The output from these sensors was then correlated to the thrust force produced by the flapping motion.

Because no data had previously been accumulated to guide solar cell placement to minimally affect the compliance of the wings, the solar cells were selected based on the location of the spars and the shapes of the different cells. The goal was to minimize the increase in stiffness of the Mylar-based film, so that the wing could still form the ideal bowl-like shape, to capture as much volume and direct the flow of air out of the wing in a direction to maintain good lift and thrust. For that reason, the solar cells were placed near the spars in areas that seemed to minimize an increase in stiffness (Fig. 12.2).

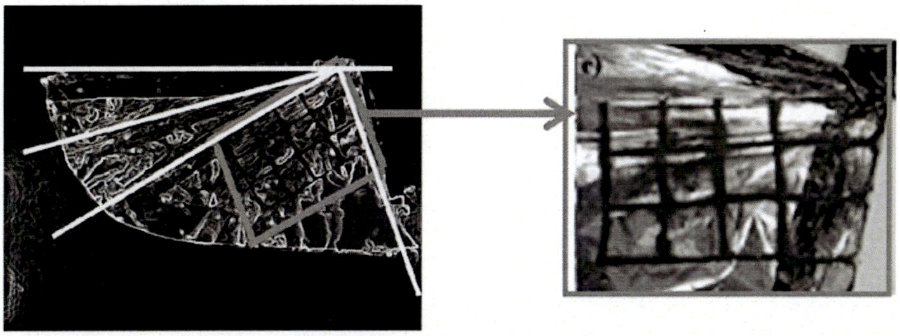

Fig. 12.1 Compliant strain sensors on a compliant wing for real-time strain sensing with carbon fiber spar locations designated in *yellow* and the strain sensing grid located in *red*

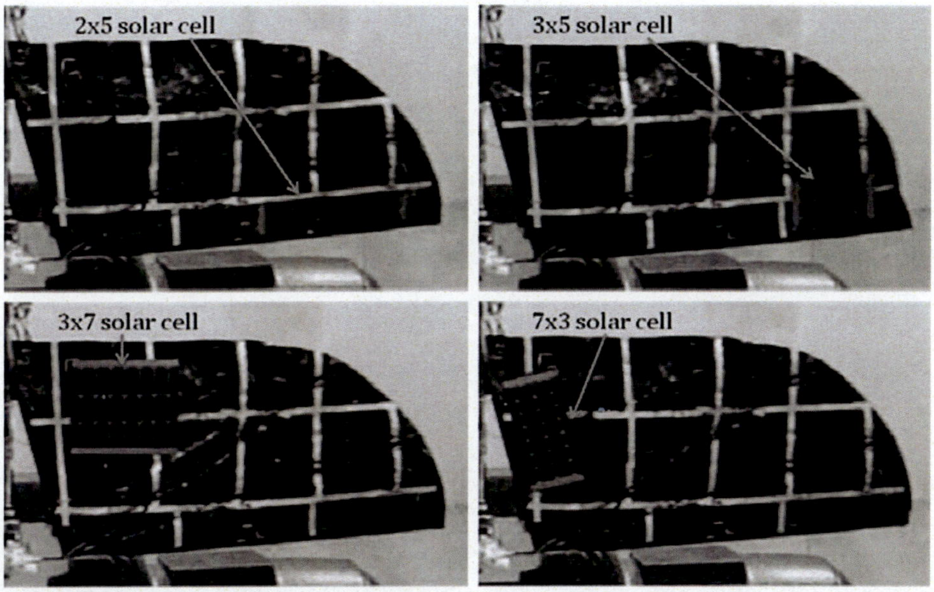

Fig. 12.2 Location of solar cells of varying size on compliant wing

12.3 Measuring Thrust Force

To measure the thrust forces generated by the flapping wings we used a test stand that we previously developed to characterize the lift and thrust forces associated with the shapes the wing takes during flight [2]. The main component of the test stand is a RABIS rigid linear air bearing form Nelson Air. The air bearing has a steel slider in it and is restricts the motion to one degree of freedom. At the end of the slider was an LCFD-1KG miniature load cell which converted the loads generated by the wing flapping into a voltage signal for processing. The voltage response from the load cell was collected by a National Instruments (NI) PXI-6040E cardbus data acquisition system using the NI Measurement & Automation Explorer software to record voltage as a function of time (Fig. 12.3).

12.4 Measuring 3D Wing Deformations Using DIC

To completely determine the effects the compliant strain sensors and solar cells have on the wings, the deformations of the wings must be measured. Using these deformations, the strains measured on the wing via the compliant strain sensors can be validated. 3D Digital Image Correlation (DIC) was used to measure the deformations of the wings.

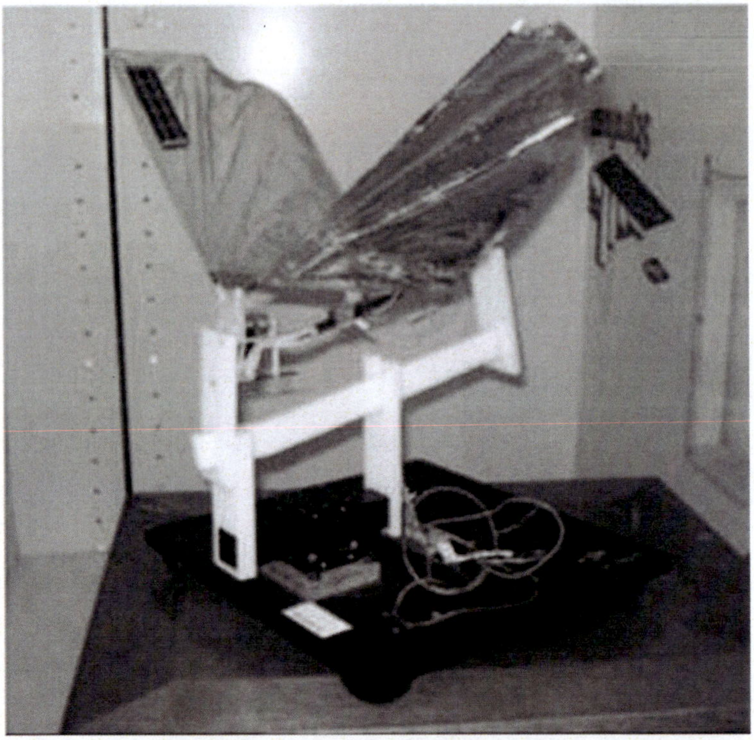

Fig. 12.3 Flapping wing MAV designed at UMD on a customized test stand for measuring thrust generated by the multifunctional compliant wings

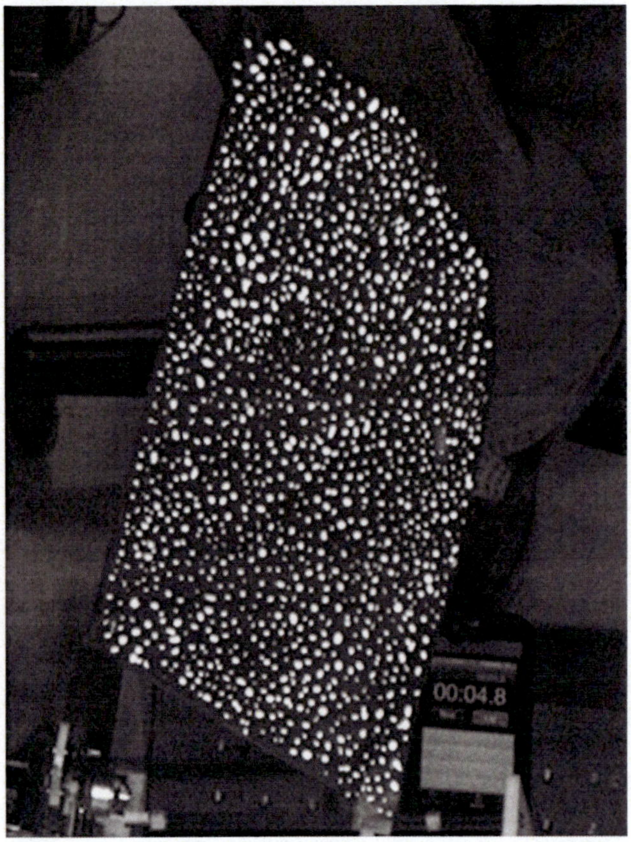

Fig. 12.4 Painted compliant wing used for DIC

Digital Image Correlation is an optical method to measure deformation on an object surface under real loading conditions. The method tracks the gray value pattern in small neighborhoods during deformation. Thus, the wings were painted with a speckle pattern on the top side for cameras to register these neighborhoods (Fig. 12.4).

The same test stand form the lift and thrust measurements was used to observe the wing deformations. In order to get a complete measurement of deformation only one wing was attached to the MAV. Using two cameras the wing was observed while the MAV was flapping at a constant frequency. The cameras were snapping pictures at 30 frames per second. Using the VIC 3D software from Correlated Solutions, the pictures were analyzed. First, the deformation of the wing was observed in the out of plane direction. Then the strains throughout the wing were observed. The deformation and strain measurements were also plotted as a function of time.

12.5 Correlations Between Strain Measurements and Test Stand Measurements

Since all of the measurements were plotted as functions of time, they could all be compared directly. In all of the data, the frequency of the flapping bird was very clear. The strains showed a large peak, whereas the force measurements showed a large valley. Thus, it appears that there is some degree of correlation between the two signals (Fig. 12.5).

To obtain a better comparison of the signals, the Shock Response Spectra (SRS) were also obtained to determine the frequency response due to the transient loading event with no damping. Figure 12.6 shows a comparison of the SRS for a compliant strain sensor and the load cell at a 3 Hz flapping frequency. From this comparison, it can be seen that the strain sensors (green) appear to be more sensitive to the lower frequencies associated with the integral modes of the flapping frequency. However, the load cell tends to have a stronger response at higher frequency integral modes. This may be due to the strain sensors giving a local response at a given point on the wing, where the load cell is measuring a global response of the wing. Thus, it may be that the transient characteristics of the strain sensors are such that they will filter out higher frequency components that contribute to generate thrust, which must be accounted for when correlating the strain sensor to thrust generation for autonomous flapping control.

In addition to the strain sensors, we have also been able to directly compare the measurements obtained using 3D DIC on the wing at the same location as the compliant strain sensor. These results can be seen in Fig. 12.7 for the normal strain transverse to the spar. These measurements indicate that the generated thrusts correlate directly with changes in the transverse strain, more so than the compliant strain sensor. Furthermore, these strain levels are within the elastic range of the wing material as well as the sensing range of the compliant strain sensor. Thus, it would appear that both the compliant strain sensors and 3D DIC are both capable of characterizing different characteristics of mechanics of the compliant wing associated with its thrust performance.

12.6 Characterization of 3D Wing Shapes with Integrated Solar Cells

It was also possible to characterize changes in the 3D wing shape due to the integration of solar cells. Figure 12.8 compares the deformation before and after solar cells were integrated onto the surface of the wing. It can be seen that it was possible to ascertain the volume changes in the upflap and downflap portions of the flapping cycle at discrete positions during the flapping

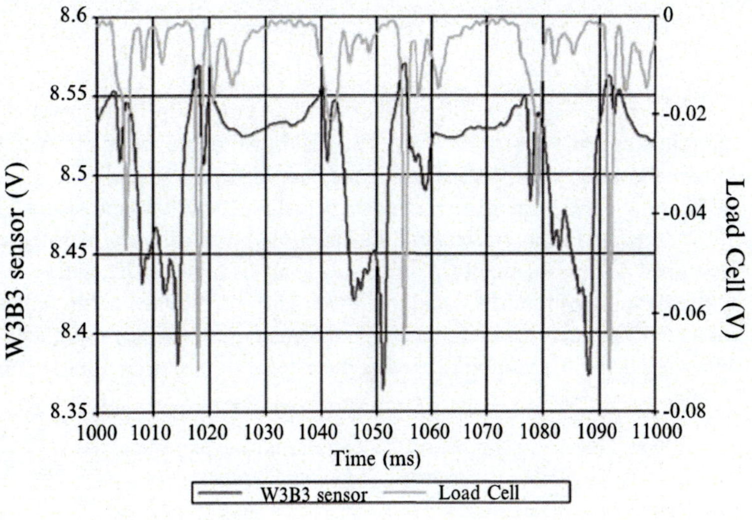

Fig. 12.5 Strain detected by the compliant strain sensor and the corresponding thrust at 3 Hz

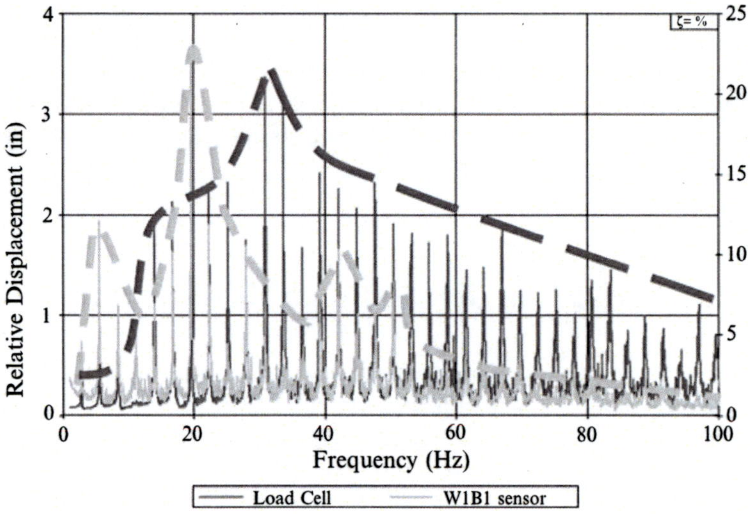

Fig. 12.6 Comparison of SRS for compliant strain sensor and the load cell measurement (3 Hz)

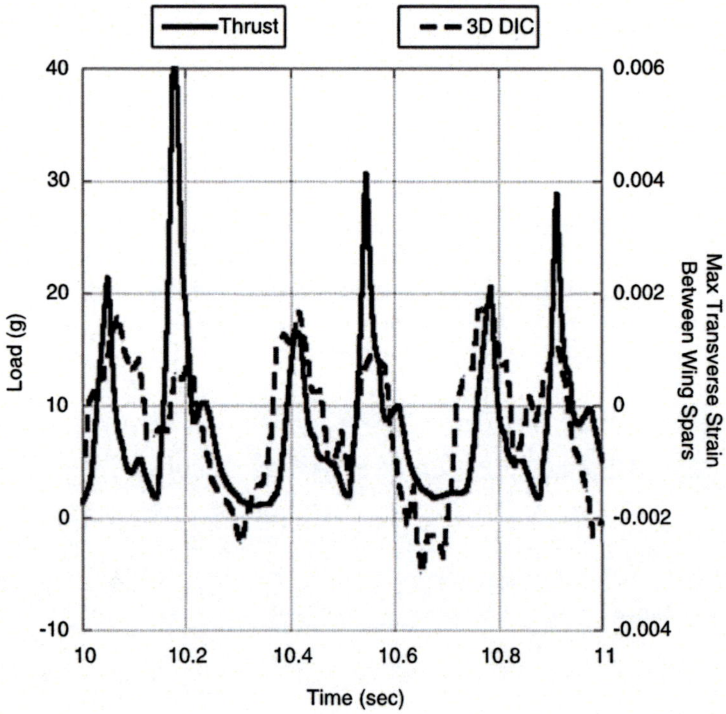

Fig. 12.7 Comparison of thrust measurements with normal strain measurements from 3D DIC obtained transverse to the spars at the same location as the compliant strain sensor

cycle [3]. From Fig. 12.8 it is evident that the wing shape is somewhat flatter than the original shape when solar cells are integrated onto the compliant wing. Thus, adding the solar cells stiffened the wing. This translates into a slight decrease in the volume the wing encompasses during the flapping cycle. In particular, it appears to reduce the volume at the apex and nadir of the flapping cycle, which are the most critical positions for generating lift due to a phenomenon known as "blowback".

To understand the impact on multifunctional performance of integrating solar cells into flexible MAV wings, we also developed a "time-of-flight (t_f)" trade-off analysis relating the energy stored in the battery (U_{bat}) with the power consumed by the flapping of the MAV and the power harvested by the solar cells (P_{SC}). U_{bat} is a function of the capacitance (C) and voltage (V) of each battery, and P_{SC} is a function of efficiency (η), solar flux (S_s), and area (A_{sc}). We have determined the following general equation to predict the effects of integrating solar cells and batteries on t_f [3].

$$t_f = \frac{U_{bat}}{kP_{structure} - P_{SC}} = \frac{n_{battery}CV^2}{U_{structure}(EI_{bat}, EI_{SC}, A_{bat}, A_{SC}, n_{bat}, n_{sc}, D)\, f - n_{SC}\eta S_s A_{sc}}$$

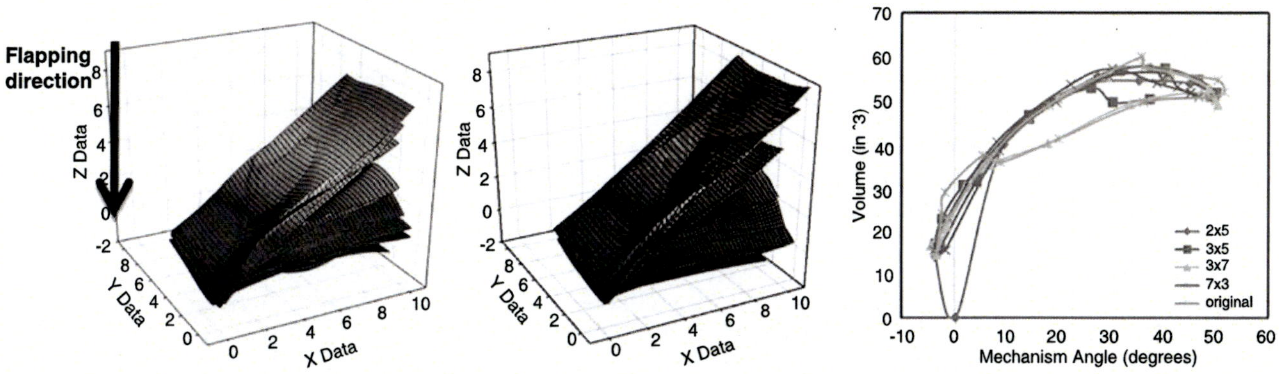

Fig. 12.8 (*left*) Shape of flexible MAV wing in Fig. 12.2 with and without solar cells during down flap cycle, and (*right*) corresponding volume measurements for various solar cell configurations [3]

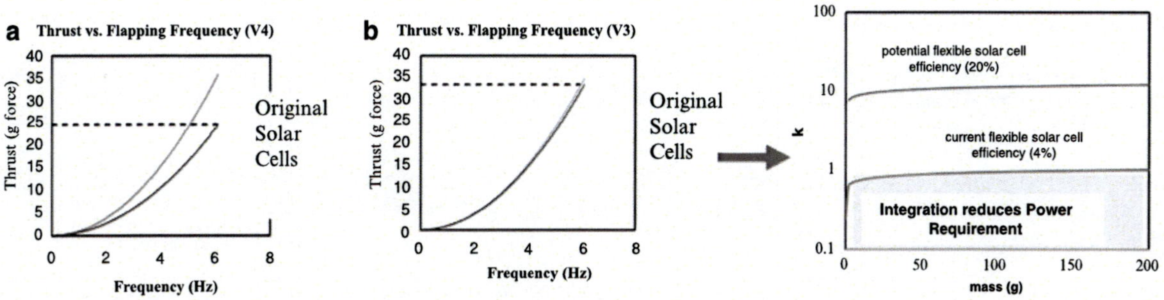

Fig. 12.9 (*left*) Thrust versus frequency profiles for two wing structures, (**a**) and (**b**), with different compliances with (*red*) and without (*blue*) SCs. Measurements were made to determine the change in multifunctional performance through time-of-flight, and (*right*) the resulting effect solar cell efficiency will have with mass through variation in multifunctional performance design parameter k [3]

In this generalized analysis, power consumption by the structure ($P_{structure}$) is related to the operational frequency (f) and the operational energy of the structure ($U_{structure}$), which generally increase with battery and SC flexural stiffness (EI_{bat} and EI_{SC}) and the number of batteries and solar cells (η_{bat}, and η_{SC}), while decreasing with fatigue damage (D) and the area of the individual batteries and SCs (A_{bat} and A_{SC}). The constant k is a "multifunctional performance index" describing the relative effects of integrating SCs and batteries on power consumption.

We used this analysis to compare t_f for MAV wings with and without SCs by determining the flapping frequency necessary to achieve the same thrust force with and without SCs (Fig. 12.9). Two different wing compliances, A and B, were examined. Results showed that B could fly twice as long when it did not have SCs, but that A and B performed similarly when the SCs were integrated. By using variations of wing area and power with mass for our MAV platform, it was possible to determine the value of k for which infinite t_f could be achieved, depending on SC efficiency (Fig. 12.9).

12.7 Conclusions

An increasing interest in MAVs has driven researchers to find ways to improve the design of these vehicles without compromising performance. Bu introducing multifunctional structures into the MAV design, we can improve the design. Integrating compliant strain sensors to compliant wings of flapping MAVs allow the vehicle to sense how much the wing is deforming to enhance performance through autonomous flapping control. The strain measurements made using the compliant strain sensor have been correlated to the thrust force produced by the wing while flapping, and a SRS analysis indicated that the strain sensor was more sensitive to lower integral modes of the flapping frequency than the thrust measurements. Strain measurements made using 3D DIC of the strain transverse to the spars at the same location as the compliant strain sensor correlated much more directly with the thrust. Thus, it should be possible to modify the output of the strain sensor to sense forces produced by the compliant wings when using this element to sense in real time, permitting more autonomous flight control for the MAV in the future.

The concept of multifunctional wing was further developed by integrating flexible SCs onto the wings. The SCs can harvest energy while the MAV is in fight, thus prolonging the flight duration and enhancing autonomy. The effects of these SCs on the compliant wing were seen through the change in the 3D wing shape. The volume the wing encompasses during flapping was measured with and without SCs. There was a slight difference between the two, especially at the apex and nadir of the flapping cycle where thrust is generated through blowback, which results in a change in thrust performance whose affect on the time-of-flight was accounted for through a new multifunctional analysis. Through the characterization data and multifunctional analysis, it is possible to determine how to minimize the effect of SCs on flapping performance, and by also increasing the efficiency of solar cells it may one day lead to MAVs completely powered by the sun.

Acknowledgments Funding for this work was provided by the US Army Research Laboratory under the MAST CTA program in the Center for Microsystem Mechanics, as well as a seed grant from the University of Maryland Robotics Center.

References

1. Kujawski M, Pearse JD, Smela E (2010) Elastomers filled with exfoliated graphite as compliant electrodes. Carbon 48(9):2409–2417
2. Mueller D, Bruck HA, Gupta SK (2010) Measurement of thrust and lift forces associated with drag of compliant flapping wing for micro air vehicles using a new test stand design. Exp Mech 50(6):725–735
3. Cellon K (2010) Characterization of flexible flapping wings and the effects of solar cells for miniature air vehicles. M.S. thesis, Department of Mechanical Engineering, University of Maryland

Chapter 13
Non-local Damage-Enhanced MFH for Multiscale Simulations of Composites

Ling Wu, Ludovic Noels, Laurent Adam, and Issam Doghri

Abstract In this work, a gradient-enhanced mean-field homogenization (MFH) procedure is proposed for fiber reinforced materials. In this approach, the fibers are assumed to remain linear elastic while the matrix material obeys an elasto-plastic behavior enhanced by a damage model. As classical finite element simulations face the problems of losing uniqueness and strain localization when strain softening of materials is involved, we develop the mean-field homogenization in a non-local way. Toward this end we use the so-called non-local implicit approach, reformulated in an anisotropic way to describe the damage in the matrix. As a result we have a multi-scale model that can be used to study the damage process at the meso-scale, and in particular the damaging of plies in a composite stack, in an efficient computational way. As a demonstration a stack with a hole is studied and it is shown that the model predicts the damaging process in bands oriented with the fibers directions.

Keywords Mean-field homogenization • Non-local • Composites • Damage • Anisotropy

13.1 Introduction

As direct numerical simulations of composite structures explicitly accounting for the individual phases are too complex to handle and as the computation expenses are unaffordable, homogenized properties are usually sought to perform structural analyzes. Homogenization techniques are known to be efficient tools to derive those homogenized material properties analytically and/or numerically, see [1, 2] for exhaustive overviews.

Among those different methods, the mean-field homogenization (MFH) approach [3] provides predictions for the macroscopic behavior of heterogeneous materials and is thus of particular interest for the modeling of particle or fiber reinforced composites. MFH methods were first developed for linear elastic structures by extending the Eshelby single inclusion solution [4] to multiple inclusions interacting in an average way in the composite phase, as the Mori-Tanaka

L. Wu (✉)
University of Liege, LTAS-CM3, Chemin des Chevreuils 1, Liège B4000, Belgium

Northwestern Polytechnical University, School of Aeronautics, Xi'an 710072, China
e-mail: L.Wu@ulg.ac.be

L. Noels
University of Liege, LTAS-CM3, Chemin des Chevreuils 1, Liège B4000, Belgium
e-mail: L.Noels@ulg.ac.be

L. Adam
e-Xstream Engineering, Rue du Bosquet, 7, Louvain-la-Neuve 1348, Belgium
e-mail: laurent.adam@e-xstream.com

I. Doghri
e-Xstream Engineering, Rue du Bosquet, 7, Louvain-la-Neuve 1348, Belgium

Université Catholique de Louvain, Bâtiment Euler, 1348, Louvain-la-Neuve, Belgium
e-mail: issam.doghri@uclouvain.be

E. Patterson et al. (eds.), *Composite Materials and Joining Technologies for Composites, Volume 7: Proceedings of the 2012 Annual Conference on Experimental and Applied Mechanics,* Conference Proceedings of the Society for Experimental Mechanics Series 44,
DOI 10.1007/978-1-4614-4553-1_13,

scheme [5, 6] and the self-consistent scheme [7, 8]. When extending MFH methods to the non-linear regime, the so-called incremental formulation rewrites the constitutive equations in a pseudo-linear form relating the stress and strain rates [9–12], allowing the use of the linear techniques.

Although homogenization schemes have achieved a high level of accuracy to capture the non-linear behavior of composites, accounting for material degradation remains highly challenging [2]. Besides the complexity of formulating the multiscale method, the governing partial differential equations lose ellipticity and finite element (FE) solution uniqueness at strain-softening onset. Recently, the authors have proposed an incremental non-local MFH accounting for damage happening in the matrix-material at the micro-scale [13]. In order to avoid the strain/damage localization caused by matrix material softening, a gradient-enhanced formulation [14] is adopted during the homogenization process. In this formulation, the non-local accumulated plastic strain of the matrix is defined and depends on the local accumulated plastic strain and on its derivatives through the resolution of a new boundary value problem [14, 15]. This formulation avoids the need to develop higher-order elements, although the elements have now one additional degree of freedom per node. As a result this new formulation allows simulating the ply-level response under quasi-static loading conditions resulting from the coupled plasticity-damage model considered at the micro-scale.

This paper extends the non-local formulation of MFH with damage [13] to the analysis of composite laminates, which are meshed by considering in each ply finite elements whose constitutive behaviors are predicted by the MFH. The elements in each ply are integrated using a material constitutive behavior depending on the orientation of the ply fibers. This extension requires reformulating the non-local approach in an anisotropic way in order to introduce different characteristic sizes in the directions parallel and perpendicular to the fibers. Also, as the non-local formulation implies new nodal degrees of freedom and the resolution of new equations, the finite element approach presented introduces discontinuities in the non-local accumulated plastic strain field at plies interfaces to satisfy the new boundary conditions at material interfaces which are associated to the non-local variables. To illustrate the efficiency of this framework, a laminate with a hole is studied, and it is shown that the model predicts the damaging process in bands oriented with the fibers directions in the different plies.

13.2 Anisotropic Non-local Gradient Model

A non-local formulation results from replacing an internal variable, as the accumulated plastic strain p, by a weighted average on a characteristic volume V_C:

$$\tilde{p}(X) = \frac{1}{V_C} \int_{V_C} p(y)\phi(y;X)dV \tag{13.1}$$

where $\phi(\boldsymbol{y}; \boldsymbol{X})$ are normalized weight functions. An elegant way to avoid the explicit integral evaluation during a finite element implementation is to reformulate this equation in an implicit way [14]

$$\tilde{p}(\mathrm{X}) - c\nabla \cdot \nabla\tilde{p}(\mathrm{X}) = p(\mathrm{X}) \tag{13.2}$$

where c comes from the integration of Eq. 13.1, and has the unit of the squared length. This new Helmholtz-type equation is completed by an appropriate boundary condition

$$\nabla\tilde{p} \cdot \boldsymbol{n} = 0 \tag{13.3}$$

on the whole body surface, where $\boldsymbol{n}$ is the outward unit normal. This model allows defining a non-local form of the accumulated plastic strain, which can be used to evaluate a damage evolution law without inducing the loss of solution uniqueness at strain softening onset, see next section.

However, the set of Eqs. 13.2 and 13.3 is characterized by a unique length scale of the material in all the directions. As we intend to combine this model with a MFH framework for which characteristic dimensions are advantageously linked to the fibers orientation, this non-local implicit model is reformulated in an anisotropic framework following our developments in [16]. Thus the governing equation reads

$$\tilde{p} - \nabla \cdot \left(\boldsymbol{c}_g \cdot \nabla\tilde{p}\right) = p \tag{13.4}$$

with the new boundary condition

$$\left(\boldsymbol{c}_g \cdot \nabla\tilde{p}\right) \cdot \boldsymbol{n} = 0 \tag{13.5}$$

In these last two equations, the symmetric second order tensor $\boldsymbol{c}_g$ represents the distribution of characteristic length sizes in the global axes where the finite element problem is formulated. If in the axes linked to the composite ply, for example with one axis along the fibers direction, the characteristic lengths are l^1 along the fibers, l^2 and l^3 in the two directions perpendicular to the fibers, and if $\boldsymbol{R}$ is the rotation matrix representing the change of orthonormal coordinates from global to local, then this tensor reads

$$\boldsymbol{c}_g = \boldsymbol{R}^T \cdot \mathrm{diag}\left(\left(l^1\right)^2, \left(l^2\right)^2, \left(l^3\right)^2 \right) \cdot \boldsymbol{R} \tag{13.6}$$

The boundary condition Eq. 13.5 states that the non-local accumulated plastic strain gradient should be aligned with the fiber directions.

13.3 MFH with Gradient-Enhanced Damage Model

This section summarizes the extension of the non-local formulation of MFH with damage [13] to the finite-element analyzes of laminated composite.

13.3.1 Incremental MFH

In the multiscale framework developed, at each macro-point $\boldsymbol{X}$ of the structure, the macro-strain tensor $\bar{\boldsymbol{\varepsilon}}$ is known, and the macro-stress tensor $\bar{\boldsymbol{\sigma}}$ is sought from the resolution of a micro-scale boundary value problem (BVP) as illustrated on Fig. 13.1.

The Hill-Mandell condition, expressing the equality between energies at both scales, transforms the relation between macro-strains and stresses into the relation between average strains and stresses over the RVE. For a two-phase isothermal composite with the volume fractions $v_0 + v_\mathrm{I} = 1$ (subscript 0 refers to the matrix and I to the inclusions), the average quantities are expressed in terms of the phase averages as

$$\bar{\boldsymbol{\varepsilon}} = v_0 \langle \boldsymbol{\varepsilon} \rangle_0 + v_\mathrm{I} \langle \boldsymbol{\varepsilon} \rangle_\mathrm{I} \tag{13.7}$$

$$\bar{\boldsymbol{\sigma}} = v_0 \langle \boldsymbol{\sigma} \rangle_0 + v_\mathrm{I} \langle \boldsymbol{\sigma} \rangle_\mathrm{I} \tag{13.8}$$

In the non-linear range no direct relation between the macro-strain tensor $\bar{\boldsymbol{\varepsilon}}$ and the macro-stress tensor $\bar{\boldsymbol{\sigma}}$ can be derived, and a linear comparison composite (LCC) is introduced by linearizing the behaviors of the two phases around the current strain state. Thus, for a given time step $[t_\mathrm{n}, t_\mathrm{n+1}]$, relations Eqs. 13.7 and 13.8 are rewritten in an incremental form

$$\Delta\bar{\boldsymbol{\varepsilon}} = v_0 \langle \Delta\boldsymbol{\varepsilon} \rangle_0 + v_\mathrm{I} \langle \Delta\boldsymbol{\varepsilon} \rangle_\mathrm{I} \tag{13.9}$$

$$\Delta\bar{\boldsymbol{\sigma}} = v_0 \langle \Delta\boldsymbol{\sigma} \rangle_0 + v_\mathrm{I} \langle \Delta\boldsymbol{\sigma} \rangle_\mathrm{I} \tag{13.10}$$

The Mori-Tanaka (M-T) assumption allows relating the strain increments in the different phases from

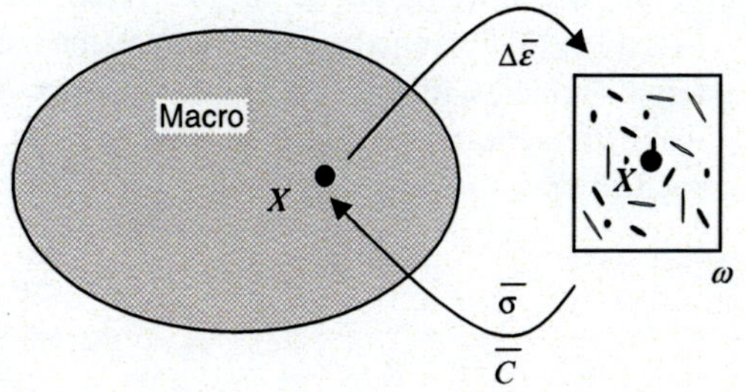

Fig. 13.1 Multiscale method

$$\langle \Delta \boldsymbol{\varepsilon} \rangle_{\mathrm{I}} = \boldsymbol{B}^{\varepsilon}\left(I, \bar{C}_0^{\mathrm{alg}}, \bar{C}_{\mathrm{I}}^{\mathrm{alg}}\right) : \langle \Delta \boldsymbol{\varepsilon} \rangle_0 \tag{13.11}$$

with the average values of the consistent algorithmic tangent modulii of the inclusions $\bar{C}_{\mathrm{I}}^{\mathrm{alg}}$ and of the matrix phase $\bar{C}_0^{\mathrm{alg}}$, and with $\boldsymbol{B}^{\varepsilon}\left(I, \bar{C}_0^{\mathrm{alg}}, \bar{C}_{\mathrm{I}}^{\mathrm{alg}}\right)$ the strain concentration tensor computed using the Eshelby tensor [4]. Thus, the system of Eqs. 13.9, 13.10 and 13.11 simplifies into

$$\Delta\bar{\sigma} = \underbrace{\left[v_{\mathrm{I}}\bar{C}_{\mathrm{I}}^{\mathrm{alg}} : \boldsymbol{B}^{\varepsilon} + v_0\bar{C}_0^{\mathrm{alg}}\right] : \left[v_{\mathrm{I}}\boldsymbol{B}^{\varepsilon} + v_0\boldsymbol{I}\right]^{-1}}_{\bar{C}} \Delta\bar{\varepsilon} \tag{13.12}$$

Finally, if the resolution scheme involves Newton–Raphson iterations, the Jacobian form of Eq. 13.12 is developed under the form

$$\delta\bar{\sigma} = \bar{C}^{\mathrm{alg}} : \delta\bar{\varepsilon} \tag{13.13}$$

13.3.2 MFH with Gradient-Enhanced Damage

In this section, the < > symbols are omitted for clarity, but the values are used in the mean sense. The damage model of Lemaître-Chaboche [17] is considered for the matrix phase. This defines an effective stress

$$\hat{\sigma}_0 = \frac{\sigma_0}{(1-D)} \tag{13.14}$$

where D is a representation of the damage in the matrix. The damage evolution reformulated in the non-local way obeys the law

$$\dot{D} = \begin{cases} 0 & \text{if} \quad \tilde{p} \le p_C \\ \left(\dfrac{\boldsymbol{\varepsilon}_0^{\mathrm{el}} : \boldsymbol{C}_0^{\mathrm{el}} : \boldsymbol{\varepsilon}_0^{\mathrm{el}}}{2S_0}\right)^s \dot{\tilde{p}} & \text{if} \quad \tilde{p} > p_C \end{cases} \tag{13.15}$$

where s, S_0 and p_C are model parameters. The damage evolution depends on the non-local accumulated plastic strain in the matrix, which is governed by relations Eqs. 13.4 and 13.5.

From relations Eqs. 13.14 and 13.15, it appears that the linearization of $\boldsymbol{\sigma}_0$ depends on $\boldsymbol{\varepsilon}_0$ and on $\tilde{p}$. Thus, the linearization of Eq. 13.10 reads

$$\delta\bar{\boldsymbol{\sigma}} = v_{\mathrm{I}}\bar{C}_{\mathrm{I}}^{\mathrm{alg}} : \delta\boldsymbol{\varepsilon}_{\mathrm{I}} + v_0\underbrace{(1-D)\bar{C}_0^{\mathrm{alg}}}_{\bar{C}_0^{\mathrm{algD}}} : \delta\boldsymbol{\varepsilon}_0 - v_0\left(\hat{\boldsymbol{\sigma}}_0 \otimes \frac{\partial D}{\partial \boldsymbol{\varepsilon}_0}\right) : \delta\boldsymbol{\varepsilon}_0 - v_0\hat{\boldsymbol{\sigma}}_0\frac{\partial D}{\partial \tilde{p}} : \delta\tilde{p} \tag{13.16}$$

The first two terms at the right hand side of this linearization represent the stress variation of a composite with a fictitious matrix material of constant non-local damage. In this work we assume that the incremental Mori-Tanaka process can be applied on these first two terms. The last two terms are related to the softening of the matrix due to the damage in the matrix, and are not considered in the Mori-Tanaka process. Indeed M-T is only defined when the two tangent moduli are definite positive, which would not be the case in the softening part, see [13].

Thus, the MFH process with gradient enhanced damage during a time step $[t_n, t_{n+1}]$ is defined as follows. Knowing the internal variables, macro-strain tensor $\bar{\boldsymbol{\varepsilon}}_n$, macro-stress tensor $\bar{\boldsymbol{\sigma}}_n$ and non-local accumulated plastic strain $\tilde{p}_n$ at time t_n, for given increments in the macro-strain tensor $\Delta\bar{\boldsymbol{\varepsilon}}_{n+1}$, and in the non-local accumulated plastic strain$\Delta\tilde{p}_{n+1}$, the system to be solved reads

$$\Delta\bar{\boldsymbol{\varepsilon}}_{n+1} = v_0\Delta\boldsymbol{\varepsilon}_{0n+1} + v_{\mathrm{I}}\Delta\boldsymbol{\varepsilon}_{\mathrm{I}n+1} \tag{13.17}$$

$$\bar{\boldsymbol{\sigma}}_{n+1} = v_0\boldsymbol{\sigma}_{0n+1} + v_{\mathrm{I}}\boldsymbol{\sigma}_{\mathrm{I}n+1} \tag{13.18}$$

$$\Delta\boldsymbol{\varepsilon}_{\mathrm{I}n+1} = \boldsymbol{B}^{\varepsilon}\left(I, \bar{C}_0^{\mathrm{algD}}, \bar{C}_{\mathrm{I}}^{\mathrm{alg}}\right) : \Delta\boldsymbol{\varepsilon}_{0n+1} \tag{13.19}$$

The resolution of this system is detailed in [13].

13.4 Finite Element Implementation

The strong form of the problem is stated by the governing equations in the body Ω

$$\nabla \cdot \boldsymbol{\sigma}^T + \rho \boldsymbol{f} = 0 \tag{13.20}$$

$$\tilde{p} - \nabla \cdot \left(\boldsymbol{c}_{\mathrm{g}} \cdot \nabla \tilde{p}\right) = p \tag{13.21}$$

with the boundary conditions

$$U = \bar{U} \text{ on } \Gamma_D \tag{13.22}$$

$$\boldsymbol{\sigma} \cdot \boldsymbol{n} = \bar{T} \text{ on } \Gamma_T \tag{13.23}$$

$$\left(\boldsymbol{c}_{\mathrm{g}} \cdot \nabla \tilde{p}\right) \cdot \boldsymbol{n} = 0 \text{ on } \Gamma_i \tag{13.24}$$

In these last equations Γ_D is the Dirichlet part of the whole body boundary and Γ_T its Neumann part. However, the boundary condition Eq. 13.24 should be enforced on the body surface Γ, but can also be enforced on each interface between different material behaviors , see [13] for a discussion on this boundary conditions between different materials. In the case of a laminated structure, the different domains with different material behaviors correspond to the different plies with different fibers orientations see Fig. 13.2.

The finite element weak formulation of the problem is detailed in [16] and results to a system with four degrees of freedom per node: three displacements and one non-local accumulated plastic strain. However, condition Eq. 13.24 has to be satisfied on each interface between different material behaviors. This is accomplished by considering the field corresponding to $\tilde{p}$ discontinuous at domain interfaces. This is achieved simply by defining for each node belonging to the interface, besides the three displacement degrees of freedom, 2 degrees of freedom corresponding to $\tilde{p}$, one for the domain 1, one for the domain 2 (n in case of an interface between n domains).

13.5 Numerical Applications

In this section we consider composite structures made of continuous carbon fibers reinforced epoxy. The material is made of prepreg Hexply M10.1/38 %/UD300/HS (R), which results in a fiber volume fraction of 52.5%. Each ply has a thickness of 0.3 mm. The properties of the epoxy matrix are reported in Table 13.1, with a hardening law following an exponential law

$$R(p) = h_0(1 - e^{-mp}) \tag{13.25}$$

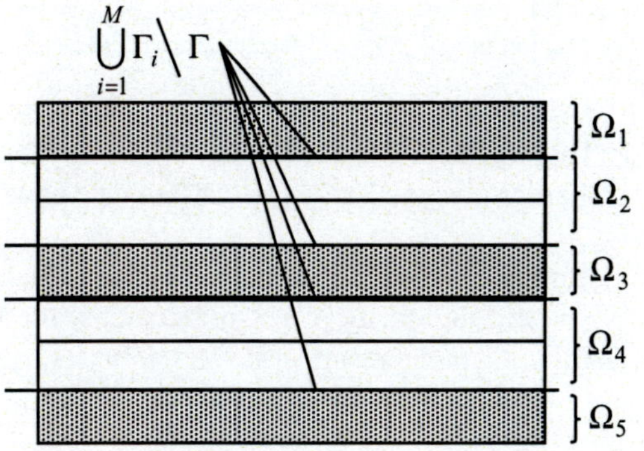

Fig. 13.2 Description of a laminate

Table 13.1 Properties of the composite constituents

Expoxy		Carbon fibers	
E_0 [GPa]	2.89	E_L [GPa]	230
ν [−]	0.3	E_T [GPa]	40
σ_Y [MPa]	35	ν_{TT} [−]	0.2
h_0 [MPa]	73	ν_{LT} [GPa]	0.256
m [−]	60	G_{TT} [GPa]	16.7
S_0 [MPa]	2	G_{LT} [GPa]	24
s [−]	0.5		
p_C [−]	0		

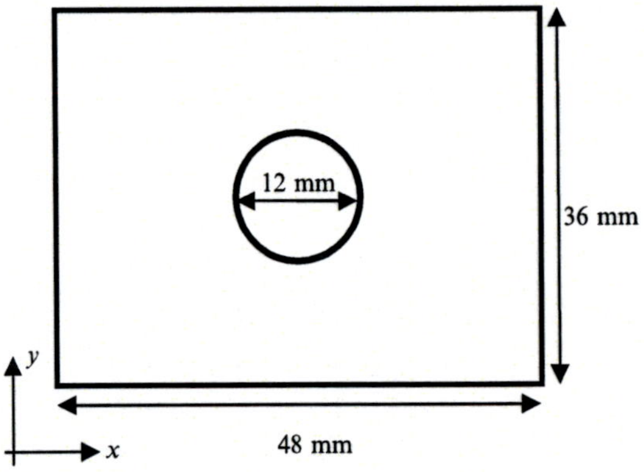

Fig. 13.3 Geometry of the laminate

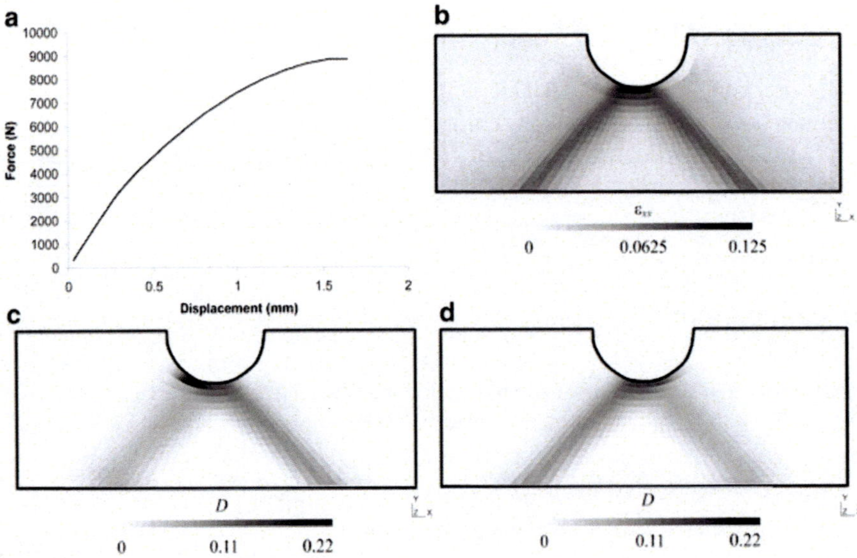

Fig. 13.4 Numerical results for the tension of a laminate with hole **(a)** Displacement vs. force curve, **(b)** Strain distribution in -45° ply, **(c)**Damage distribution in -45° ply, **(d)** Damage distribution in 45° ply

and a damage law following Eq. 13.15. The carbon fibers are linear elastic and transversely isotropic with the mechanical properties given in Table 13.1. A laminate plate with the stacking sequence $(-45)_2/(45)_4/(-45)_2$ has a hole in its center, see Fig. 13.3.

Numerical results of a tensile test on the laminate with a hole are presented in Fig. 13.4. Although the plate is not symmetric due to the fibers orientations, the balanced distribution of the plies induces a quasi-symmetric stress/strain distribution at the macro-scale and only one half of the plate is represented. The deformation distribution at maximal force, in Fig. 13.4a, is presented in Fig. 13.4b. Strain concentration occurs in bands parallel to the fiber orientations. Similarly, the damage distributions in the −45°, Fig. 13.4c, and in the +45°, Fig. 13.4d, plies exhibit bands, which initiate at the hole, and which propagate with the fibers orientations, respectively −45° and 45°. Due to the balanced stacking sequence, there is a second band in each ply, of lower intensity, which is parallel to fiber orientation of the other plies, respectively along 45° and −45°. It also appears that the location of the band initiation at the hole differs in the −45°, Fig. 13.4c and in the +45°, Fig. 13.4d, plies.

13.6 Conclusions

In order to analyze failure of laminated composite structures, a gradient-enhanced mean-field homogenization (MFH) procedure has been developed. In this approach, the fibers are assumed to remain linear elastic while the matrix material obeys an elasto-plastic behavior enhanced by a damage model. The loss of solution uniqueness at onset of strain softening has been avoided by using a non-local implicit approach, reformulated in an anisotropic way, to describe the damage in the matrix. As a result we have a multi-scale model that can model the damage evolution in each ply of a composite stack. As a demonstration a stack with a hole is studied, and it is shown that the model predicts the damaging process in bands oriented with the fibers directions.

Acknowledgement The research has been funded by the Walloon Region under the agreement SIMUCOMP n° 1017232 (CT-EUC 2010-10-12) in the context of the ERA-NET +, Matera + framework.

References

1. Kanouté P, Boso D, Chaboche J, Schrefler B (2009) Multiscale methods for composites: a review. Arch Comput Method Eng 16:31–75
2. Geers M, Kouznetsova V, Brekelmans WAM (2010) Multi-scale computational homogenization: trends and challenges. J Comput Appl Math 234:2175–2182
3. Doghri I, Brassart L, Adam L, Gérard JS (2011) A second-moment incremental formulation for the mean-field homogenization of elasto-plastic composites. Int J Plast 27(3):352–371
4. Eshelby JD (1957) The determination of the elastic field of an ellipsoidal inclusion, and related problems. Proc R Soc Lond Ser A Math Phys Sci 241(1226):376–396
5. Mori T, Tanaka K (1973) Average stress in matrix and average elastic energy of materials with misfitting inclusions. Acta Metall 21 (5):571–574
6. Benveniste Y (1987) A new approach to the application of Mori-Tanaka's theory in composite materials. Mech Mater 6(2):147–157
7. Kröner E (1958) Berechnung der elastischen Konstanten des Vielkristalls aus den Konstanten des Einkristalls. Zeitschrift f¨ur Phys A Hadrons Nucl 151:504–518
8. Hill R (1965) A self-consistent mechanics of composite materials. J Mech Phys Solids 13(4):213–222
9. Hill R (1965) Continuum micro-mechanics of elastoplastic polycrystals. J Mech Phys Solids 13(2):89–101
10. Pettermann HE, Plankensteiner AF, Böhm HJ, Rammerstorfer FG (1999) A thermo-elasto-plastic constitutive law for inhomogeneous materials based on an incremental Mori-Tanaka approach. Comput Struct 71(2):197–214
11. Doghri I, Ouaar A (2003) Homogenization of two-phase elasto-plastic composite materials and structures: study of tangent operators, cyclic plasticity and numerical algorithms. Int J Solids Struct 40(7):1681–1712
12. Chaboche J, Kanouté P, Roos A (2005) On the capabilities of mean-field approaches for the description of plasticity in metal matrix composites. Int J Plast 21(7):1409–1434
13. Wu L, Noels L, Adam L, Doghri I (Submitted) Multiscale mean-field homogenization method for fiber-reinforced composites with gradient-enhanced damage model. Comput Method Appl Mech Eng
14. Peerlings R, Geers M, de Borst R, Brekelmans W (2001) A critical comparison of nonlocal and gradient-enhanced softening continua. Int J Solids Struct 38:7723–7746
15. Geers M (1997) Experimental analysis and computational modelling of damage and fracture, Ph.D. thesis. University of Technology, Eindhoven (Netherlands)
16. Wu L, Noels L, Adam L, Doghri I (in preparation)Anisotropic gradient-enhanced damage mean-field homogenization for multiscale analysis of composite laminates
17. Lemaitre J, Desmorat R (2005) Engineering damage mechanics: ductile, creep, fatigue and brittle failures. Springer, Berlin. ISBN 3540215034

Chapter 14
Composite Damage Detection with Self-Sensing Fibers and Thermal Sprayed Electrodes

Toshio Nakamura and Masaru Ogawa

Abstract Production of novel damage sensor system for carbon fiber-reinforced composite structures is presented. It utilizes carbon fibers as self-sensing sensors. Two key components are the fabrications of electrodes required to measure electrical resistivity changes due to damage, and the development of robust data interpretation scheme to diagnose the damage stage. The former takes the advantage of new technology based on the thermal spray process in which molten copper are directly deposited onto composite laminate surfaces to fabricate network of electrodes. The latter task relies on the simulation study and the optimization/inverse analysis to estimate damage state through processing of measured data. The initial experimental measurements on composite laminates with artificially made damage indicate increased resistivity among thermal sprayed electrodes.

Keywords Carbon fiber reinforced polymer (CFRP) • Structural health monitoring • Self-sensing fibers

14.1 Introduction

Carbon fiber reinforced polymer (CFRP) is being increasingly used as primary components of aerospace structures. However, there is yet to be an effective tool to monitor structural integrity and assess possible damage. In the past, to detect material defects during in-service inspection, non-destructive evaluation (NDE) techniques such as ultrasonic inspection, vibration response, infrared thermal images and eddy current test have been commonly used. However, these conventional monitoring methods are relatively expensive and time-consuming and also less effective to perform in-situ monitoring. Therefore, a novel structural health monitoring approach that can overcome these difficulties and maintain structural reliability of composites is needed.

Utilizing the electrical conductivity of carbon fibers in CFRP laminates, an alternate concept in health monitoring of composite was emerged [1, 2]. The idea is to conduct electricity through fibers and measure potential or resistivity changes between multiple points (i.e., electrodes). Since damage or straining of composite can impede current pathways through some fibers, it results in the increased resistance of a circuit made of carbon fiber network (distributed throughout the structure). Hence, this monitoring concept relies on the fibers themselves to act as the sensing elements which should minimize any structural attrition due to sensor incorporations. An advantage of this method is that any fiber loss/breakage, which lowers structural integrity, is closely tied to the resistance changes of fibers. In other words, the damage is directly captured as an increased electrical resistance in the fiber network. In addition, since a resistance is taken over two locations, a damage located away from electrodes is still detectable as long as the electrical passage between the electrodes is affected. Recently, methods based on self-sensing fibers have been refined by several investigators. For examples, Todoroki and his co-workers have studied both the electrical potential change method (EPCM) and the electrical resistance change method (ERCM) to search for potential embedded flaws in composites [3, 4]. Their efforts concentrated on identification of interlaminar flaw shape and size in multi-ply CFRP laminates. Other similar studies utilized the circumferential lead wires with silver paste to probe damage in the unidirectional CFRP [5].

T. Nakamura (✉) • M. Ogawa
Department of Mechanical Engineering, State University of New York at Stony Brook, Stony Brook, NY 11794, USA
e-mail: toshio.nakamura@sunysb.edu

E. Patterson et al. (eds.), *Composite Materials and Joining Technologies for Composites, Volume 7: Proceedings of the 2012 Annual Conference on Experimental and Applied Mechanics*, Conference Proceedings of the Society for Experimental Mechanics Series 44,
DOI 10.1007/978-1-4614-4553-1_14,

Although some successful results were reported in these works, major obstacles remain in adopting the conductive carbon fibers to identify damage outside the laboratory environment. First, the fabrication of high-quality electrical input/output terminals (i.e., electrodes) onto CFRP laminates is laborious if not practical in large number. For each electrode, matrix phase encapsulating carbon fibers must be removed to ensure good contacts. Second, many studies considered only 2D models where the electrical flow is essentially along one direction in laminates (except near the damage). In actual applications, dominant electrical flow is over a plane (in shell-like structures) which requires 3D models. The purpose in this study is to develop an in-situ detection device using ERCM and to offer robust identification of damage. In order to process measured resistance changes, an efficient procedure based on inverse analysis is proposed. To measure electrical resistances along fibers, many electrodes must be attached to composite laminates. Currently available methods require laborious steps [6]. In order to develop more effective technique, we are utilizing thermal spray (TS) techniques to deposit electrodes directly onto laminates. In a thermal spray process, copper powder is fed through a spray gun which instantly melts the powder at high temperature and sprays them directly onto the composite surface. The molten copper particles rapidly solidify and form bonding with the substrate. In this paper, a proposed procedure to detect damage in CFRP laminate is presented and a simulation study is carried out to verify the method. Then a preliminary work on actual laminates with thermal sprayed electrodes is described to illustrate the capability of self-sensing fibers.

14.2 Damage Identification Process

The proposed sensor system measures resistances among electrodes attached to composite structures and estimate its size and location within each grid formed by 4 (or 9) electrodes. These electrodes serve as input/output of electrical current, and also identify the approximate location of local damage. The process is repeated grid by grid throughout the electrode network. The damage is evaluated according to the following steps. First, the relationship between damage state and electrical resistance changes is established as the reference solution. Second, changes in electrical resistances due to damage are measured. Third, the measurements are processed by an inverse analysis to obtain estimated damage parameters. In the present study, the downhill simplex method, which is a multi-dimensional minimization algorithm, is used to find the best estimates.

14.2.1 Model

In order to establish the reference or so-called forward solutions relating the resistance changes to various damage states, 3D finite element model is constructed. Here, an eight-ply $[0/90]_{2s}$ composite laminate with 4 electrodes as shown in Fig. 14.1 is considered. Each ply has 2 element layers with the total element layer along the thickness (z'-axis) to be 16. The in-plane or the x'-y' plane is divided into 100 by 100 elements. Thus the total of 160,000 8-noded electrical linear brick elements is used to construct the model. The electrical properties of CFRP are highly anisotropic and they also depend on the fiber volume fractions. In the analysis, the electrical conductivities along three directions of composite ply are chosen as $\sigma_0 = 5{,}500$ S/m, $\sigma_{90} = 204$ S/m and $\sigma_T = 20.7$ S/m along fiber, transverse and thickness directions, respectively [7]. These values correspond to a composite laminate with 62% fiber volume fraction (that of tested laminate in Section 4). Note that other values used in the simulations and they showed very small differences in the estimated accuracy. The electrical conductivity of copper electrode is $\sigma_{Cu} = 58.8 \times 10^6$ S/m. The thickness of the model is set $t = 1.6$ mm. Though the results are not presented here, other thicknesses ranging from $t = 0.8$ mm to 2.0 mm are also investigated. The distance between the two neighboring electrodes is denoted as a and set as $a = 0.8\ l$ where l is the side length of panel. The spacing a is varied to investigate the effects of electrode density. A surface damage is located on the opposite side of laminate and assumed to be in square shape. The damage size is expressed via the side length d and the depth δ. The coordinates of damage center are denoted by x and y, respectively. A total of four parameters, d, δ, x and y, represents the size and location of damage. The simple damage configuration is considered here but it may be also used as a representative model of more complex damage in actual composites. As shown in Fig. 14.1, there are altogether six electrode pairs where resistances can be measured; *AB*, *AC*, *AD*, *BC*, *BD* and *CD*. In actual CFRP laminate, there also exist contact resistance between an electrode and carbon fibers, which is not modeled in the numerical analysis. The contact resistances do not play a direct role in identifying the damage since they are subtracted when resistance changes are measured. However, they may influence the accuracy of damage estimates.

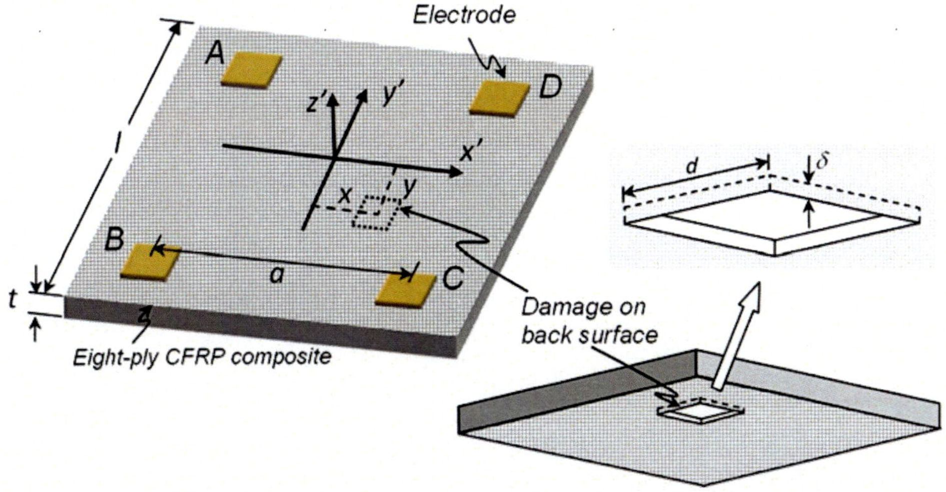

Fig. 14.1 Square composite model with 4 electrodes on the front surface containing damage on the back surface. The coordinate and the damage parameters are noted

14.2.2 Estimation of Damage Size and Location

An existence of damage influences the electrical behavior of the laminate, and this relation is utilized to assess the damage state. As described in the previous section, the finite element model is used to establish such relations for data processing. In an iterative type of post process, generating the relations at every updating step is impractical since the total number of calculations can be very large. In order to alleviate the time-consuming process, a reference or forward solution set is established a priori. Furthermore, to reduce the total number of computations (for various combinations of 4 damage variables), a scheme to utilize cubic Lagrangian interpolation functions is adopted. In order to extract the most probable damage parameters, the difference between the measured electrical resistances and the estimated ones is minimized. In this study, the error object function for *n* electrical potential measurements is defined as,

$$\varphi(x, y, d, \delta) = \frac{1}{n}\sum_{\alpha=1}^{n}\left(\frac{R_\alpha(x^{est}, y^{est}, d^{est}, \delta^{est}) - R_\alpha^{meas}}{R_\alpha^{ref}}\right)^2 \qquad (14.1)$$

Here, x^{est} and y^{est} are the estimated damage locations, d^{est} and δ^{est} are the estimated damage size and depth, respectively. Estimated resistance values $R_\alpha(x^{est}, y^{est}, d^{est}, \delta^{est})$ are calculated from the interpolation. The measured resistance in the damage model is denoted as $R_\alpha{}^{meas}$, and $R_\alpha{}^{ref}$ is taken as the reference resistance damage. The total number of electrode pair is given as *n*. Through the minimization of the error objective function, suitable damage parameters are estimated. The minimum of error objective function is obtained by the 'downhill simplex method'. This method is a popular method in the multi-dimensional optimization process, especially when derivatives of objective function are either discontinuous or difficult to obtain [8, 9]. It is essentially a search process based on the concept of a simplex, which is a polytope of $N + 1$ vertices in *N* dimensions. Since there are four unknown parameters in the present work, *N* is 4 with a polytope of 5.

14.3 Computational Verification

To verify the proposed approach, various numerical simulations are carried out with different damaged models. Initially, to reduce the number of computations, the depth is set constant at $\delta = 0.4$ mm, and three unknown damage parameters *x*, *y* and *d* are sought. The damage spacing of $a = 64$ mm is chosen here. After the reference solutions are established, simulated resistance measurements are obtained by arbitrary choosing the values for *x*, *y* and *d*. For an example, in the first case, the damage state described by $x = 0$ mm, $y = 0$ mm (center of 4 electrodes) and $d = 14.0$ mm is imposed on the finite element model.

Table 14.1 Estimated damage parameters compared with exact values under different electrode spacing. Large spacing corresponds to lower density of electrodes

Electrode spacing a		x (mm)	y (mm)	d (mm)
40 mm	Exact	10.5	−2.5	10
	Estimated	10.8	−2.8	10.4 (4.3%)
64 mm	Exact	16.8	−4.0	9.6
	Estimated	15.2	−3.0	10.4 (8.2%)
80 mm	Exact	21.0	−5.0	10
	Estimated	19.3	−7.4	11.1 (11.3%)
96 mm	Exact	25.2	−6.0	9.6
	Estimated	21.1	−5.3	11.4 (18.4%)
120 mm	Exact	31.5	−7.5	10.5
	Estimated	27.2	−5.5	12.8 (21.7%)

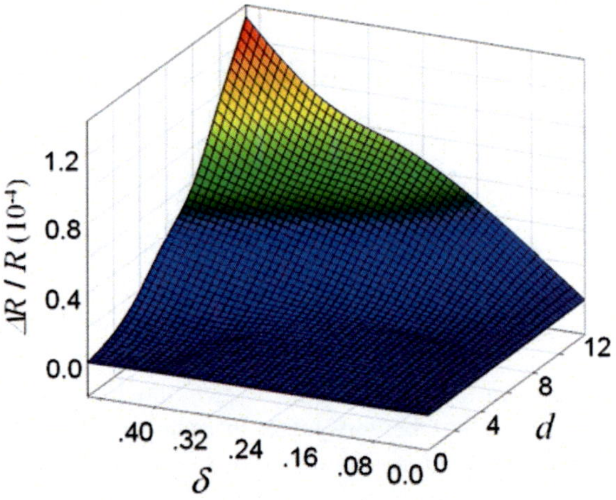

Fig. 14.2 Estimated damage size error is shown as a function of electrode density

With these conditions, the six electrical resistance measurements among 4 electrodes are computed. Using these values as inputs, the downhill simplex method is carried out to estimate the damage parameters (Table 14.1).

It is expected that for a given size of damage, its detection becomes more difficult as the spacing of electrodes increases or the total number of electrode reduces for a given size of panel. Here the effects of electrode density are evaluated by considering various distances between electrodes while the other damage parameters are kept similar. In this analysis, the model thickness, the damage size and depth are chosen as $t = 1.6$ mm, $d \simeq 10$ mm and $\delta = 0.4$ mm, respectively, while various values of a are assigned. In addition, the relative positions of damage are set the same with respect to the electrode positions ($x = 0.2625a$, $y = -0.0625a$) in all cases (note a varies in each case). The estimated errors are shown as a function of the electrode density in Fig. 14.2. Here the electrode density is expressed in terms of $(d/a)^2$ where d is the damage size.

14.4 Fabrication of Thermal Spray Electrodes

14.4.1 Specimen Preparation

To further demonstrate the capability of proposed system, actual test specimens with electrodes deposited by the thermal spray (TS) process are fabricated. Then resistances between the electrodes are measured with and without damage. Eight-ply composite laminates with $[0/90]_{2s}$ arrangements and ~2 mm thickness were cut to 12″ by 9″ dimensions. Their surfaces were initially grit-blasted to remove excess epoxy layer (~50 μm). Then, copper powder was fed to Atmospheric Plasma Spray (APS) and sprayed over a mask with holes for electrodes. Here, the distance between the panel and the injection tip of the APS was 150 mm, and the nozzle raster speed was set at 700 mm/s. Total of five passes were made to fabricate 75 μm thick copper electrodes. Argon was used as a shield gas with the electric current at 300 A. In this experiment, the 12 circular shaped

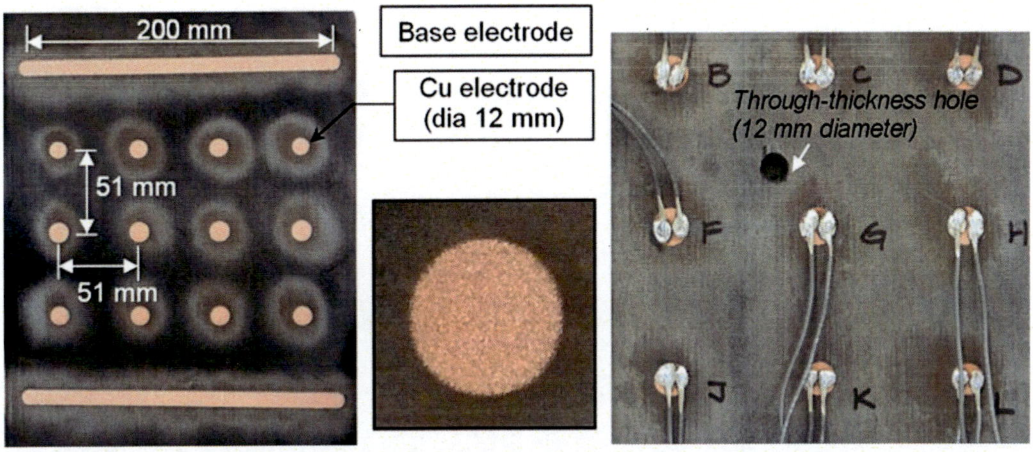

Fig. 14.3 Grid of electrodes on composite laminate fabricated by plasma spray. Magnified view of an electrode and wired plate with a hole are also shown

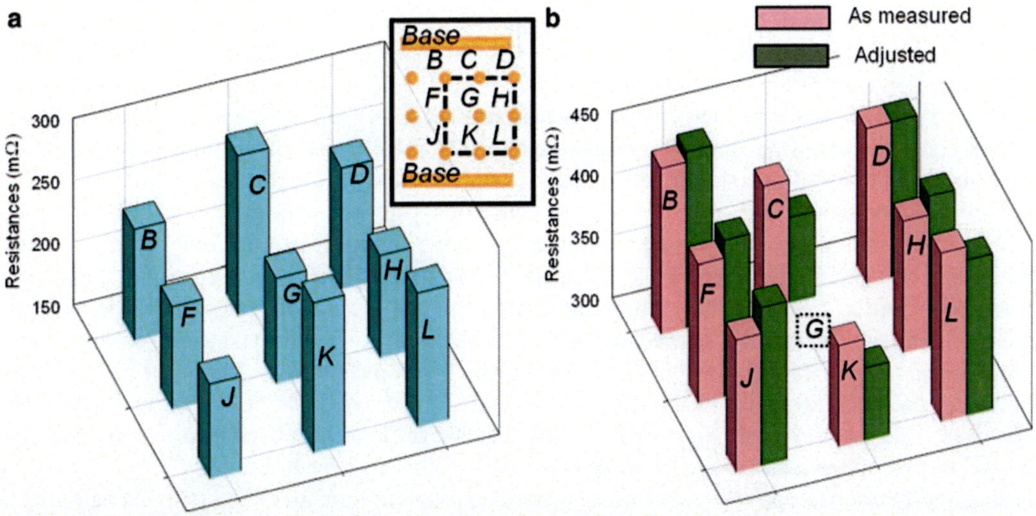

Fig. 14.4 (**a**) Averaged resistances between each circular electrode and the two base electrodes without damage. (**b**) Resistances between the central electrode G and the neighboring electrodes. As measured and adjusted with estimated contact resistances are shown

electrodes with 12 mm diameter as well as two base electrodes as shown in Fig. 14.3 were fabricated. Note unlike the square electrode shape assumed in the simulation study, the circular electrodes were fabricated in the actual tests to reduce the possibility of corner delamination. On every electrode, two electric wires were soft-soldered to measure resistances by the 'four-point probe' method, which offsets instrumental contact resistances. The spacing between electrodes is $a = 51$ mm.

14.4.2 Measured Resistances

Resistances between electrodes were measured using a low current meter (Keithley, Inc.) with the four-point probe method. First, resistances between circular electrodes and base electrodes at the top and bottom (2 locations each) are measured to determine resistance variations. Since the effective distances from each electrode to wiring locations of the base electrodes are approximately same, the averaged resistance should offer some information on the contact resistances between electrode and carbon fibers. It is expected the resistances between two electrodes to be comprised of the carbon fiber resistance and the interface resistance between carbon fibers and copper electrodes. Although the latter effects can be subtracted when the difference in resistances due to damage are computed, large contact resistances can increase measurement error or noise. Raw measurements of averaged resistances among electrodes are shown Fig. 14.4a. Then, in order to examine the resistances

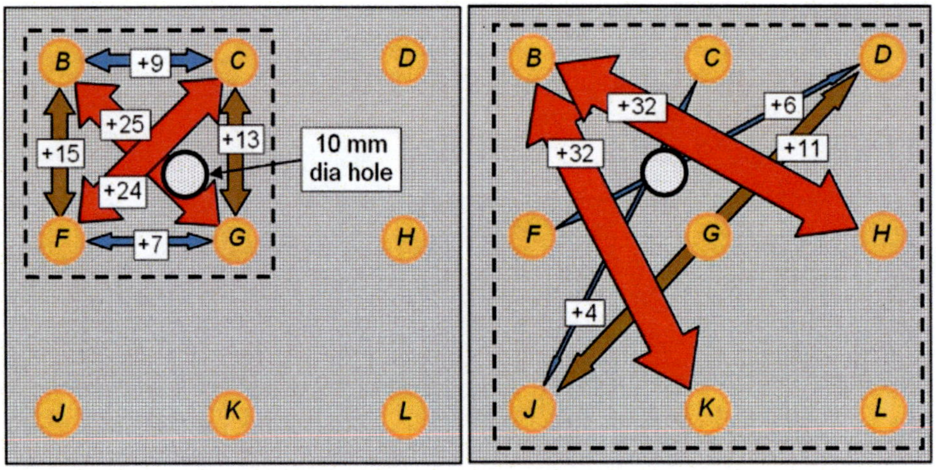

Fig. 14.5 Changes in resistances due to damage (10 mm diameter hole) between electrode pairs are illustrated in different *arrow* thicknesses and different colors. For clarity, the measurements from 4 to 9-electrode grid are shown separately

within an electrode grid, the electrode *G* was chosen as the central electrode. Figure 14.4b shows the resistances between the central electrode *G* and its neighboring electrodes. Here the pink bars correspond to "as measured" results while the green bars represent adjusted results. The each adjustment was made by adding difference of contact resistance (not directly the contact resistance), which was estimated from the measurements shown in Fig. 14.4a, to measured value. With the adjustments, more consistent results (e.g., similar resistances for *GB*, *GJ*, *GD* and *GL* pairs) were obtained. Although not shown here, the similar analyses of the other panels including unidirectional $[0]_8$ laminates.

In an initial step toward monitoring actual damage in composite laminate with thermal sprayed electrodes, 10 mm diameter hole was drilled through-thickness to represent damage in the current feasibility study. The size and depth of the hole were chosen arbitrary but made large enough to detect resistance measurements beyond measurement errors. Thus through-thickness hole, unlike a part-through damage as described in the simulation study was made. Note that the post-process to estimate the damage was not carried out in this preliminary experiment since it would require additional information on the resistivity of each ply within the laminate which was available. As shown in Fig. 14.4, the variations of contact resistances (between fibers and electrodes) are relatively large compared to expected resistance changed due to part-through damage. Thus the present aim is to simply verify the changes in resistances due to an existence of a hole in this preliminary phase, and the refinement to produce more consistent electrodes is underway.

Prior to drilling, the 11 electrode pairs were selected and their resistances were measured (without damage). Their values ranged 401 ~ 500 mΩ. Then the hole was drilled as shown in Fig. 14.5, and the resistance measurements were again made. Since the electrical flow was obstructed with the hole, the resistances between electrode increased by 4 ~ 32 mΩ. Note the estimated measurement errors were ±5 ~ 8 mΩ in these test. The changes in resistances are graphically shown in Fig. 14.5. Here the values were slightly calibrated with the earlier measurements. As expected, large resistances were observed in electrode pairs whose paths cross or run near the drilled hole (*BG*, *CF*, *BH*, *BK* pairs). However some inconsistencies were also observed. For examples, the *CJ* and *DF* pairs showed only marginal increases even though their paths pass through the hole. Further adjustments in these electrode fabrications and resistance measurements are necessary in the future studies.

14.5 Conclusions

A novel health monitoring approach, which utilizes the electrical conductivities of carbon fibers, is proposed to identify damage on the CFRP laminate. Two key features of the procedure are the robust data processing scheme based on an inverse analysis technique to identify damage state, and the thermal spray technique to fabricate multiple electrodes. In the inverse analysis, the downhill simplex technique is adopted to determine the best estimate of damage parameter. Here an error objective function is introduced to search its global minimum in the four parameter space representing damage location and size. Several damage models are constructed for the finite element simulations to test the accuracy of estimated parameters. The results indicate better estimates to be obtained when the damage is located near or on the path connected two electrode pair. The accuracy of the method is further examined in the error sensitivity analysis. Here random errors/noises are added to the simulated resistance changes and processed in the inverse analysis. The accuracy is significantly worsened but the damage estimate can be improved by adding extra resistance measurements from the nine-electrode grid. With five additional resistance measurements, the estimated errors are reduced by about half. In summary, the present simulation study indicate that to identify the damage size within a reasonable accuracy (±20%), the electrode density or spacing to be about 10 times the expected size of damage. A smaller damage may be still detected but with less accuracy.

In order to prove the feasibility of proposed technique, a network of copper electrodes were deposited on actual composite laminate. First, resistance measurements were made without damage to verify the electrical conductance among electrodes. From various electrode pairs, somewhat scattered results were observed. This may be attributed to inconsistent contact conditions between carbon fibers and electrodes. Next, as a preliminary step toward establishing damage monitoring system with thermal sprayed electrodes, a 10 mm diameter hole was drilled through-thickness. Since relatively large measurements errors and noises were expected in this initial phase of test, a through-thickness hole, and not part-through damage as analyzed in the simulation study was selected. The size and depth of the hole was chosen arbitrary but chose to be large enough to detect resistance changes. The measured results indicated promising results with the increased resistances due to the hole. Although much refinement is needed before reliable estimates in damage size can be established, these measurements support the capability of self-sensing fibers with thermal spray electrodes to detect damage.

Acknowledgements The authors gratefully acknowledge the Army Research Office for their support under grant W911NF1010317 and the Center for Thermal Spray Research at Stony Brook University.

References

1. Moriya K, Endo T (1988) Trans Japan Soc Aeronaut Space Sci 36:139–146
2. Schulte K, Baron C (1989) Compos Sci Technol 36:63–76
3. Todoroki A, Tanaka M, Shimamura Y (2005) Compos Sci Technol 65:37–46
4. Ueda M, Todoroki A (2008) Eng Fract Mech 75:2737–2750
5. Wang X, Chung DDL (1997) Polym Compos 18:692–700
6. Todoroki A (2008) J Intel Mater Syst Struct 19:291–298
7. Todoroki A, Tanaka M, Shimamura Y (2002) Compos Sci Technol 62:619–628
8. Nelder JA, Mead R (1965) Comput J 7:308–313
9. Ramanujam N, Nakamura T, Urago M (2005) Int J Fract 132:153–173

Chapter 15
Experimental Investigation of Fatigue Behavior of Carbon Fiber Composites Using Fully Reversed Four Point Bending Test

A. Amiri and M.N. Cavalli

Abstract Carbon fiber reinforced polymer (CFRP) has become an increasingly notable material for use in structural engineering applications. Some of its advantages include high strength-to-weight ratio, high stiffness-to-weight ratio, and good moldability. Prediction of the fatigue life of composite laminates has been the subject of various studies due to the cyclic loading experienced in many applications. Both theoretical studies [1–3] and experimental tests [4–6] have been performed to estimate the endurance limit and fatigue life of composite plates. One of the main methods to predict fatigue life is the four point bending test. In most previous works the tests have been done in one direction (load ratio, R, > 0). In the current work, we have designed and manufactured a special fixture to perform a fully reversed bending test ($R = -1$). Four point bending tests are carried out on samples of carbon fibers ($(0°/90°)_{15}$) and polyester resin. The results provide additional insight into the fatigue behavior of polymer composites.

Keywords Carbon fiber reinforced polymer (CFRP) • Fatigue life • Four point bending test

15.1 Introduction

Due to mechanical and structural advantages, carbon fiber reinforced polymers (CFRP) have become increasingly notable materials used in many engineering applications from aerospace to infrastructure to alternative energy. Some of their advantages include high strength-to-weight ratio, high stiffness-to-weight ratio, and good moldability. Efficient and effective use of CFRP in design requires understanding their behavior and properties [3]. Materials in engineering applications are subjected to long-term loading and often go through repeated loading and unloading conditions. All materials, when subjected to fatigue loading, are prone to fail after certain number of cycles; therefore, one of the main aspects of structural design is to study their resistance to cyclic stresses [4, 5, 7]. Fatigue in materials takes place when they are subjected to alternating cyclic stresses, with the service life depending on the applied stress level, loading frequency, and other factors. This behavior is typically collected in the form of an S-N curve, shown schematically in Fig. 15.1 [6, 8, 9]. Some materials exhibit a stress below which fatigue failure is not likely to occur (endurance limit). However, an endurance limit has not been observed in most composites.

As a result of the inhomogeneous and anisotropic nature of composite materials, their fatigue behavior is more complicated than for more homogeneous and nominally isotropic materials such as metals. Composite fatigue failure modes include combinations of damage such as matrix failure, fiber cracking, debonding, delamination and ply failure [10]. Unlike the most isotropic brittle materials, fatigue failure in composite materials is usually accompanied with broad damage through the specimen volume [11]. Many theoretical studies and experimental tests have been performed to estimate the endurance limit and fatigue behavior of composite plates under axial tension and compression [12, 13]. These tests typically follow international standards [14] and provide the stress-life curves for tested materials. However, the actual loading condition of composite materials in service is more complex, potentially involving a combination of multiaxial tension/compression, bending and torsion. The majority of laboratory fatigue studies are conducted under the conditions of constant frequency and constant amplitude, e.g. a constant ratio of minimum and maximum load is applied [15]. Figure 15.2 shows examples of cycle types for stress σ [15].

A. Amiri (✉) • M.N. Cavalli
Department of Mechanical Engineering, University of North Dakota, 243 Centennial Drive, Stop 8359, Grand Forks, ND 58202-8359, USA
e-mail: ali.amiri@my.und.edu

E. Patterson et al. (eds.), *Composite Materials and Joining Technologies for Composites, Volume 7: Proceedings of the 2012 Annual Conference on Experimental and Applied Mechanics,* Conference Proceedings of the Society for Experimental Mechanics Series 44,
DOI 10.1007/978-1-4614-4553-1_15,

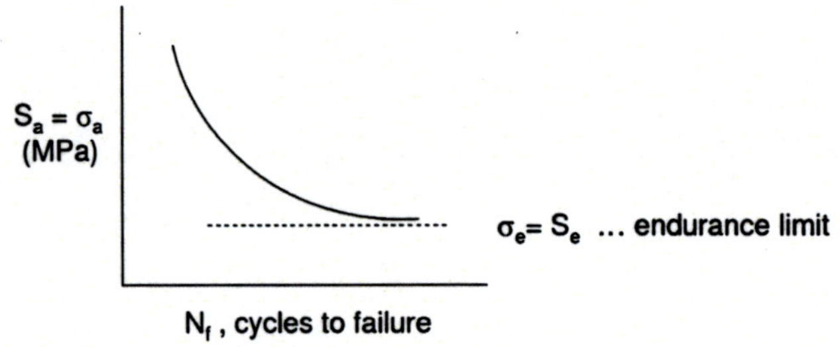

Fig. 15.1 Typical S-N curve, average stress (σ_a) versus cycles to failure (N_f) [6, 8, 9]

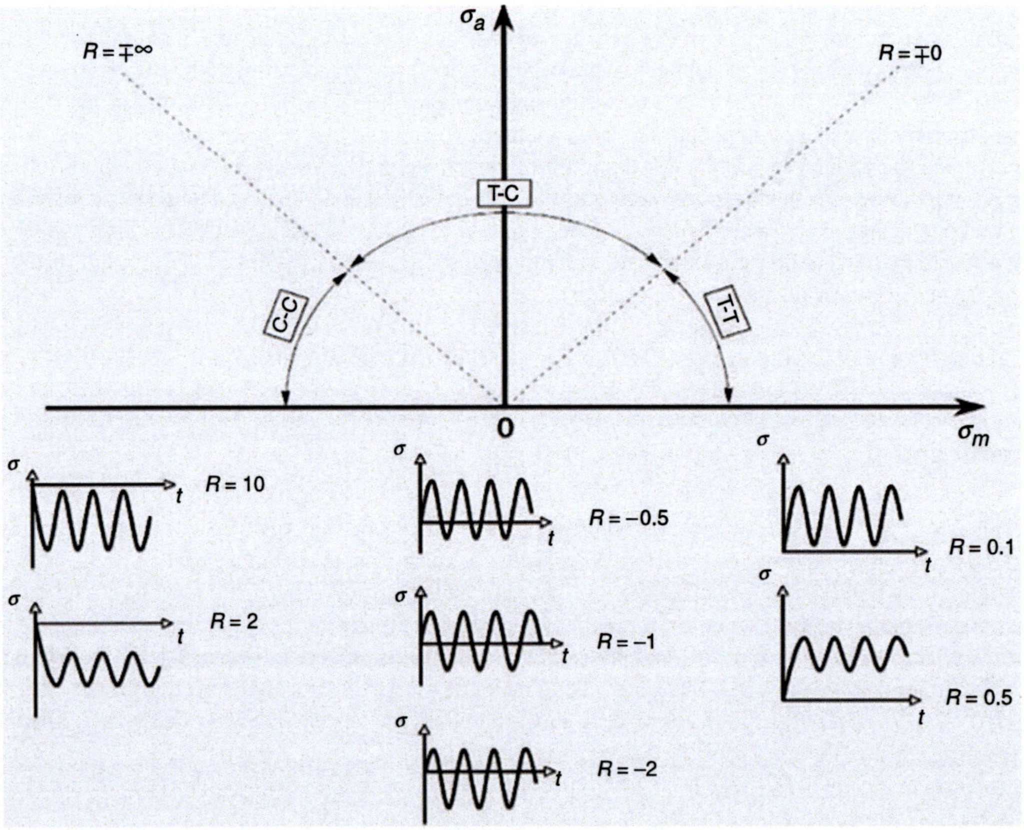

Fig. 15.2 Constant amplitude stress-time patterns [15]

In this study, the mechanical fatigue in bending of CFRP samples has been investigated. Bending tests are widely used for research purposes [15–19] and they have some advantages including: (1) in-service load conditions often include bending components, (2) the danger of Euler buckling is removed, and (3) the forces required to achieve the required stresses are typically much lower than for uniaxial loading [20]. In order to characterize fatigue behavior in bending, we have designed and manufactured a special fixture in our lab to perform a fully-reversed four-point bending test. Four-point bending tests are carried out on samples of carbon fibers ($(0/90)_{15}$) and polyester resin at constant frequency of 5 Hz. In this study all tests were done with a load ratio, R, of −1 where R = Load Min/Load Max.

15.2 Materials and Methods

15.2.1 Sample Preparation

A plate 600 mm long, 300 mm wide with a thickness of 4 mm was made out of 15 layers of carbon fiber plies with fibers in 0°/90° directions using VARTM (Vacuum Assisted Resin Transfer Molding). Plate layup and curing details are given in Table 15.1.

A VARTM set-up is shown in Fig. 15.3. After gelling of the resin, the plate was cured at room temperature for 24 h, and then post-cured at 40°C for 16 h. In order to determine the volume fraction of the fibers, a burn-off test was conducted in accordance with ISO 14127: 2008 [21] and the volume fraction of fibers in the plate was calculated to be 0.47. Test samples were cut out of the manufactured plate in accordance with ASTM D6272-02 [22] as shown in Fig. 15.4.

Table 15.1 CFRP specimen lay-up, method and curing details

Lay-up	Method	Details
Biaxial Carbon Plies (0°/90°)$_{15}$	VARTM	Curing 24 h at room temperature Post curing 16 h at 40°C

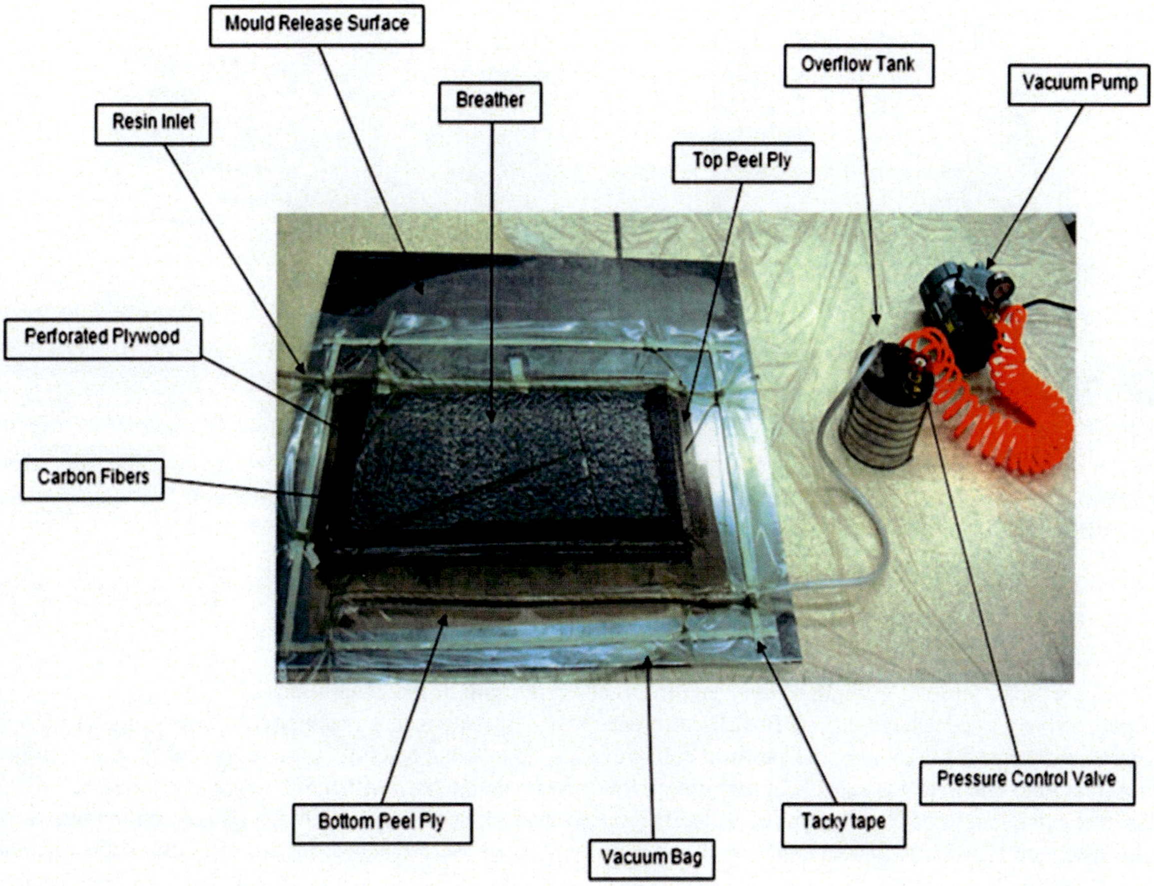

Fig. 15.3 VARTM set-up during resin infusion

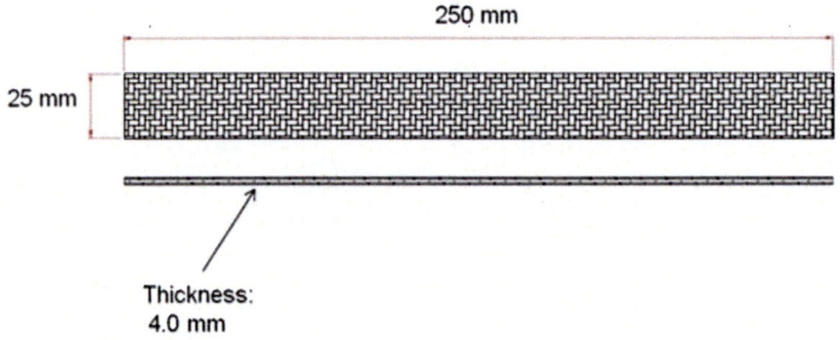

Fig. 15.4 Sample geometry, in accordance with ASTM D6272-02

Fig. 15.5 Test apparatus and four-point bending test fixture

15.2.2 Testing Methods

Four-point static bending test were conducted to determine yield and ultimate stress of the samples. The tests were done using a Shimadzu AG-IS 50 KN tensile test machine with rate of crosshead motion of 10 mm/min. The maximum stress in the outer fibers was calculated using Eq. 15.1 [22]:

$$S = PL/bd^2 \tag{15.1}$$

Where S is stress in the outer fiber throughout the load span (MPa), P is load at a given point on the load-deflection curve (N), L is the support span (mm), b is width of the beam (mm) and d is depth of the beam (mm).

Fatigue testing was conducted using a Bose Electroforce 3510 test system. The machine has a load range of ±7.5 kN and a displacement range of ±25 mm. To perform fully-reversed four-point bending tests, a special fixture was designed and manufactured. The fixture is capable of performing four-point bending tests with maximum applied load of 7 KN. A sine waveform at a frequency of 5 Hz was applied. All tests were done with a stress ratio, R, of −1. Fatigue testing was carried out until the specimen failed with visible cracks in it or 1.7 million cycles were reached. Figure 15.5 shows the apparatus and four-point bending test fixture.

Table 15.2 Mechanical properties of composite samples

Property	Measured average	CV[a] (%)
Maximum load (N)	747	1.8
Maximum stress (MPa)	315	3.4
Maximum strain (%)	0.58	5.7
Modulus of elasticity in bending (GPa)	62	4.2

[a]CV = coefficient of variation = (standard deviation/average) × 100%

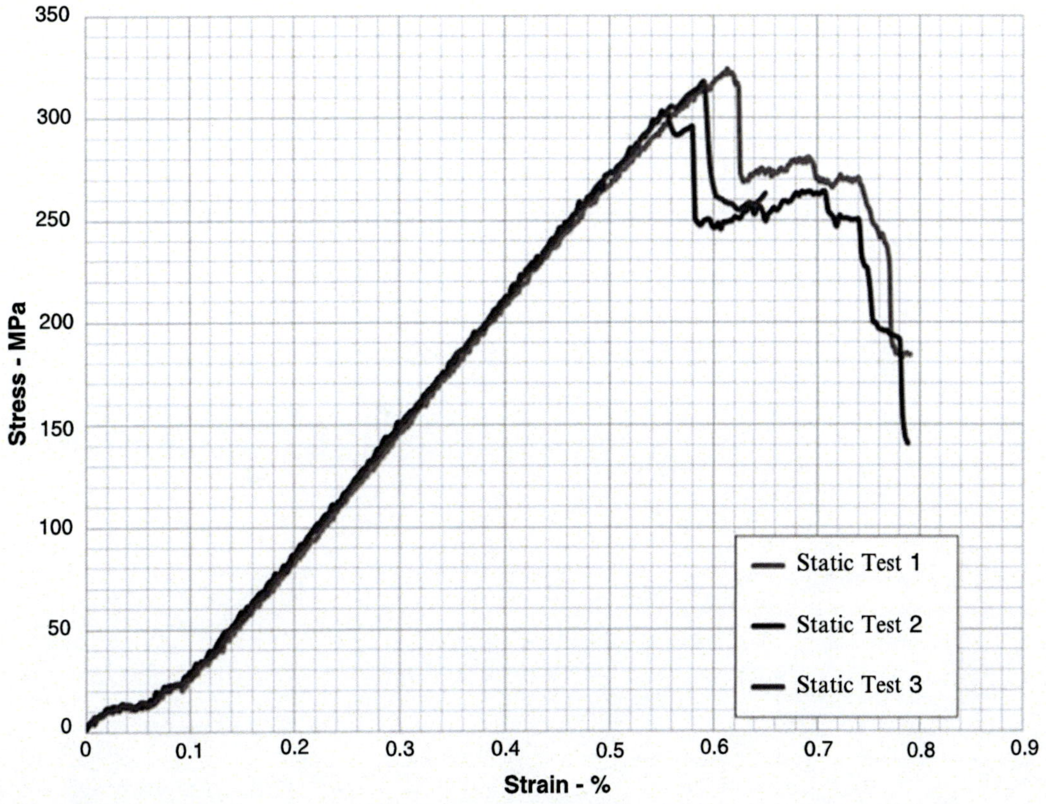

Fig. 15.6 Stress-deformation curves for four-point bending static tests

15.2.3 Experimental Results

Static four-point bending tests were carried out to measure mechanical properties of CFRP samples. Load was applied at the rate of 10 mm/min until failure. In all cases, samples failed by cracking/buckling on the compressive side of the sample. Table 15.2 shows the result of these tests.

The modulus of elasticity in bending was calculated using Eq. 15.2 [22]:

$$E_B = 0.21L^3m/bd^3 \tag{15.2}$$

Where E_B is modulus of elasticity in bending (MPa), L is support span (mm), b is width of the beam (mm), d is depth of the beam (mm) and m is slope of tangent line to the initial straight line in load-deflection curve for tested samples. The resulting stress–strain curves for tested samples are plotted in Fig. 15.6. The strain was calculated using Eq. 15.3 [22]:

$$r = 4.70Dd/L^2 \tag{15.3}$$

Where r is strain in the outer fiber (mm/mm) at deflection D (mm) of the center of the beam, d is depth of the beam (mm) and L is support span (mm). The maximum strain occurs at the midspan when deflection of the center of the beam is maximum.

Table 15.3 Fatigue lives for CFRP samples with $(0°/90°)_{15}$ layup

Sample	Test condition	R	σ_{max}/σ_y	Number of cycles
2	σ_{max} = 140 MPa	−1	0.45	1.72×10^6
14				1.76×10^6
1	σ_{max} = 175 MPa	−1	0.56	1.71×10^6
3				1.69×10^6
4				1.71×10^6
6	σ_{max} = 210 MPa	−1	0.67	1.91×10^5
11				4.66×10^5
12				4.73×10^5
8	σ_{max} = 225 MPa	−1	0.72	1.31×10^5
9				1.15×10^5
10				1.21×10^5
5	σ_{max} = 240 MPa	−1	0.76	3.20×10^4
13				2.40×10^4
15				4.04×10^4

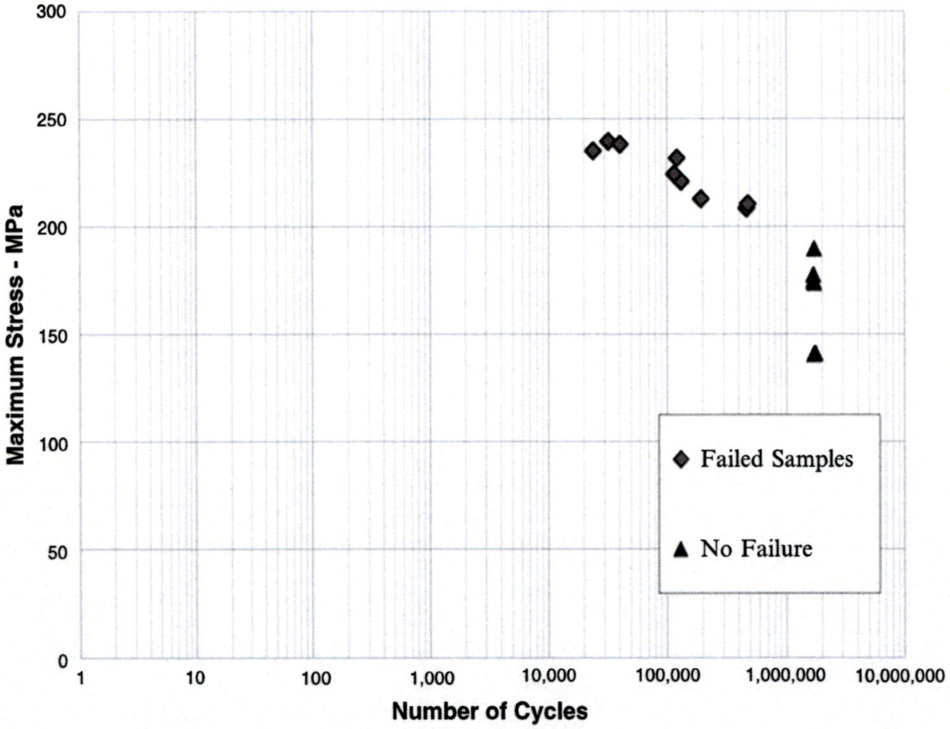

Fig. 15.7 S-N data for CFRP samples with $(0°/90°)_{15}$ layup

Four-point bending fatigue tests were carried out with maximum stresses of approximately 45%, 56%, 67%, 72% and 76% of the measured yield stress. The resulting maximum stresses were nominally 140, 175, 210, 225 and 240 MPa. Fatigue tests results are listed in Table 15.3. For maximum stresses of 140 and 175 MPa there were no failure detected therefore tests were stopped after 1.7×10^6 cycles. In all samples that failed during cyclic loading, there was visible crack through the thickness of the samples.

15.3 Discussion and Conclusions

The fatigue results shown in Table 15.3 and Fig. 15.7 show the expected downward trend in fatigue life with increasing maximum applied stress. There appears to be a threshold for 'infinite' life (defined as 1.7 million cycles in the current work) at a maximum stress of about 200 MPa. No change in mechanism was observed for specimens that did exhibit failure.

Tomita et al. examined the bending fatigue behavior of a variety of different carbon fibers and layups in fully-reversed bending at a frequency of 30 Hz [23]. For a cross-ply specimens and fatigue life of 10,000 cycles, they found stress amplitudes ranging from 400 to 475 MPa could be applied. This would correspond to maximum stresses of 200–237 MPa, very consistent with the values observed in the current work.

Based on previous results in the literature, e.g. [24], fully-reversed fatigue loading is expected to significantly reduce the fatigue life of carbon composites compared to positive values of R between 0 and 1. Comparing the current results with those of Kawai and Maki [25], we see that this is indeed the case. Under room temperature bending fatigue at $R = 0.1$ and 10 HZ, those authors found that a maximum stress of about 700 MPa resulted in a fatigue life of one million cycles. In the current study, the maximum stress resulting in failure at one million cycles was only 200 MPa. For fatigue lives of 10,000–1,000,000 cycles, the allowable maximum stress in the work of Kawai and Maki decreased from about 1,000 to 700 MPa (a decrease of 30%). In the current work, for the same range of fatigue lives, the allowable maximum stress decreased from about 250 MPa to about 200 MPA (a decrease of 20%), perhaps indicating a fundamental difference in the damage accumulation in the material during testing. Additional characterization work is needed to determine if this is the case.

References

1. Lee J, Harris B, Almond DP, Hammett F (1997) Fibre composite fatigue life determination. Composites A 28:5–15
2. Tai NH, Ma CCM, Wu SH (1995) Fatigue behaviour of carbon fiber/PEEK laminate composites. Composites 26:551–559
3. Fatigue life prediction in composite materials, achieving confidence in simulation using multiscale analysis, firehole composites white paper series, 2010
4. Belingardi G, Cavatorta MP, Frasca C (2006) Bending fatigue behavior of glass–carbon/epoxy hybrid composites. Compos Sci Technol 66:222–232
5. Amiri M, Khonsari MM (2010) Rapid Determination of fatigue failure based on temperature evolution: Fully reversed bending load. Int J Fatigue 32:382–389
6. Couillard RAA, Schwartz P (1997) Bending fatigue of carbon-fiber-reinforced epoxy composite strands. Compos Sci Technol 57:229–235
7. Smith TR, Owens MJ (1969) The progressive nature of fatigue damage in glass reinforced plastics. In: Sixth international reinforced plastics conference, London
8. Schackelford JF (1988) Introduction to materials science for engineers. Macmillan, New York
9. Timoshenko S, Young DH (1962) Elements of strength of materials. D. Van Nostrand, Princeton
10. Bedi R, Chandra R (2009) Fatigue-life distributions and failure probability for glass-fiber reinforced polymeric composites. Compos Sci Technol 69:1381–1387
11. Tsai SW (1988) Composite design, 4th edn. Think Composites, Dayton, USA
12. Peijis AAJM (1993) Hybrid composites based on polyethylene and carbon fibres. Composites 24:19–32
13. Swanson SR (1993) Compression fatigue response for carbon fiber with conventional and toughened epoxy matrices with damage. Eng Mater Technol 115:116–121
14. ASTM D4762 – 08, Standard guide for testing polymer matrix composite materials
15. Fatigue in composites, Bryan Hariss, Woodhead Publishing Ltd, 2003
16. Kawai M, Murata T (2010) A three-segment anisomorphic constant life diagram for the fatigue of symmetric angle-ply carbon/epoxy laminates at room temperature. Compos: Part A Appl Sci Manuf 10(10):1498–1510
17. Sedrakian A, Ben Zineb T, Billoet JL, Sicot N, Lardeur P (1997) A numerical model of fatigue behaviour for composite plates: application to a three point bending test. In: Degallaix S, Bathias C, Fougères R(eds) International conference on fatigue of composites. Proceedings, Paris, France, La Société Française de Métallurgie et de Matériaux, pp 415–423
18. Sidoroff F, Subagio B (1987) Fatigue damage modelling of composite materials from bending tests. In: Matthews FL, Buskell NCR, Hodgkinson JM, Morton J (eds) Sixth international conference on composite materials (ICCM-VI) & second European conference on composite materials (ECCM-II): proceedings, vol 4. Elsevier, London, pp 4.32–4.39
19. Caprino G, D'Amore A (1998) Flexural fatigue behaviour of random continuous-fibre-reinforced thermoplastic composites. Compos Sci Technol 58:957–965
20. De Baere I, Van Paepegem W, Degrieck J (2009) Comparison of different setups for fatigue testing of thin composite laminates in bending. Int J Fatigue 31:1095–1101
21. ISO 14127: 2008 Determination of the resin, fiber and void contents, Carbon-fiber-reinforced composites
22. ASTM D6272-02 Flexural properties of unreinforced and reinforced plastics and electrical insulating materials by four-point bending
23. Tomita Y, Morioka K et al (2001) Bending fatigue of long carbon fiber-reinforced epoxy composites. Mater Sci Eng 319–321:679–682
24. Himmel N, Bach C (2006) Cyclic fatigue behavior of carbon fiber reinforced vinylester resin composites manufactured by RTM and VARI. Int J Fatigue 28:1263–1269
25. Kawai M, Maki N (2006) Fatigue strengths of cross-ply CFRP laminates at room and high temperatures and its phenomenological modeling. Int J Fatigue 28:1297–1306

Chapter 16
Spark Plasma Sintering and Characterization of Graphene Reinforced Silicon Carbide Nanocomposites

Arif Rahman, Ashish Singh, Sandip P. Harimkar, and Raman P. Singh

Abstract This study investigates the possibility of incorporating graphene nanoplatelets as a nanofiller in silicon carbide matrix composites for improvement in mechanical properties. Silicon carbide, an advanced ceramic material, is considered suitable for structural applications due to its properties that include high hardness, toughness at high temperatures, high thermal conductivity, and chemical resistance. However, there has been a drive to improve strength and toughness to make these materials more reliable for engineering applications. Reinforcing these materials with appropriate nanofillers has the potential to significantly improve strength and toughness. These nanofillers can provide a variety of extrinsic toughening mechanisms. Nonetheless, the dispersion of nanoscale fillers in ceramic matrix composites remains challenging. In this study, dispersion of graphene nanoplatelets in precermic polymer using ball milling is done followed by polymer pyrolysis to obtain near stoichiometric SiC–graphene nanoplatelet powder. Subsequently, spark plasma sintering (SPS) is used to consolidate SiC-graphene nanoplatelet powders to form bulk samples. Different processing conditions are used, including various temperatures, and pressures while limiting the grain growth. Density and porosity of SiC-graphene nanoplatelet nanocomposites are determined as a function of processing temperature. XRD is used to determine changes in phase at different processing temperature and to estimate grain size variation. Ring-on-ring tests are performed to observe the effect of nanofiller on biaxial strength of bulk samples. Microhardness of the samples are also determined to understand the effect of graphene on hardness. Increasing the sintering temperature to 2100°C has resulted in relative density of 90% with significant improvement in mechanical properties.

Keywords Silicon carbide • Polymer pyrolysis • X-ray diffraction • Spark plasma sintering • Graphene nanoplatelet • Ceramic composite

16.1 Introduction

Significant improvements in mechanical properties are expected for nanostructured silicon carbide (SiC) compared to the conventional microcrystalline (micro-grained) microstructure, primarily due to the very high grain boundary area-to-volume ratio observed in nanostructured materials [1–3]. Nonetheless, the bulk processing of nanostructured SiC is challenging. Reducing ceramic grain size to nanoscale along with nanoscale reinforcement in ceramic matrix results in improvement in mechanical properties [4–7]. Nanoscale fillers also have potential for improving other properties of ceramic matrix composites (CMCs). For instance, carbon nanotubes (CNTs) significantly affect electrical and thermal properties of CMCs along with mechanical properties [8–10]. However, processing of nanograined CMCs is difficult. Processing issues include grain growth during sintering process [11–13] and dispersion of nanofillers (such as CNTs) in the ceramic matrices [14–16]. Hence, it is critically important to overcome these processing issues to understand the potential of nanograined CMCs to full extent in advanced applications.

In the current investigation, a novel approach involving polymer precursor processing and spark plasma sintering (SPS) is developed to process nanostructured SiC and nanostructured CMC with possibilities of retaining SiC grain size down to sub-100 nm range while being able to fabricate net-shape dense forms. Spark plasma sintering is a novel technique, where

A. Rahman • A. Singh • S.P. Harimkar • R.P. Singh (✉)
School of Mechanical and Aerospace Engineering, Oklahoma State University, Stillwater, OK 74075, USA
e-mail: raman.singh@okstate.edu

E. Patterson et al. (eds.), *Composite Materials and Joining Technologies for Composites, Volume 7: Proceedings of the 2012 Annual Conference on Experimental and Applied Mechanics*, Conference Proceedings of the Society for Experimental Mechanics Series 44, DOI 10.1007/978-1-4614-4553-1_16,

combined effect of pulse direct current, pressure and temperature is used to sinter the material at comparatively low sintering temperatures ($200-300°C$ lower than conventional sintering) and minimal sintering time (few minutes). Since very high heating rates are achievable using SPS due to the passage of high amount of pulsed direct current through graphite dies and punches, grain growth of nanostructured bulk SiC can be inhibited due to exposure to high temperatures for very little time.

Recently discovered graphene is a one atom thick 2-D layer of sp^2 carbon arranged in a honeycomb lattice [17, 18]. The properties of graphene, such as Young's modulus of 1TPa and ultimate strength of 130 GPa, make it one of the strongest materials available [17, 19]. Moreover, graphene is considered to have super-electrical properties and very high thermal properties [18, 20]. All these unique properties of graphene make it an attractive nanofiller. Recent studies have demonstrated that graphene can significantly improve mechanical properties of polymer matrices for very low loading [21, 22]. It is evident that the graphene–polymer nanocomposite system is well studied for quite some time now. However, graphene–ceramic composite system has not been well studied till now to understand the potential of graphene in ceramic matrix system.

In the current investigation, graphene nanoplatelet is used as nanofiller for different weight percentages to understand the effect of graphene nanoplatelet on properties of SiC matrix. Samples with or without graphene were fabricated using different processing temperatures and the phase and mechanical properties were characterized.

16.2 Material Fabrication

As a first step amorphousnanocrystalline SiC powder was prepared from a commercially available preceramic polymer, allylhydridopolycarbosilane (AHPCS) (Starfire Systems Inc., Malta, NY). The powder preparation process was started by heating the liquid polymer precursor to 650°C at 1°C/min, under an inert atmosphere and then holding it at 650°C for 10 min. This initiated the cross-linking of the polymer precursor. For complete conversion to amorphous-nanocrystalline SiC, the heating was continued till 1400°C at 3°C/min. All material was held at the final temperature for 1 h to ensure thermal equilibrium and complete processing. Finally, the material was cooled down to room temperature, at a rate of 5°C/min.

Due to the release of hydrogen gas during the polymer to ceramic conversion, the final ceramic product contained large voids. This was then ground using a hand grinder until the particles passed through a colander of mesh size 12 followed by subsequent milling into fine powder (~ 0.9μm) using a high energy ball mill (Pulverisette, Fritsch GmbH). Subsequently, spark plasma sintering (SPS) technique was used to consolidate nanostructured SiC obtained through pyrolysis of pre ceramic precursor using an SPS system (Model 10-3, Thermal Technology, LLC., Santa Rosa, California, USA). The ball milled SiC powder was loaded in between graphite dies and punches and sintered at various temperatures ranging from 1600 to 1900°C, at 100°C/min. Sintering was performed at 70 MPa of pressure for a soak time of 10 min. Based on the density and porosity data, a different set of samples were sintered with 150°C/min heating rate to a final temperature of 2100°C. Pressure and soak time were kept the same.

Exfoliated graphene nanoplatelets, xGnP®-M-5 grade (99.5% carbon) with an average diameter of 5μm were obtained in dry powder from XG Sciences, Inc. (East Lansing, MI). For preparing SiC-graphene powder mixture, controlled weight fraction of graphene nanoplatelets was mixed with AHPCS polymer and ball milled using a planetary ball mill (PM-100, Retsch GmbH, Haan, Germany) for the first set of samples. Subsequently, this mixture was pyrolyzed using the similar procedure mentioned earlier. The pyrolyzed amorphous SiC reinforced with graphene nanoplates was carefully milled such that uniform mixture without significant damage of graphene nanoplatelets can be achieved. Finally, this uniform mixture of powder was sintered using SPS to different final temperatures ranging from 1600 to 1900°C. Another set of samples was prepared using a heating rate of 150°C/min and a sintering temperature of 2100°C. Samples prepared by mixing graphene nanoplatelets with AHPCS polymer by ball milling are referred as AHPCS–%C in subsequent sections. Samples prepared by mixing graphene nanoplatelets to pyrolyzed amorphous SiC using ball milling are referred as SiC–%C. Figure 16.1 shows a schematic of the fabrication process.

16.3 Experimental Procedure

Phase analysis of the sintered compacts was performed using X-ray diffraction (XRD). The XRD analysis of the starting SiC and SiC–graphene powder was carried out using Philips Norelco X-ray diffractometer operating with Cu Kα ($\lambda = 1.54178$ Å) radiation at 45 kV and 40 mA. The diffraction angle was varied between 10° and 90° 2θ at a step

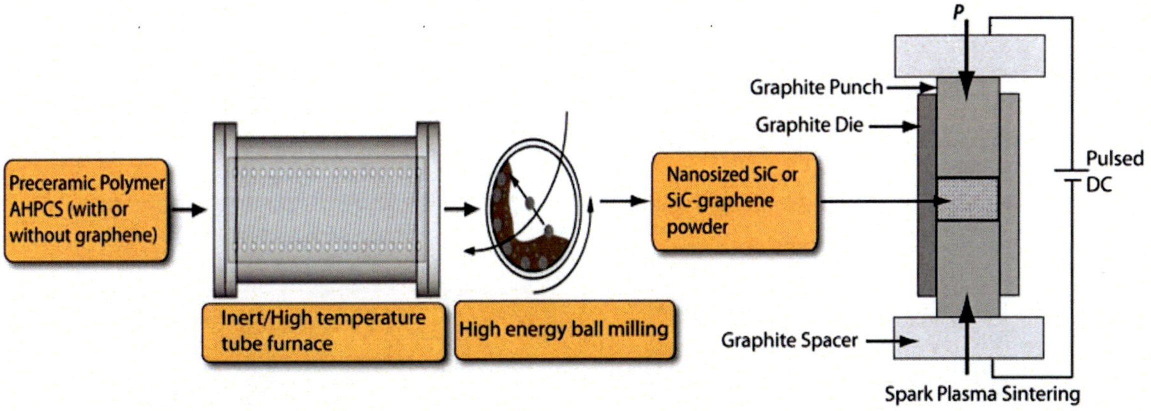

Fig. 16.1 Schematic of the fabrication process

increment of 0.02° 2θ with a count time of 1 s. In addition, the buoyancy method [23] was used for determining bulk density and porosity of the samples using a density measurement kit along with a high-resolution analytical balance.

Ring-on-ring (RoR) biaxial flexure testing was used to determine the flexural strength for SiC and SiC-graphene composite samples. The configuration used in this test exposes the maximum area under a constant maximum stress. This configuration employed a support ring diameter of 15 mm and the loading ring diameter of 5 mm, which ensures valid fracture mode of the samples [24]. The discs were loaded in the RoR fixture using a table-top test frame (Instron® 5567, Instron Corporation, Norwood, Massachusetts, USA). Adhesive tape was applied on the compressive side of the discs as per ASTM C1499–05 [25]. Displacement controlled loading at a rate of 0.5 mm/min. was used and the peak load at failure was recorded. For each kind of sample, at least five discs were subjected to the test to provide a statistical distribution of data. The flexure strength was then determined from the peak load at failure as per the equation used by Ovri et al. [26].

Microhardness of the polished samples were obtained using a microhardness tester (Clark Instrument, Inc., Novi, Michigan, USA) at a load of 9.8 N applied for 15 s. Twenty readings were collected for each sample and an average value is reported here.

16.4 Experimental Results

16.4.1 Microstructure Characterization

Based on available literature, allylhydridopolycarbosilane (AHPCS) completely converts to amorphous SiC at or above 1150°C. Usually crosslinking occurs between 250 to 400°C due to the loss of hydrogen at Si sites [27]. At higher temperatures, rearrangement of the Si–C network structure occurs. For the current work, SiC powder prepared using polymer pyrolysis at 1400°C was characterized using X-ray diffraction. Figure 16.2a shows XRD patterns of starting amorphous powder as well as samples compacted using spark plasma sintering technique.

The XRD pattern of the initial SiC powder, obtained through pyrolysis at 1400°C, exhibited characteristic broad halo peaks with diffused intensity showing a mostly amorphous structure. Spark plasma sintering of amorphous powder at various temperatures ranging from 1600 to 1900°C resulted in partial crystallization of amorphous powder during sintering. Peaks at 2θ value of 35.7°, 60.2°, and 72.0° correspond to β-crystalline SiC and suggest ordering of the starting amorphous SiC structure as processing temperature was increased. Peaks of WC were also present in the XRD data that is mainly due to the WC balls used for ball milling. Crystallite sizes were estimated from peak broadening using the Debye-Scherrer equation as shown in Table 16.1. Calculated crystallite sizes showed increasing trend reaching highest value of 35.4 nm for sample sintered at 1900°C. This shows, first processing of amorphous SiC powder using pyrolysis of preceramic precursor and then sintering using SPS can be used effectively in order to control the grain size or to make amorphous/crystalline nanocomposites where in-situ nanosized grains are reinforced in the amorphous matrix during spark plasma sintering.

Samples prepared with 5 and 25 wt% graphene were also studied with XRD as shown in Figure 16.2b, c. In each case, characteristic graphene-2H peak at 26.6° was prominent and other peaks correspond to β-SiC and grinding media WC as mentioned earlier. No change in peak position for graphene indicated insufficient exfoliation of graphene nanoplatelets.

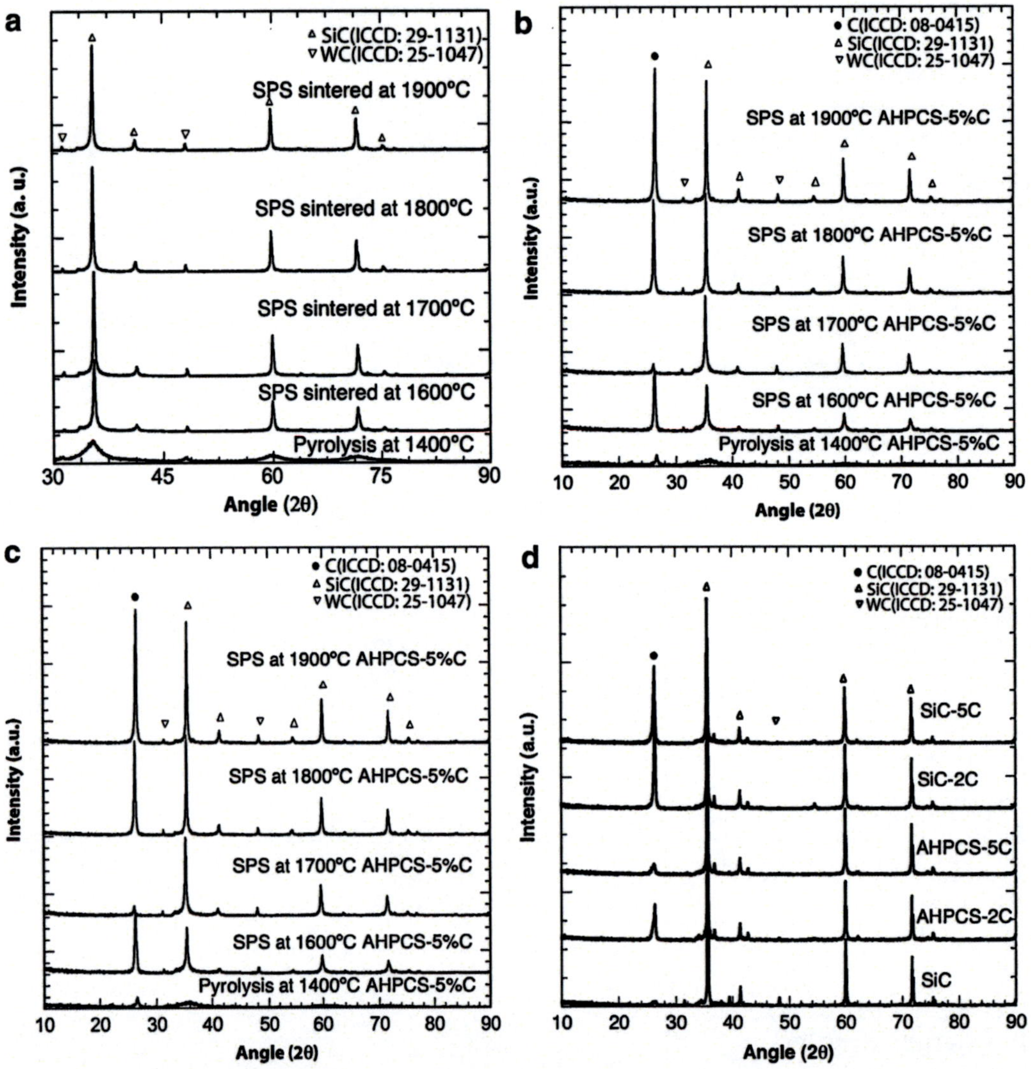

Fig. 16.2 XRD patterns from (**a**) amorphous SiC powder and spark plasma sintered SiC, (**b**) AHPCS–5 wt%graphene compacts, (**c**) AHPCS–25 wt%graphene compacts processed at temperatures ranging from 1600 to 1900°C, and (**d**) SiC–% graphene compacts processed at 2100°C

Table 16.1 Crystallite size of SPS sintered SiC and AHPCS-C as function of processing temperature

Sample constituent	SiC grain size (nm) at processing temperature			
	1600(°C)	1700(°C)	1800(°C)	1900(°C)
SiC	30.1	31.7	34.3	35.4
AHPCS-5 wt%C	24.53	29.3	36.5	37.1
AHPCS-25 wt%C	17.4	22.2	37.2	38.8

Crystallite size data for samples prepared with 5 and 25 wt% graphene also showed an increasing trend with increasing processing temperature. However, it was interesting to note that inclusion of graphene have restricted grain growth of SiC at lower processing temperature as observed by Fan et al. for Al_2O_3–graphene system [28]. Increasing graphene content to 25 wt% reduced grain size to as low as 17 nm for samples processed at 1600°C. This reduction in grain size was not observed for material processed at 1800 and 1900°C.

Table 16.2 Crystallite size of SPS sintered SiC, C–AHPCS, and C–SiC at 2100°C

Sample constituent	Grain size (nm) at 2100°C	
	SiC	Graphene
SiC	59.6	–
2 wt%C–AHPCS	48.0	21.5
5 wt%C–AHPCS	44.0	17.0
2 wt%C–SiC	49.5	31.5
5 wt%C–SiC	47.5	28.0

Table 16.3 Density and porosity of SPS sintered SiC and AHPCS–C as function of processing temperature

Measured property	Sample constituent	Processing temperature (°C)			
		1600	1700	1800	1900
Bulk density (g/cc)	SiC	2.0	2.1	2.3	2.42
	AHPCS–5 wt%C	1.96	2.0	2.17	2.38
	AHPCS–25 wt%C	2.0	2.13	2.14	2.47
Open porosity (%)	SiC	28.0	27.9	20.8	5
	AHPCS–5 wt%C	27.5	27.8	23.12	3
	AHPCS–25 wt%C	22.2	20.3	20.6	9.0
Relative density (%)	SiC	62.4	68.3	73.6	75.3
	AHPCS–5 wt%C	62.6	63.8	69.3	75.6
	AHPCS–25 wt%C	70.0	73.8	68.3	85.6

Samples prepared with 2 and 5 wt% graphene at 2100°C were also studied as shown in Fig. 16.2d. There was an increase in crystal size of SiC to ~ 60 nm as sintering temperature was increased from 1900 to 2100°C. However, addition of graphene restrained grain growth of SiC grain to ~ 50 nm for all the samples as shown in Table 16.2.

16.4.2 *Density and Porosity*

Generally, density and porosity of a bulk sample indicate how effectively the sample has been processed. Bulk mechanical properties are usually affected by sample density and porosity. Denser samples are desired for ceramic material as pores in them act as defects and lowers strength. For the current work, samples were processed to different temperatures with a pressure of 70 MPa. Bulk densities and porosity are listed in Table 16.3 for processed samples with or without graphene nanoplatelet. Relative densities were calculated assuming absolute densities for each case. For SiC samples, density of β-SiC (~ 3.2 g/cc) was considered to be absolute density for that material. For AHPCS–graphene samples absolute densities were calculated using rule of mixture equation.

Material density increased with processing temperature, which is in accordance with the amorphous to crystalline conversion with higher processing temperature. The density of polymer derived amorphous SiC (typically 2.0–2.8 g/cc) is lower than that of crystalline SiC [36] because at lower processing temperature there are bonds between Si and residual H, other than Si–C bonds, that reduce the density. At higher processing temperature, the amount of residual hydrogen was reduced in the structure and density increased. Also, with increasing processing temperature the volume diffusion enhances resulting in further increase in density. Although SiC processed at 1900°C had the highest observed bulk density (~ 2.42 g/cc), this value is still lower than the theoretical density of SiC (~ 3.2 g/cc). This could be the result of several factors. First, bulk density of the material is affected by the porosity in the material. Since material sintered at 1900°C had ~ 18.5% open porosity, the relative density of the material could have been ~ 81.5%. As can be seen from Table 16.2, the relative density of sintered SiC at 1900°C was ~ 75.4% which indicated other factors might have played a role in lowering the bulk density of the material. Second, material processed at 1900°C was not completely crystalline, and contained amorphous SiC that has a lower density. Moreover, the presence of excess amorphous carbon could also lower the effective material density, as amorphous carbon is significantly less dense than SiC [29].

Inclusion of graphene in SiC matrix also showed a similar trend of increasing density with increase in processing temperature. Although bulk densities of 5 wt% graphene–AHPCS were lower compared to bulk SiC samples, however, this is reasonable as long as the open porosity remained the same for both set of samples. Since, density of graphene is less

Table 16.4 Density and porosity of SPS sintered SiC, C–AHPCS, and C–SiC at 2100°C

Sample constituent	Density (g/cc)	Open porosity (%)	Relative density (%)
SiC	2.86	2.4	89.3
2wt %C–AHPCS	2.81	4.2	88.2
5wt %C–AHPCS	2.85	1.7	90.5
2wt %C–SiC	2.85	3.8	89.6
5wt %C–SiC	2.6	10.2	82.5

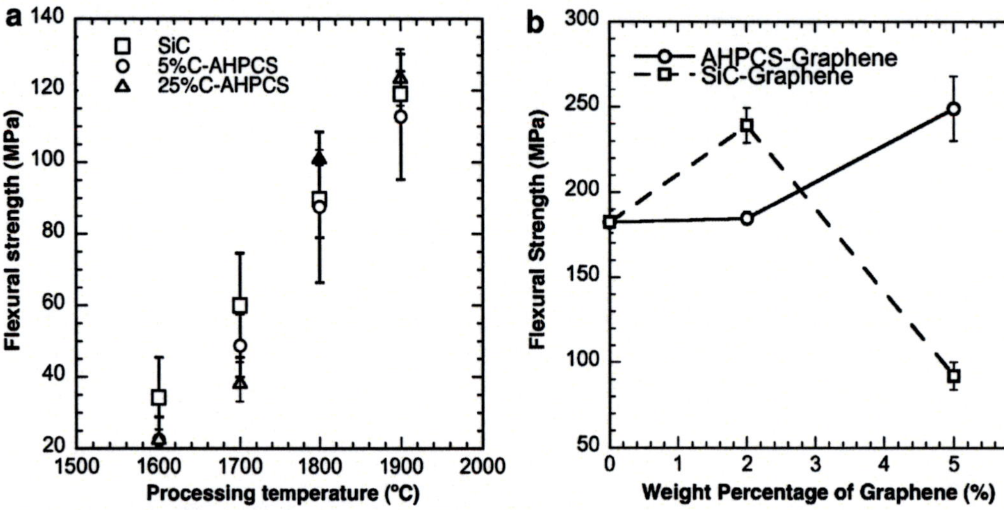

Fig. 16.3 Variation of biaxial strength of bulk SiC and AHPCS–C samples (**a**) sintered at various processing temperatures from 1600 to 1900°C and (**b**) sintered at 2100°C as a function of graphene weight percentage

compared to crystalline SiC, overall bulk density of AHPCS–C samples should be less. It is interesting to note that bulk densities of 25 wt% graphene–AHPCS were almost similar to bulk SiC samples. This is mainly due to the reduction of porosity in 25 wt% graphene–AHPCS bulk samples.

Table 16.3 shows decrease in open porosity with increasing processing temperature for all set of samples. This decrease in open porosity could be attributed to enhanced diffusion mechanism of particles during sintering at higher processing temperature.

Change of the processing parameters resulted in improvement in density as shown in Table 16.4. Higher heating rate favors densification over particle coarsening, hence this improvement in density. In addition, higher processing temperature also favors volume diffusion resulting in further increase in density. 5 wt%C–SiC sample had the highest open porosity of 10.2%, which could be due to the fact that dry ball milling of graphene nanoplatelets in SiC powder produced non-uniform dispersion and overall densification mechanism was affected.

16.4.3 Flexural Strength

The biaxial flexure strengths obtained for the bulk samples composed purely of SiC derived from AHPCS and AHPCS–graphene fabricated using SPS process are shown in Fig. 16.3a. The samples followed a trend of increasing strength with increasing processing temperature. SiC processed at 1600°C showed biaxial strength of ~ 34 MPa which is lower compared to strength obtained for other materials. This is reasonable since material processed at 1600°C had the lowest bulk density, and possibly contained amorphous SiC.

It should be noted that there was a ~ 75% increase in strength as processing temperature was increased from 1600 to 1700°C. This increase in strength could be attributed to two factors. One of the factors is increase in density. The other one is significant conversion of amorphous SiC to nano-crystalline SiC at higher processing temperature. This trend was observed

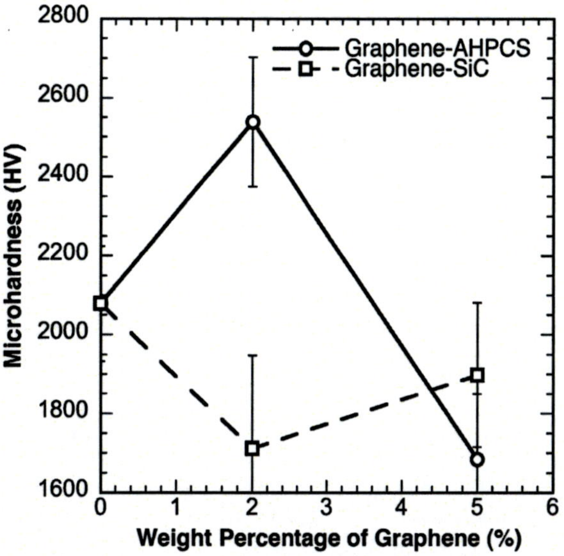

Fig. 16.4 Microhardness of 2100° C samples as a function of graphene weight fraction

for material processed at higher processing temperatures. This increase could be attributed to enhanced nanocrystallization at higher processing temperature.

Inclusion of graphene in the SiC system was expected to enhance materials ability to resist fracture. However, for materials processed at 1600 and 1700°C showed sudden decrease in strength when compared to monolithic SiC samples processed at the same temperature. The reason for this sudden decrease in strength at these temperatures could be due to insufficient densification of the system. Strength of AHPCS–C samples sintered at 1800 and 1900°C was almost similar to that of sintered monolithic SiC.

Biaxial-strength of samples prepared using higher heating rate and sintering temperature is shown in Fig. 16.3b. There was a ~ 40 increase in strength from monolithic SiC to 5 wt%–AHPCS sample sintered at 2100°C. As both the samples had similar relative density, increase in strength could be attributed to toughening by graphene nanoplatelets. Addition of graphene nanoplatelets by dry ball milling also improved strength of SiC matrix by ~ 30% for 2 wt%–SiC sample. However, non-uniform dispersion of graphene nanoplatelets in 5 wt%–SiC resulted in sudden decrease in strength.

16.4.4 Microhardness

To understand the effect of graphene inclusion on micro-scale property, samples prepared only using higher heating rate and sintering temperature were tested. As all the samples fabricated at 2100°C had about the same relative density, either similar microhardness values or some improvement were expected. Figure 16.4 shows 25% increase in microhardness of AHPCS–2 wt% graphene sample from monolithic SiC sample. However, other samples showed decrease in microhardness. Further investigation is underway to understand this phenomena.

16.5 Conclusion

Polymer infiltration and pyrolysis technique followed by spark plasma sintering technique was used successfully to fabricate nano-crystalline bulk SiC samples by in-situ crystallization of amorphous-SiC obtained from pyrolysis of polymer precursor. Density and porosity of the compacts were affected by heating rate and sintering temperature. Relative density was increased to ~ 90% with increase in heating rate to 150°C/min and sintering temperature to 2100°C for samples with or without graphene inclusion. From x-ray diffraction it was observed that with increase in processing temperature SiC grain size increased to ~ 60 nm at 2100°C; however, inclusion of graphene nanoplatelets restricted grain growth resulting in an average SiC grain size of ~ 50 nm. Bulk mechanical property was found to be affected by graphene inclusion. There was a ~ 40% increase in strength for AHPCS–5 wt%C sample as compared to monolithic SiC sintered at 2100°C. This increase in strength was mainly due to toughening by graphene nanoplatelets. There was a ~ 25% increase in microhardness of AHPCS–2 wt%C sample as compared to monolithic SiC sintered at 2100°C. Thus, effect of graphene nanoplatelets as nanofiller in SiC matrix was studied in terms of microstructural and mechanical properties.

References

1. Gubicza J, Nauyoks S, Balogh L, Labar J, Zerda TW, Ungar T (2007) Influence of sintering temperature and pressure on crystallite size and lattice defect structure in nanocrystalline SiC. J Mater Res 22(5):1314–1321
2. Szlufarska I, Nakano A, Vashishta P (2005) A crossover in the mechanical response of nanocrystalline ceramics. Science 309(5736):911–914
3. Kearand B, Mukherjee A (2007) Far-from-equilibrium processing of nanostructured ceramics. In: Materials processing handbook. CRC Press, Boca Raton, pp 7–1–7–18
4. Hofler H, Averback R (1990) Grain-growth in nanocrystalline $TiO_2$and its relation to vickers hardness and fracture-toughness. Scr Metall Mater 24(12):2401–2406
5. Karch J Birringer R (1990) Nanocrystalline ceramics – possible candidates for net-shape forming. Ceram Int 16(5):291–294
6. Zhan G, Kuntz J, Garay J, Mukherjee A (2003) Electrical properties of nanoceramics reinforced with ropes of single-walled carbon nanotubes. Appl Phys Lett 83(6):1228–1230
7. Peigney A (2003) Composite materials: tougher ceramics with nanotubes. Nat Mater 2(1):15–16
8. Thostenson E, Karandikar P, Chou T (2005) Fabrication and characterization of reaction bonded silicon carbide/carbon nanotube composites. J Phys D Appl Phys 38(21):3962–3965
9. Zhan G, Mukherjee A (2004) Carbon nanotube reinforced alumina-based ceramics with novel mechanical, electrical, and thermal properties. Int J Appl Ceram Technol 1(2):161–171
10. Zhan G, Kuntz J, Garay J, Mukherjee A (2003) Electrical properties of nanoceramics reinforced with ropes of single-walled carbon nanotubes. Appl Phys Lett 83(6):1228–1230
11. Chen I, Wang X (2000) Sintering dense nanocrystalline ceramics without final-stage grain growth. Nature 404(6774):168–171
12. Mayo M, Hague D, Chen D (1993) Processing nanocrystalline ceramics for applications in superplasticity. Mater Sci Eng A Struct Mater Prop Microstruct Process 166(1–2):145–159
13. Averback R, Hofler H, Tao R (1993) Processing of nano-grained materials. Mater Sci Eng A Struct Mater Prop Microstruct Process 166(1–2):169–177
14. Peigney A, Laurent C, Flahaut E, Rousset A (2000) Carbon nanotubes in novel ceramic matrix nanocomposites. Ceram Int 26(6):677–683
15. Ionescu E, Francis A, Riedel R (2009) Dispersion assessment and studies on AC percolative conductivity in polymer-derived Si-C-N/CNT ceramic nanocomposites. J Mater Sci 44(8):2055–2062
16. Gao L, Jiang L, Sun J, (2006) Carbon nanotube-ceramic composites. J Electroceram 17(1):51–55
17. Geim A, Novoselov K (2007) The rise of graphene. Nat Mater 6(3):183–191
18. Kim H, Abdala A, Macosko C (2010) Graphene/polymer nanocomposites. Macromolecules 43(16):6515–6530
19. Lee C, Wei X, Kysar J, Hone J (2008) Measurement of the elastic properties and intrinsic strength of monolayer graphene. Science 321(5887):385–388
20. Cai D, Song M (2010) Recent advance in functionalized graphene/polymer nanocomposites. J Mater Chem 20(37):7906–7915
21. Ramanathan T, Abdala A, Stankovich S, Dikin D, Herrera-Alonso M, Piner R, Adamson D, Schniepp H, Chen X, Ruoff R, Nguyen S, Aksay I, Prud'homme R, Brinson L (2008) Functionalized graphene sheets for polymer nanocomposites. Nature Nanotechnol 3(6):327–331
22. Yasmin A, Luo J, Daniel I, (2006) Processing of expanded graphite reinforced polymer nanocomposites. Compos Sci Technol 66(9):1182–1189
23. ASTM-C830-00 (2006) Standard test methods for apparent porosity, liquid absorption, apparent specific gravity, and bulk density of refractory shapes by vacuum pressure. American Society for Testing and Materials, West Conshohocken
24. Wereszczak AA, Swab JJ, Kraft RH (2003) Effects of machining on the uniaxial and equibiaxial flexure strength of cap3 ad-995 al_2o_3. ARL Technical report
25. ASTM-C1499-05 (2006) Standard test method for monotonic equibiaxial flexural strength of advanced ceramics at ambient temperature. American Society for Testing and Materials, West Conshohocken
26. Ovri JE (2000) A parametric study of the biaxial strength test for brittle materials. Mater Chem Phys 66(1):1–5
27. Zunjarrao SC (2008) Ploymer derived ceramics: processing–structure–property relationships. Ph.D. thesis, Oklahoma State University
28. Fan Y, Wang L, Li J, Li J, Sun S, Chen F, Chen L, Jiang W (2010) Preparation and electrical properties of graphene nanosheet/Al_2O_3 composites. Carbon 48(6):1743–1749
29. Moraes KV (2000) The densification, crystallization and mechanical porperties of allylhydridoploycarbosilane-derived silicon carbide. Ph.D. thesis, Rensselaer Polytechnic Institute

Chapter 17
Processing, Microstructure, and Properties of Carbon Nanotube Reinforced Silicon Carbide

Thomas A. Carlson, Charles P. Marsh, Waltraud M. Kriven, Peter B. Stynoski, and Charles R. Welch

Abstract The addition of multi-walled carbon nanotube reinforcements to a ceramic matrix has been suggested to improve the fracture toughness. The hypothesized improvement is thought to be the result of crack bridging and other toughening mechanisms. However, no such improvement in toughness has not been achieved to date for a multi-walled carbon nanotube and silicon carbide composite. However, there are several processing techniques, compositions, and methods for producing said composite, which may inhibit or foster success. Here, we report the processing, microstructure, and properties of a multi-walled carbon nanotube and silicon carbide composite material. The processing required careful mixing of the carbon nanotubes within the matrix in order to maximize dispersion and minimize carbon nanotube damage. The sintering required careful control of specific parameters to produce the desired microstructure and maximum density. The spark plasma sintering technique used was. These processing methods resulted in unique microstructures which in turn affected the material properties. The effect on the mechanical strength was evaluated using three-point flexural testing.

Keywords Silicon carbide (SiC) • Carbon nanotubes (CNTs) • Spark plasma sintering (SPS) • Flexural strength

17.1 Introduction

Silicon carbide (SiC) is a ceramic material with high strength [1]. However, it is brittle and has a low resistance to fracture [1]. Methods to improve the fracture toughness of SiC have been identified, including changing the SiC grain size and shape [2] as well as by incorporating additional phases [3]. One method is to introduce carbon nanotubes (CNTs) as a reinforcement material [4]. To successfully improve the toughness of the SiC, the CNTs should be homogeneously dispersed throughout the SiC [5], retain structural integrity during processing [6] and sintering, and interact well with the SiC [7]. These are the main challenges associated with SiC-CNT composites. Here we describe the dispersion, sintering, and properties of a SiC-CNT composite. The objective of our experiment was to synthesize a SiC composite containing dispersed CNTs that resulted in more desirable properties than either component on its own. We used commercial SiC and multi-walled carbon nanotubes (MWCNTs) as the component materials. The sintering method was spark plasma sintering (SPS). The mechanical property of interest was flexural strength. In this paper, we explain the methods of our material characterization, processing, composite formation, and flexural testing.

T.A. Carlson (✉) • C.P. Marsh • P.B. Stynoski
Engineer Research and Development Center, Construction Engineering Research Laboratory, 2902 Newmark Drive, Champaign, IL 61822, USA
e-mail: thomas.a.carlson@usace.army.mil

W.M. Kriven
Department of Materials Science and Engineering, University of Illinois at Urbana-Champaign, 1304 W. Green St, Urbana, IL 61801, USA

C.R. Welch
Advanced Materials Initiative, Engineer Research and Development Center, Information Technology Laboratory, 3909 Halls Ferry Road, Vicksburg, MS 39180, USA

E. Patterson et al. (eds.), *Composite Materials and Joining Technologies for Composites, Volume 7: Proceedings of the 2012 Annual Conference on Experimental and Applied Mechanics*, Conference Proceedings of the Society for Experimental Mechanics Series 44, DOI 10.1007/978-1-4614-4553-1_17,

17.2 Methods

17.2.1 Material Characterization

The MWCNTs (Helix Material Solutions, Richardson, TX) manufacturer's stated diameter was 20–40 nm, the length was 0.5–40 um, and the purity was >95%. The MWCNTs were used in the as-received state and after a silica functionalization treatment [8]. The as-received MWCNTs were characterized using x-ray diffraction (XRD), scanning electron microscopy (SEM), and transmission electron microscopy (TEM). The functionalized CNTs were characterized by TEM.

The SiC powder (NanoAmor, Inc., Houston, TX) manufacturer's stated particle size was 45–55 nm. The stated phase was β-SiC. The as-received SiC was characterized by XRD, SEM, and TEM.

17.2.2 Powder Processing

Mixtures containing 5 wt% MWCNTs in SiC were ball milled together for 24 h using a 100 ml HDPE bottle with 20 g of 5 mm yttria-doped zirconia grinding medium (Union Process, Akron, OH). Two samples for each mixture were made so that one could be ball milled dry and the other ball milled in ethanol. The five MWCNT-SiC mixtures discussed in this paper are given in Table 17.1.

Upon mixing of the MWCNTs and the SiC, additional characterization was completed by SEM and TEM. SEM was used to qualify the degree of mixing, and TEM was used to determine the degree of damage to the CNTs and the degree of clumping of the MWCNTs and the SiC.

17.2.3 Composite Formation

A spark plasma sintering unit located at the Army Research Laboratory, Aberdeen, MD, was used to densify the processed powders. The unit consists of a large vacuum chamber located within a load frame such that the rams pass through the walls of the chamber and apply the compressive load to the sample. The rams also deliver the pulsed DC electric current; however, the load system is independent of the electric current system. The processed powder was placed within a graphite die and punch assembly placed between the upper and lower rams, as shown in Fig. 17.1.

After filling the die, placing it in the chamber, and applying the desired load to the rams, argon gas was introduced into the chamber while the pulsed DC current was applied. The current was increased at a rate of 300 A/min until the final temperature was reached, resulting in 4–5 min ramp times. The temperature of the die wall was measured using an optical pyrometer. When the desired temperature was reached, either a hold time began, or the current and load were removed from the sample. The sample was allowed to cool to 100 ° Celsius before removal from the chamber. The parameters used for each sample are given in Table 17.2.

Bulk density measurements for the specimens were completed by using the Archimedes method. Characterization of the specimens was performed using XRD, SEM, and TEM. XRD was used to determine the presence of CNTs in the matrix. Measurement of grain size using the XRD spectra was done using peak-fitting analysis. Measurement of grain size from the SEM images was completed using the line-intercept method.

Table 17.1 Description of the powder compositions and processing for all specimens

Sample	SiC (g)	MWCNT (g)	Milling	Theoretical density (g/cm^3)
2	5.00	0.00	Wet	3.21
3a	4.75	0.25	Dry	3.12
3b	4.75	0.25	Wet	3.12
4a	4.75	0.25	Dry	3.12
4b	4.75	0.25	Wet	3.12

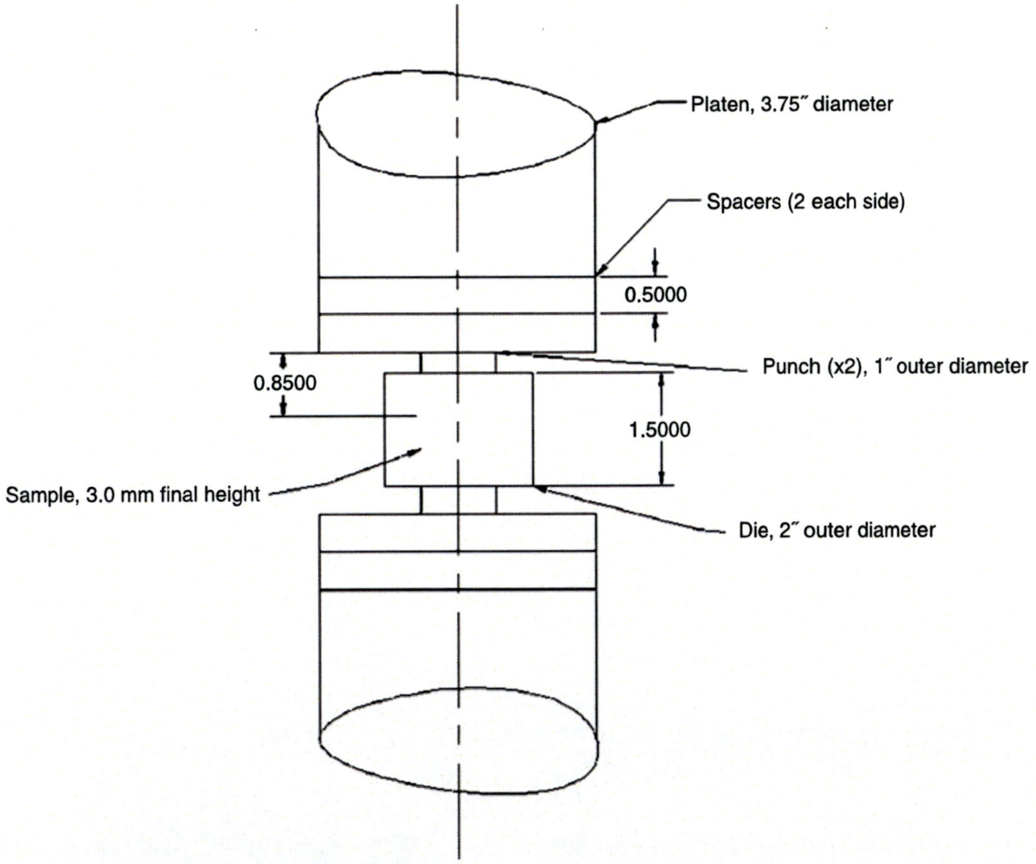

Fig. 17.1 Geometry of the graphite die and punch assembly

Table 17.2 Spark plasma sintering parameters used for each sample

Sample	Current ramp rate (A/min)	Load (MPa)	Final temperature (Celsius)	Hold at final temperature (min)
2	300	55	1,800	5
3a, 3b, 4a, 4b	300	55	1,800	0

17.2.4 Flexural Testing

Specimens were prepared for three-point flexural testing by cutting bend bars from the sintered disks. The bars were cut using a slow-speed saw with a 4 in. diamond wafering blade. The bend bars were then polished using a 15 μm and then a 6 μm diamond polishing wheel under running water. Surface uniformity and the removal of all surface features were confirmed by optical microscope. Typical bend bar dimensions were 3 mm by 3 mm square in section, with varying lengths.

A load frame (Model 5882, Instron, Norwood, MA) was used to test the bend bars. The loading configuration is given in Fig. 17.2. Crosshead displacement rate was 0.1 mm/min.

Flexural strength for the specimens was determined using Eq. 17.1, where σ is the flexural strength (Pa), F is the measured load (N), L is the support span (m), b is the specimen width (m), and d is the specimen thickness (m).

$$\sigma = \frac{3FL}{2bd^2} \tag{17.1}$$

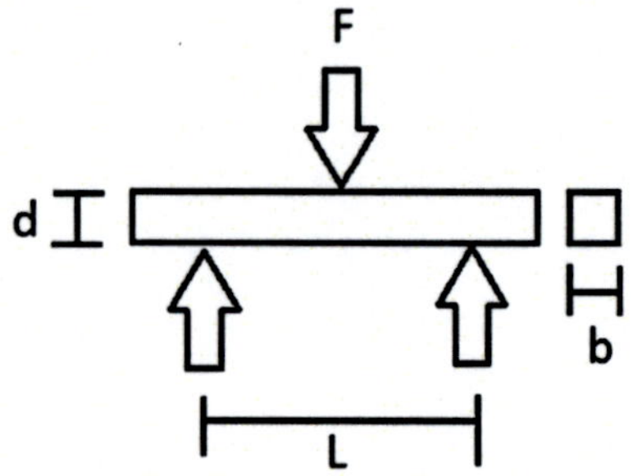

Fig. 17.2 Loading configuration for flexural test

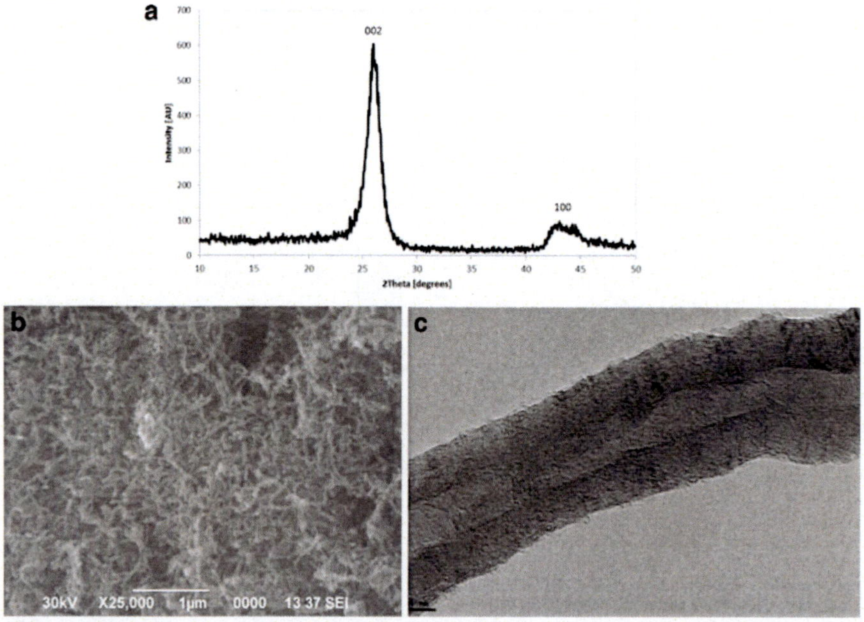

Fig. 17.3 As-received MWCNTs showing expected crystal structure, agglomeration, and multiple walls. **(a)** XRD spectra. **(b)** SEM micrograph. **(c)** TEM micrograph

17.3 Results

17.3.1 Material Characterization

The microstructure of the as-received MWCNT, as determined by XRD, SEM, and TEM, is summarized in Fig. 17.3. The microstructure of the silica-functionalized MWCNTs is given in Fig. 17.4.

The microstructure of the SiC, as determined by XRD, SEM, and TEM, is summarized in Fig. 17.5. The SiC particle size measured using these methods is given in Table 17.3.

17.3.2 Powder Processing

The microstructure of the mixed MWCNTs and SiC are given in Fig. 17.6.

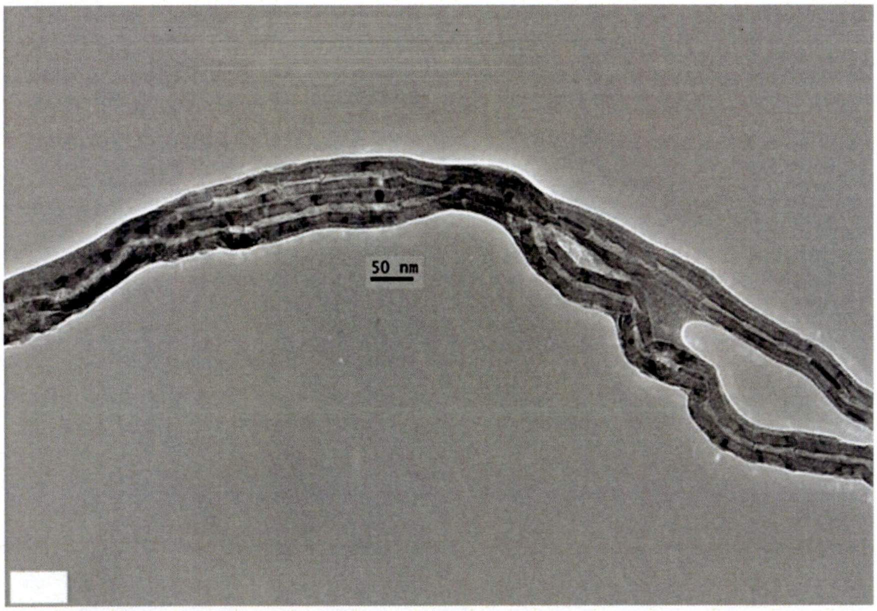

Fig. 17.4 TEM micrograph of silica-functionalized MWCNT showing silica surface coating

17.3.3 Composite Formation

Digital photographs of the spark plasma sintered specimens are shown in Fig. 17.7.

Densities and percent of the theoretical density for the spark plasma sintered specimens are given in Table 17.4.

Figures 17.8 and 17.9 are the XRD spectra and SEM micrographs, respectively, of the spark plasma sintered specimens.

The grain size in each composite specimen, determined using XRD and SEM, is given in Table 17.5. Measurement of the grain sizes for specimens 3a, 3b, 4a, and 4b could not be obtained by the line-intercept method because individual grains were not observed in the micrographs which was assumed to be a result of the high porosity.

17.3.4 Flexural Testing

A typical stress versus strain curve for 3-point flexural tests is given in Fig. 17.10, and the flexural strengths of the specimens are provided in Table 17.6. The reported values are the average of two flexural tests per specimen. Specimen 3b was not tested.

17.4 Discussion

17.4.1 Materials Characterization

The XRD spectra of the MWCNTs shows the presence of the graphite structure. The SEM micrograph of the MWCNTs reveals a dense agglomeration of MWCNTs with relatively large diameters as compared to single-walled carbon nanotubes (SWCNTs). The larger diameter of the MWCNTs allows them to be imaged easier and to withstand mechanical mixing. The TEM image of the MWCNT confirms the presence of multiple layers in the wall structure.

The XRD spectra of the SiC displays the expected reflections for the SiC structure. Analysis of the spectra confirms that the powder was the β-SiC phase and the particle size was about 18 nm. The SEM micrograph of the SiC displays small agglomerates of the SiC, which themselves are about 50–60 nm in diameter. The TEM characterization of the SiC indicates SiC particles of about 20–40 nm in diameter. There is evidence of a surface oxidation layer, assumed to be $Si0_2$.

a

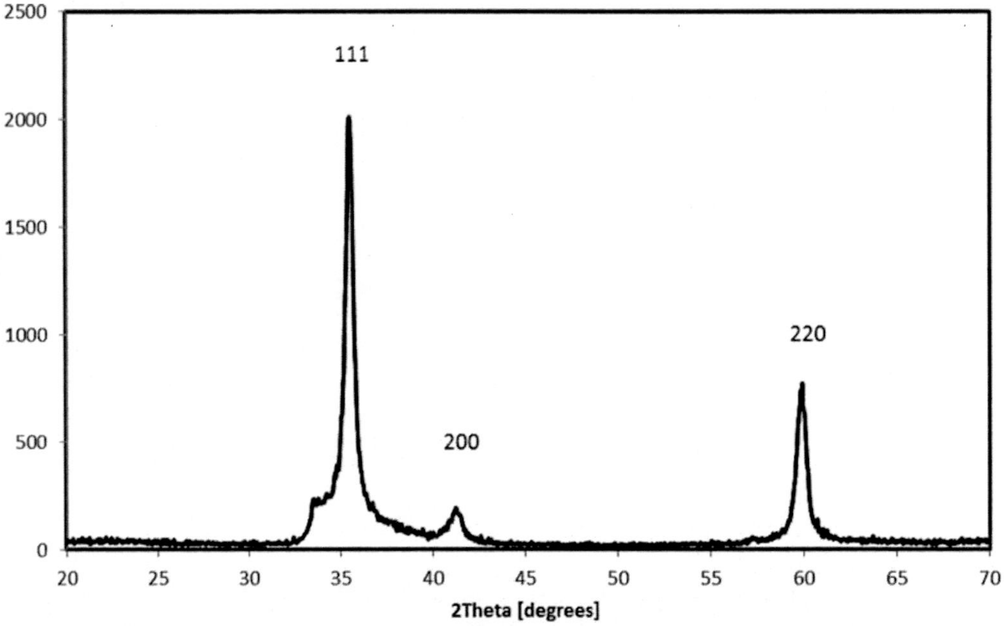

b

Fig. 17.5 As-received SiC powder showing expected crystal structure, agglomeration, and surface oxidation. (**a**) XRD spectra. (**b**) SEM micrograph. (**c**) TEM micrograph

c

Fig. 17.5 (continued)

Table 17.3 Particle size of as-received SiC

Method	Particle size (nm)
XRD	18
SEM	50–60
TEM	20–40

The particle sizes measured using the line-intercept method from the SEM and TEM micrographs are similar. However, because we are not looking at a flat cross-section, errors could be introduced. The grain sizes measured from the XRD spectra were obtained using a strain and size calculation provided by the analysis software. The calculation uses peak profiling, taking into account the three peaks shown in the spectra. The peak-profile fitting method is only valid for pure phases, which was the case for this SiC sample.

17.4.2 Powder Processing

The XRD characterization of the mixed MWCNT and SiC contains peaks from both the MWCNTs and the SiC. The MWCNT signal was not as strong as the SiC signal, but that was to be expected because the composition contained significantly less MWCNTs than SiC.

The SEM micrograph of the MWCNTs and SiC indicated homogeneous mixing. There were single de-agglomerated MWCNTs surrounded by small agglomerates of SiC.

The TEM micrograph of the mixed MWCNTs and SiC also confirmed that the two components were mixed well. The MWCNTs were not clumped in large bundles. The micrograph does show two MWCNTs aligned next to each other. The SiC particles formed agglomerates, but the agglomerates were separated by bridging MWCNTs.

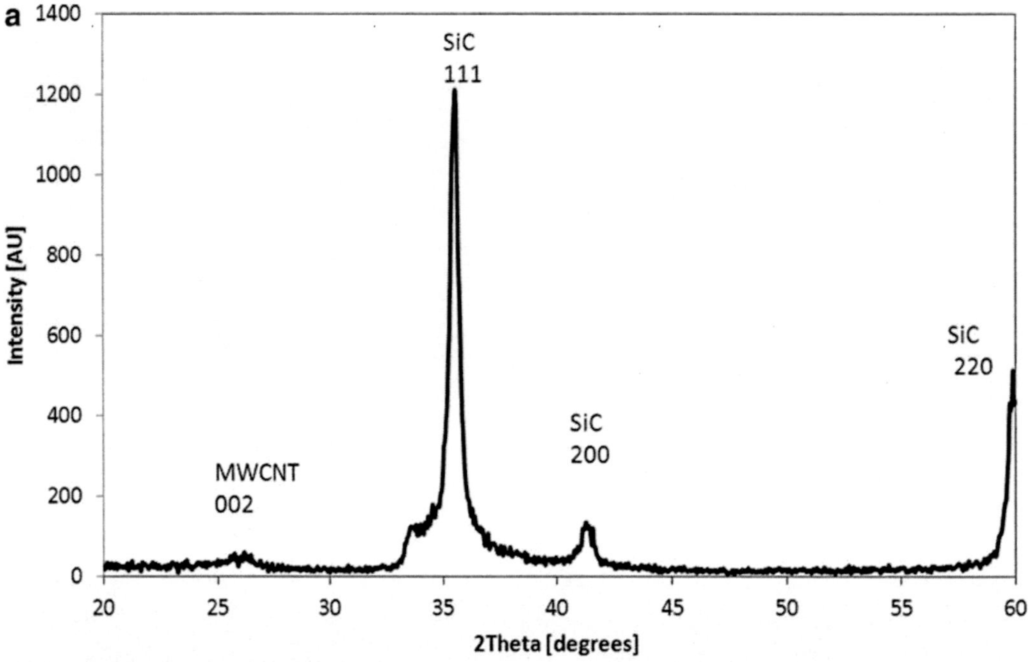

Fig. 17.6 Microstructure of MWCNTs and SiC showing combined crystal structure, homogeneous mixing, and undamaged MWCNTs. (**a**) XRD spectra. (**b**) SEM micrograph. (**c**) TEM micrograph

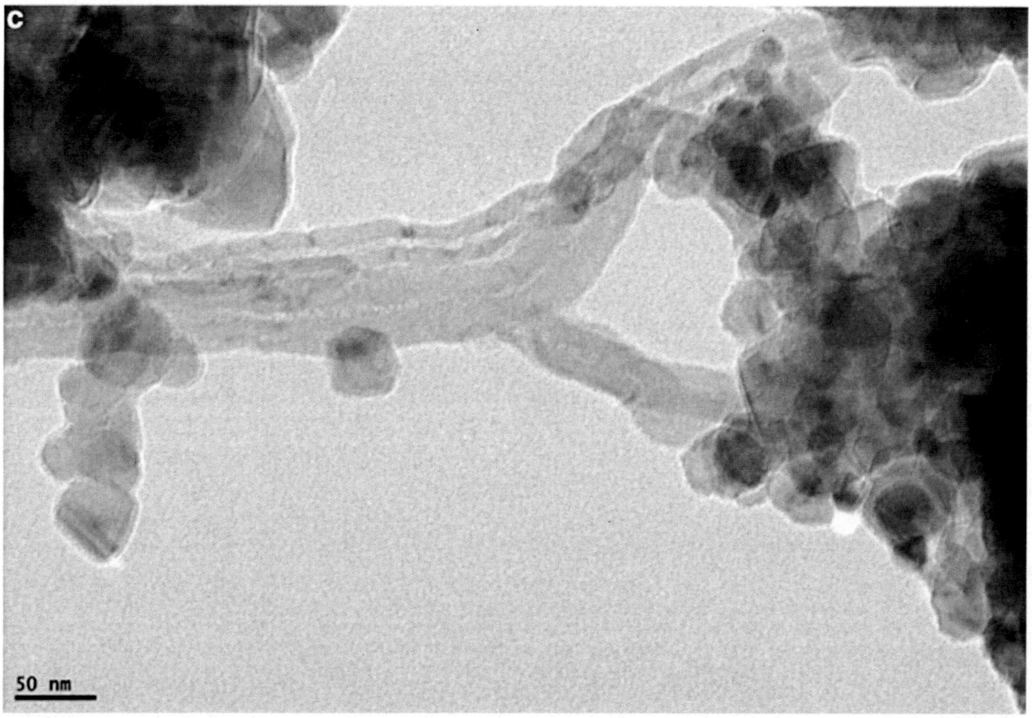

Fig. 17.6 (continued)

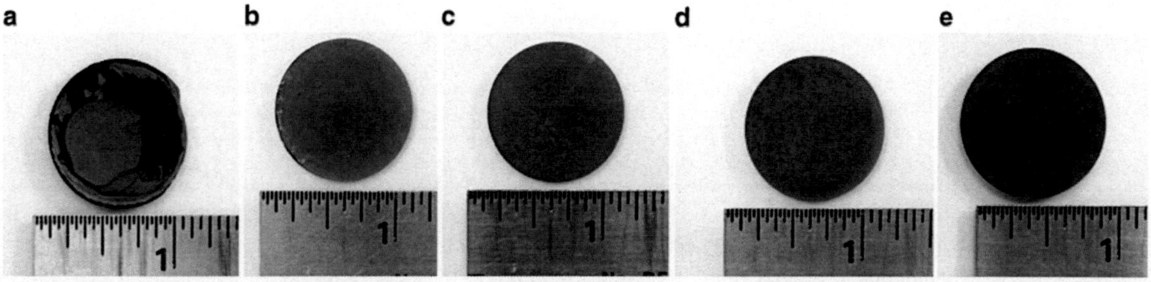

Fig. 17.7 Digital photographs of the specimens showing the effect of surface sticking. (**a**) Specimen 2. (**b**) Specimen 3a. (**c**) Specimen 3b. (**d**) Specimen 4a. (**e**) Specimen 4b

Table 17.4 Densities of the spark plasma sintered specimens

Sample	Archimedes density (g/cm^3)	%TD
2	2.79	86.8
3a	2.91	93.3
3b	2.75	87.9
4a	2.61	83.6
4b	2.58	82.7

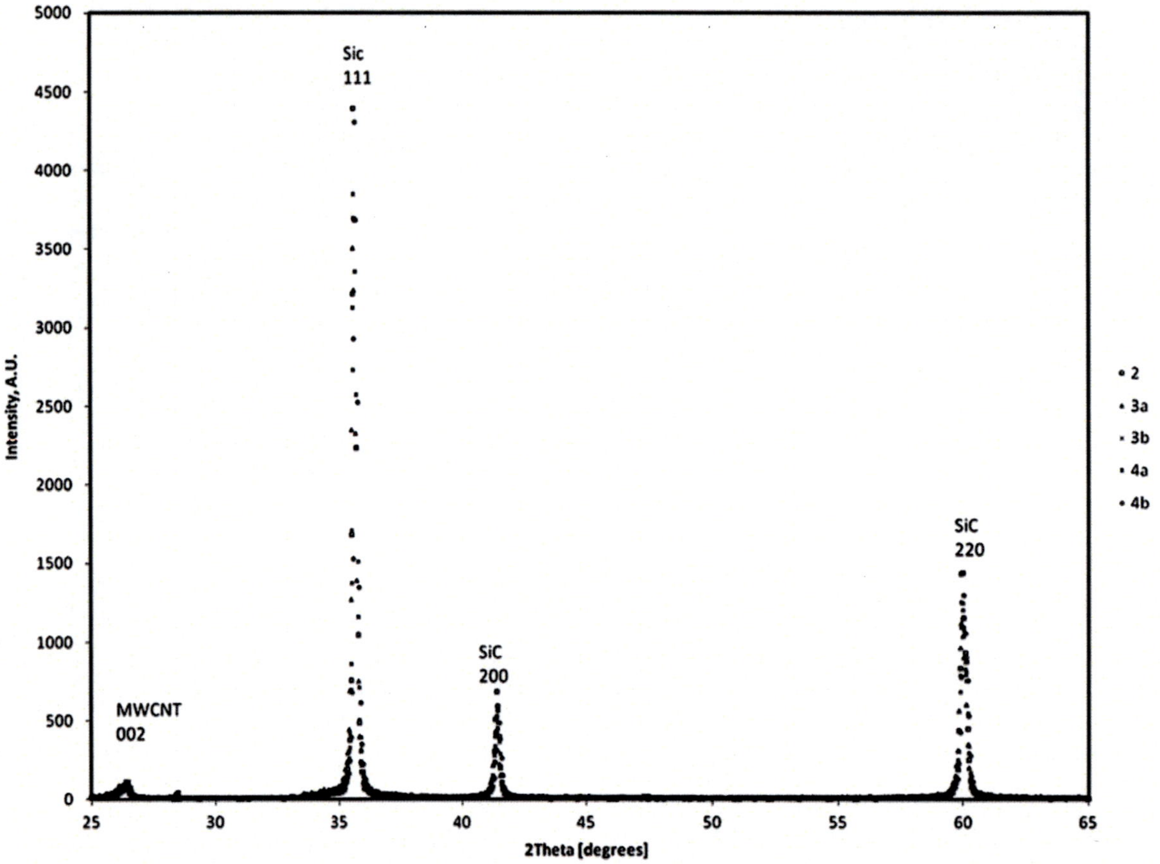

Fig. 17.8 XRD spectra of the spark plasma sintered specimens

17.4.3 Composite Formation

Digital photographs were made of those specimens that could be removed from the die and punch without being damaged. Specimen 2 had partial sticking, however, so the surface was not flat. The specimens measured about 3 mm thick. The hold time for specimen 2 was 5 min. Because of the sticking, the hold time was eliminated for the remaining specimens.

The densities of the specimens given in Table 17.4 were compared to the maximum theoretical density possible for that composition as given in Table 17.1. This comparison is denoted in Table 17.4 as %TD (theoretical density). Specimen 2, pure SiC, had a %TD of 86.8. Specimens 3a and 3b had %TDs of 93.3 and 87.9, respectively. The increase in %TD for 3a compared to specimen 2 is a result of the added MWCNTs, making the TD initially lower. The decrease between 3a and 3b is attributed to the processing in ethanol. Samples 4a and 4b had a lower density than all the others, most likely due to the presence of extra silica as a result of the functionalization. Again, there is a decrease between 4a and 4b, due to the processing in ethanol.

As indicated by the fact that none of the samples reached 100% of TD, the samples were porous. The samples were cut and polished in order to image the grains on the interior and measure the grain sizes. The SEM images of the polished surfaces show the presence of pores of various sizes. The pores appear to be closed off. Grains were not observed in specimens 3a, 3b, 4a, or 4b. This was assumed to be a result of the porosity or the carbon nanotubes within the matrix preventing the acquisition of quality micrographs. The grain size measured by the line-intercept method for specimen 2 yielded a grain size much larger than that obtained by XRD analysis.

17.4.4 Flexural Testing

Bend bars cut from specimen 2 were not long enough to perform an accurate test. The specimens broke at the support instead of at the midpoint of the bar, producing an unreliable test. All other bend tests resulted in a clean break at the center of the

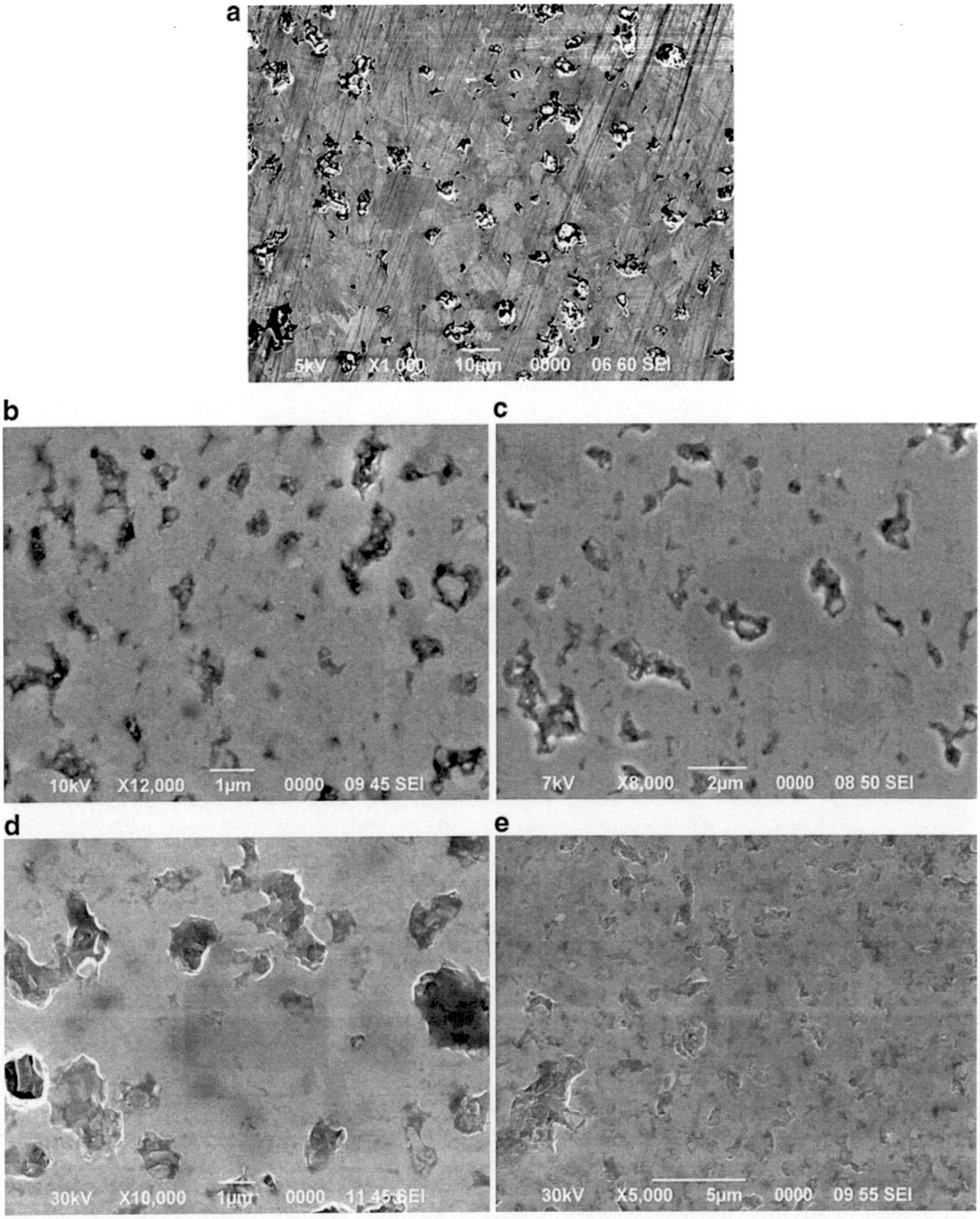

Fig. 17.9 SEM micrographs of the spark plasma sintered specimens. (**a**) Specimen 2. (**b**) Specimen 3a. (**c**) Specimen 3b. (**d**) Specimen 4a. (**e**) Specimen 4b

Table 17.5 Grain sizes of spark plasma sintered composites (units in nm)

Sample/method	2	3a	3b	4a	4b
XRD	477	168	204	207	265
SEM	4,656	–	–	–	–

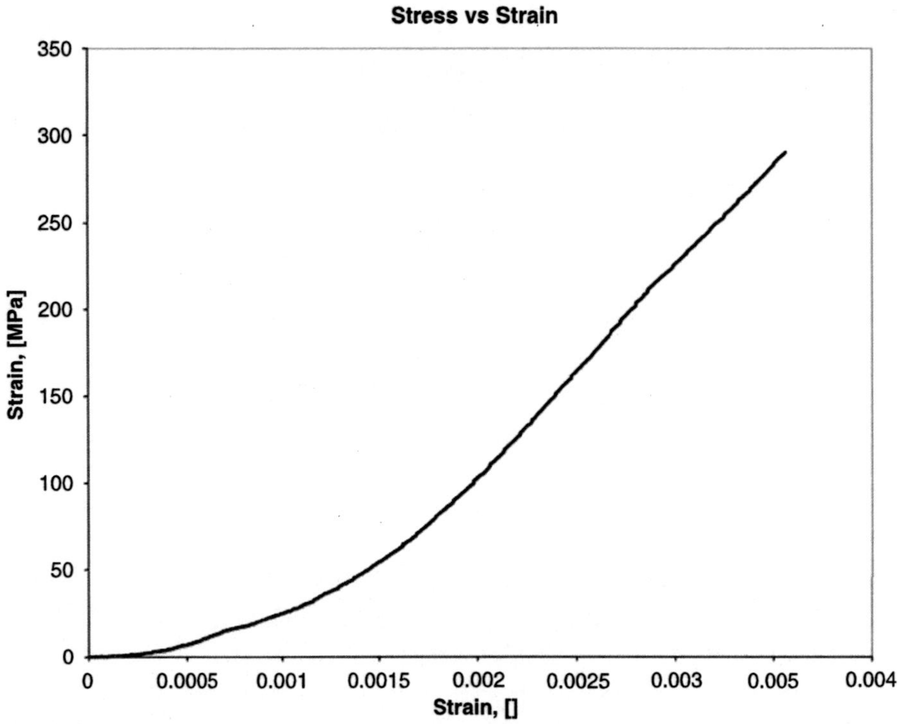

Fig. 17.10 Typical stress versus strain curve for spark plasma sintered bend bars

Table 17.6 Flexural strengths for the spark plasma sintered specimens

Specimen	Flexural strength (MPa)
2	–
3a	271
3b	–
4a	249
4b	317

specimen. Because specimen 3b was not tested, no useful comparison was available for specimen 3a. Specimen 3a demonstrated higher flexural strength than specimen 4a. This result is consistent with the higher density observed for specimen 3a. Specimen 4a demonstrated a lower flexural strength than specimen 4b. This result is not consistent with the higher density of specimen 4a. All specimens recorded lower flexural strengths than that listed in published material property tables [1]. This result is consistent with measured densities being lower than the density for the SiC listed in published material property tables, which was >98%.

17.5 Conclusion

The results of these experiments confirm that the addition of MWCNTs to a SiC matrix alters the microstructure and material properties of the resulting composite. The density and flexural strength have decreased and the porosity has increased. These changes in microstructure and properties are not an improvement as compared with pure SiC. To improve pure SiC, additional processing steps may be needed or current processing steps need to be modified. The use of MWCNTs as a reinforcement material for SiC may yet have potential, but reaching that potential will require additional experimentation. It is recommended that the composition of the MWCNT and SiC mixture be altered to deduce the optimal amount of

MWCNT reinforcement to use. Bonding between the MWCNTs and the SiC matrix needs to be further studied and optimized. The exact location of the MWCNTs within the microstructure should be determined by TEM characterization. The fracture toughness of the composite should be determined by single-edged notch beam tests.

Acknowledgments This project is funded through the Center Directed Research Program from the Engineer Research and Development Center under a project entitled " Nanoscale Studies of Polycrystalline Materials with Emphasis on Ceramics Synthesis". The authors would like to thank the Army Research Laboratory at Aberdeen Proving Ground, MD, for the use of their SPS equipment and their time in making the samples discussed in this paper.

References

1. Munro RG (1997) Material properties of sintered alpha-SiC. J Phys Chem Ref Data 26:1195–1203
2. Padture NP (1994) In Situ-toughened silicon carbide. J Am Ceram Soc 77:519–523
3. Kim S, Kriven WM (1997) Preparation, microstructure, and mechanical properties of silicon carbide dysprosia composites. J Am Ceram Soc 80:2997–3008
4. Chen YL, Liu B, Huang Y, Hwang KC (2011) Fracture toughness of carbon nanotube-reinforced metal- and ceramic-matrix composites. J Nanomaterials p.9 doi:10.1155/2011/746029
5. Inam F, Yan H, Reece MJ, Peijs T (2008) Dimethylformamide: an effective dispersant for making ceramic–carbon nanotube composites. Nanotechnology 19:195710
6. Seraphin S, Zhou D, Jiao J, Withers JC, Loutfy R (1993) Effect of processing conditions on the morphology and yield of carbon nanotubes. Carbon 31:685–689
7. Inam F, Yan H, Reece MJ, Peijs T (2010) Structural and chemical stability of multiwall carbon nanotubes in sintered ceramic nanocomposite. Adv Appl Ceram 109:240–245
8. Li H, Ha C, Kim I (2009) Fabrication of carbon nanotube/SiO2 and carbon nanotube/SiO2/Ag nanoparticles hybrids by using plasma treatment. Nanoscale Res Lett 4:1384

Chapter 18
HP/HT Hot-Wet Resistance of Thermoplastic PEEK and Its Composites

Yusheng Yuan, Jim Goodson, and Rihong Fan

Abstract Advanced thermoplastic polyetheretherketone (PEEK) resin is known to have excellent resistance to elevated-temperatures in a broad range of severe chemical environments. One of the most challenging oilfield downhole environments for most engineering polymers is the high-pressure/high-temperature (HP/HT) hot-wet environment. PEEK resin is proven to be capable of resisting steam and water environments at elevated temperatures above 400°F because of its thermoplastic polyetherketone polymer chemistry and semi-crystalline polymer structure. However, when various reinforcement phases are incorporated in the PEEK resin to form PEEK matrix composites, their HP/HT hot-wet resistant capabilities can be very different, depending on the type of reinforcement, the resin-reinforcement interface bonding and the reinforcement geometry. A clear understanding of the HP/HT hydrolysis resistant behavior of the PEEK resin and its composites and the related mechanisms is critical.

In this study, virgin PEEK resin and its glass-filled composites were investigated, including 30% glass-fiber and 30% glass-bead filled PEEK grades. The hot-wet degradations studied included hygrothermal material structural damage, moisture-induced thermal mechanical degradation and hydrolytic reactions. The experimental methods involve HP/HT hot-wet environmental exposure in NaCl brine up to 400°F and 5,000 psi, mechanical testing and dynamic mechanical thermal analysis before and after the HP/HT hot-wet exposure. Experimental results and data analysis will be presented. Deformation and failure mechanisms are also discussed.

Keywords Thermoplastic PEEK • Glass-filled PEEK composites • Hot-wet resistance • Hydrolysis • Mechanical tests • Dynamic mechanical thermal analysis • Interface debonding

18.1 Introduction

Advanced high-performance thermoplastic PEEK resin and its family were introduced in early 1980s as tough crystalline thermoplastic aromatic polyetherketone polymers for critical high-temperature applications [1]. Further mechanical and chemical resistance characterization and applications indicated that they were capable of resisting a broad range of severe chemical environments to certain elevated temperature levels [2]. With the advantages in high-temperature and chemical resistance capabilities, this thermoplastic polymer family has been used extensively in almost all the critical industrial areas including electronics, aerospace, automotive, chemical and energy, including oilfield downhole applications [3, 4]. Among commonly encountered oilfield downhole environments, a most challenging one to many engineering polymers is the HP/HT hot-wet environment involving hygrothermal material structural damage, moisture-induced thermomechanical degradation and hydrolytic reactions [5–7]. PEEK resin has been proven to be capable of resisting steam, water and water-based fluid environments at elevated temperatures above 400°F because of its thermoplastic aromatic polyetherketone polymer chemistry and semi-crystalline polymer structure [2–5]. However, when various reinforcement phases are incorporated in the PEEK resin to form PEEK matrix composites, their HP/HT hot-wet resistant capabilities may become very different depending on the type of reinforcement, the resin-reinforcement interface bonding and the reinforcement geometry.

Y. Yuan (✉) • J. Goodson • R. Fan
Baker Hughes Incorporated, 14990 Yorktown Plaza, Houston, TX 77040, USA
e-mail: yusheng.yuan@bakerhughes.com

E. Patterson et al. (eds.), *Composite Materials and Joining Technologies for Composites, Volume 7: Proceedings of the 2012 Annual Conference on Experimental and Applied Mechanics*, Conference Proceedings of the Society for Experimental Mechanics Series 44,
DOI 10.1007/978-1-4614-4553-1_18,

The engineering-grade PEEK matrix composites commercially available can be classified as continuous fiber reinforced and particulate or short fiber filled composites. The continuous fiber reinforced PEEK composites include continuous S-glass fiber reinforced and continuous carbon fiber reinforced PEEK composites with a standard ~61% fiber volume fraction. The particulate or short-fiber filled PEEK composites include glass-fiber filled, carbon-fiber filled and glass-bead filled composite grades with reinforcement contents ranged from 15% to 60% by weight for different processing and application requirements. The lengths of the chopped short fibers in the filled composites are about 1/32 in. typically optimized from processability and mechanical performance concerns. The orientation of the short fiber in the molded components depends on the molding method, mold design and flow pattern. Compared with the glass-fiber filled PEEK composites, the glass beads have less efficiency for enhancing mechanical properties of the composites according to composite micromechanics theories [8]. However, when the molding geometry stability of the molded components is critical, the glass-bead filled grade becomes preferable because there is no fiber-orientation induced effects on molding residual stresses and shrinkage.

Reinforcement fibers in composite materials play important roles providing stiffening and strengthening and taking major mechanical loads. The commonly used reinforcement fibers in advanced composite materials include glass fibers, carbon fibers and Kevlar® (aramid) fibers. These advanced reinforcement fibers are commonly small in diameter (5–15 μm) and provide high mechanical strength, enhanced stiffness and excellent temperature resistant capability in dry or ambient conditions. However, when contacted with a high-temperature hot-wet environment, some of these fibers behave very differently [9]. After exposure in a 3% NaCl brine at 204°C, 5,000 psi for 168 h, the glass fibers (E-glass and S2-glass) became very fragile with no mechanical strength because of hydrolytic scission of the inorganic glass polymer chains in the glass fibers in this high-temperature hot-wet environment.

In this study, unfilled virgin PEEK, short glass-fiber filled and glass-bead filled PEEK composites are investigated. The hot-wet environmental degradations considered include hygrothermal material structural damage, moisture-induced thermomechanical degradation and hydrolytic reactions. The experimental methods involve HP/HT hot-wet exposure in NaCl brine up to 400°F and 5,000 psi, mechanical tests at ambient and elevated temperatures and dynamic mechanical thermal analysis before and after the hot-wet environmental exposure. Experimental results and data analysis will be presented. Deformation and failure mechanisms are also discussed.

18.2 Experimental

Materials and Test Specimens: The test materials used in this study include Victrex unfilled PEEK 450 G, 30% glass-fiber filled PEEK composite 450 G L30 and 30% glass-bead filled PEEK composite 450 G B30; all are standard flow, injection molding grades. Tensile test specimens of 450 G, 450 G B30 and 450 G L30, provided by Victrex, used in this study are injection-molded standard Type 1 reduced gage-section tensile specimens with a nominal 50.8 × 12.7 × 3.2 mm gage section according to ASTM D638 tensile test standard,. Four tensile test specimens, at least, were prepared for each test condition. All the material and specimen information is listed in Table 18.1.

HP/HT Hot-wet Environmental Exposure: The HP/HT hot-wet environmental exposure of the test specimens was conducted in a 3% NaCl brine in a HP/HT autoclave at 121°C, 149°C and 204°C under a nitrogen pressure at 34.5 MPa for 240 h, respectively, simulating a class of short-term down hole environments and applications. A slow pressure release rate at ~0.14 MPa (20 psi)/min was used in the depressurization process to avoid any explosive depressurization (ED) damage to the HP/HT exposed specimens. Weight and the thickness of the gauge section of each test specimen were measured to 0.0001 g and ~0.01 mm (0.0005 in) before and after the hot-wet exposure. Measured weight increase of each exposed specimen will be used as its water absorption based on an assumption of minimal leaching of the PEEK resin during the exposure from an earlier study [3, 5] where the leaching of a carbon/PEEK composite was undetectable after exposure in 3% NaCl brine at 204°C, 34.5 MPa for 240 h.

Mechanical Tests: Tensile tests for the unfilled and filled 450 G, 450 G B30 and 450 G L30 specimens were performed at room and elevated temperatures in an MTS servo-hydraulic test frame under displacement control mode and a crosshead

Table 18.1 List of materials and specimens in study

Material designation	Matrix resin	Reinforcement	Reinforcement fraction	Fabrication method	Specimen type
450 G	Victrex PEEK 450 G	Unfilled	0	Injection molding	ASTM D638 Type 1
450 G L30	Victrex PEEK 450 G	Short glass fibers	30% by wt.	Injection molding	ASTM D638 Type 1
450 G B30	Victrex PEEK 450 G	Glass beads	30% by wt.	Injection molding	ASTM D638 Type 1

speed ~5 mm (0.2 in.) per minute per ASTM D638 before and after the HP/HT hot-wet exposure. The tensile strain during the test was measured by using a high-temperature extensometer with a 2-in. gage length and 50% extension capability. This extension capability covered most of the tests at room and lower elevated temperatures, but was not enough to cover the full elongation of the specimens when tested at a high temperature. In these cases, the extensometer was used only in the early part of the tests (< 50% strain) and then it was removed from the test specimen and the test was continued further with displacement reading. Since the maximum stroke of the MTS test frame used for this study at that time is only 6 in., in some elevated-temperature test cases, this stroke was not long enough to break the test specimens. In all the tensile tests, load, displacement and strain were recorded. Modulus, ultimate tensile strength, yield strength and strain at break for each test specimen were calculated and determined from its stress–strain curve or stress-extension curve. When the stress–strain curve showed a maximum yield stress or a ductile fracture or a yielding plateau within a low strain range (<5%), the maximum stress or the stress at the plateau was used to define the yield strength. When this method could not be used for the purpose, the 0.2% offset scheme was used to determine the yield strength [10, 11].

Thermal Mechanical Analysis: Glass transition temperatures (T_g) and the related dynamic viscoelastic behavior of the Victrex 450 G, 450 G B30 and 450 G L30 samples were analyzed and determined by dynamic mechanical analysis (DMA) using an TA AR2000 or an ARES G2 analyzer with torsional oscillation mode per ASTM D4065 before and after the specified hot-wet environmental exposure. In the DMA analysis, a 0.05% torsional strain was applied under 1 Hz frequency and a heating rate of 3–5°C per minute in an inert atmosphere. The T_g values were determined by the peak temperature of the loss modulus (G″). The torsional beam samples were about 51 × 12.7 × 3.2 mm in dimensions. This relatively large sample size is helpful to keep the absorbed moisture inside the exposed test samples during the DMA test.

Microstructure and Interface Examinations: To investigate directly the effect of the HP/HT hot-wet exposure on microstructures of the test materials under the study and support the mechanical testing and analysis results, microstructures of the injection molded 450 G, 450 G B30 and 450 G L30 specimens, including resin and reinforcement phases and their interfaces, were examined under a high-magnification optical microscope, before and after the specified hot-wet exposure. The unexposed and exposed 450 G L30 samples were cut specially in both longitudinal and transverse directions with consideration of the glass-fiber flow orientation. All the cut samples were mounted, polished and ultrasound cleaned before the micro-photographing.

18.3 Results and Discussion

Water Absorption and Geometric Stability: In study of HP/HT hot-wet resistance of the unfilled PEEK and its glass-filled composites, tensile specimens of 450 G, 450 G B30 and 450 G L30 were exposed in 3% NaCl brine at 121°C (E250), 149°C (E300) and 204°C (E400) under 34.5 MPa for 240 h, respectively. The measured averaged water absorption and the thickness changes of the gauge section of the specimens of each material group in each exposure condition are presented in Table 18.2. From the data presented we can see that: (1) Water absorption increases with the exposure temperature but decreases with the glass-filled grades since the glass fibers and beads have less water absorption than the PEEK resin matrix. (2) The water absorption rate of the unfilled PEEK 450 G ranges from 0.87% to 1.39% which is much lower than that of other resins such as high-temperature epoxy resins under the similar exposure conditions (3–4%) [3, 5]. (3) The thickness increases of all the PEEK grades in the study are minimal (0.035–0.375%) even after the E400 exposure compared with other resins and their composites [3, 5], indicating an excellent geometric stability of all the PEEK grades in the HP/HT hot-wet environment. (4) The thickness changes of the glass filled 450 G B30 and 450 G L30 specimens increase with the exposure temperature and are higher than those of the unfilled 450 G specimens when the exposure temperature reaches 149°C or higher. Considering the glass filled 450 G L30 and 450 G B30 specimens have a higher thickness increase but lower water absorption than the unfilled 450 G specimens when the exposure temperature reaches 149°C or higher,

Table 18.2 Measured water absorption and thickness changes of 450 G, 450 G B30 and 450 G L30 specimens after each exposure condition

	Exposed in 3% NaCl brine at 121°C, 34.5 MPa for 240 h (E250)		Exposed in 3% NaCl brine at 149°C, 34.5 MPa for 240 h (E300)		Exposed in 3% NaCl brine at 204°C, 34.5 MPa for 240 h (E400)	
	Water absorption (%)	Thickness change (%)	Water absorption (%)	Thickness change (%)	Water absorption (%)	Thickness change (%)
450 G	0.875	0.185	0.870	0.185	1.395	0.085
450 G B30	0.510	0.035	0.675	0.210	1.310	0.275
450 G L30	0.052	0.090	0.640	0.260	0.855	0.375

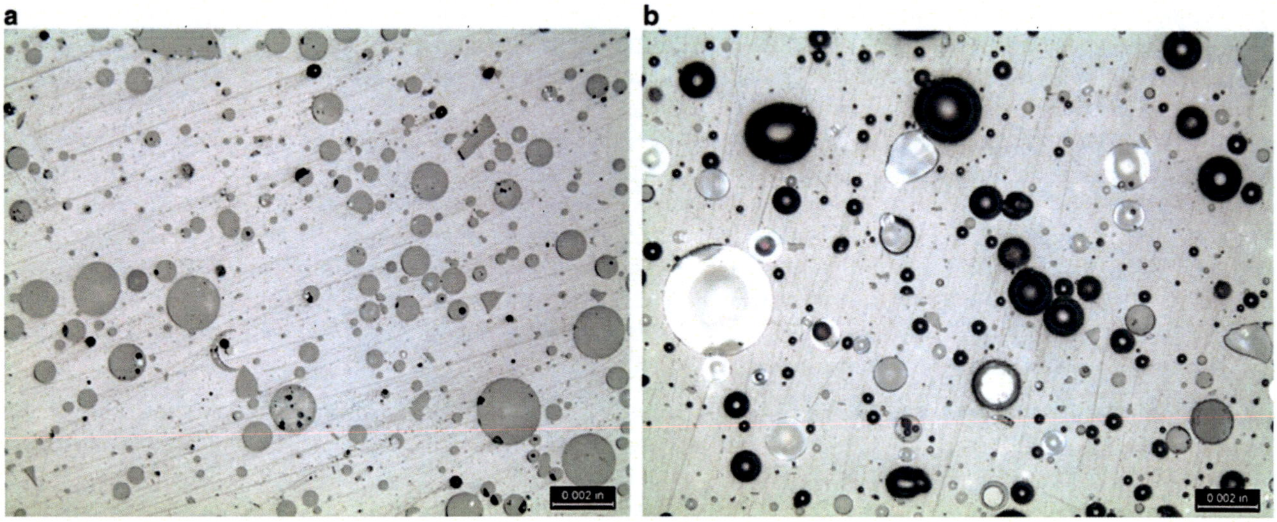

Fig. 18.1 Micrographs of cross-sections of 450 G B30 specimens: (**a**-*left*) Unexposed (UE) sample shows good glass-bead/PEEK resin interface bonding; (**b**-*right*) E400 exposed sample shows glass-bead/PEEK resin interface debonding and separations of the glass beads from the PEEK resin matrix

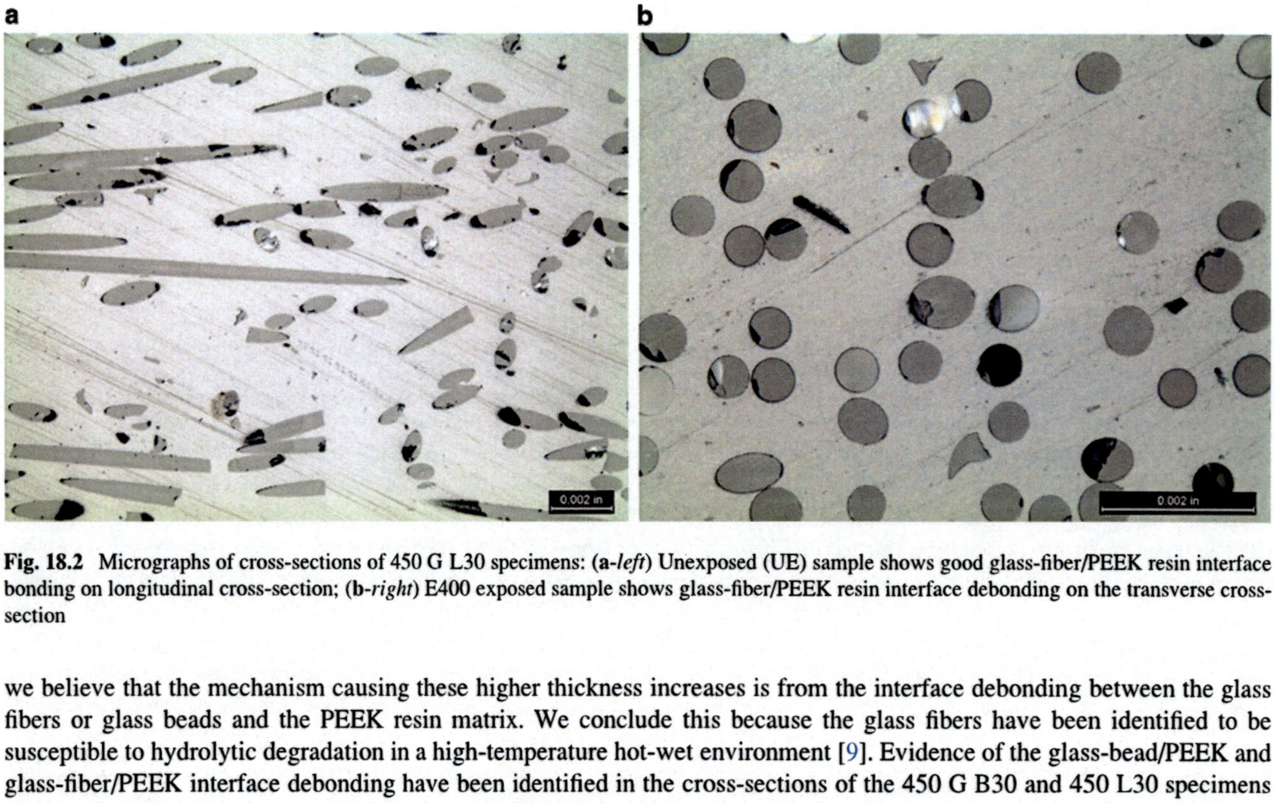

Fig. 18.2 Micrographs of cross-sections of 450 G L30 specimens: (**a**-*left*) Unexposed (UE) sample shows good glass-fiber/PEEK resin interface bonding on longitudinal cross-section; (**b**-*right*) E400 exposed sample shows glass-fiber/PEEK resin interface debonding on the transverse cross-section

we believe that the mechanism causing these higher thickness increases is from the interface debonding between the glass fibers or glass beads and the PEEK resin matrix. We conclude this because the glass fibers have been identified to be susceptible to hydrolytic degradation in a high-temperature hot-wet environment [9]. Evidence of the glass-bead/PEEK and glass-fiber/PEEK interface debonding have been identified in the cross-sections of the 450 G B30 and 450 L30 specimens after the E400 hot-wet exposure and presented in Figs. 18.1 and 18.2, respectively. (5) No structural damage observed in the unfilled, E400 exposed 450 G sample (not shown) and in the PEEK matrix phases in the E400 exposed 450 G B30 and 450 G L30 samples as shown in Figs. 18.1 and 18.2. (6) The thickness increase of the unfilled 450 G specimens after the E400 exposure is less than the ones determined from the specimens after the E250 and E300 exposure at a lower temperature. The reason causing this controversy is not clear. A further investigation on this observation is needed.

Glass-Transition and its Wet Shifting: The measured DMA curves of the unexposed (Dry) 450 G, 450 G B30 and 450 G L30 samples are presented in Fig. 18.3, where the dry T_gs of 450 G, 450 G B30 and 450 G L30 samples were determined to be 149.8°C, 149.8°C and 148.7°C, respectively. These measured T_g values are quite consistent since these

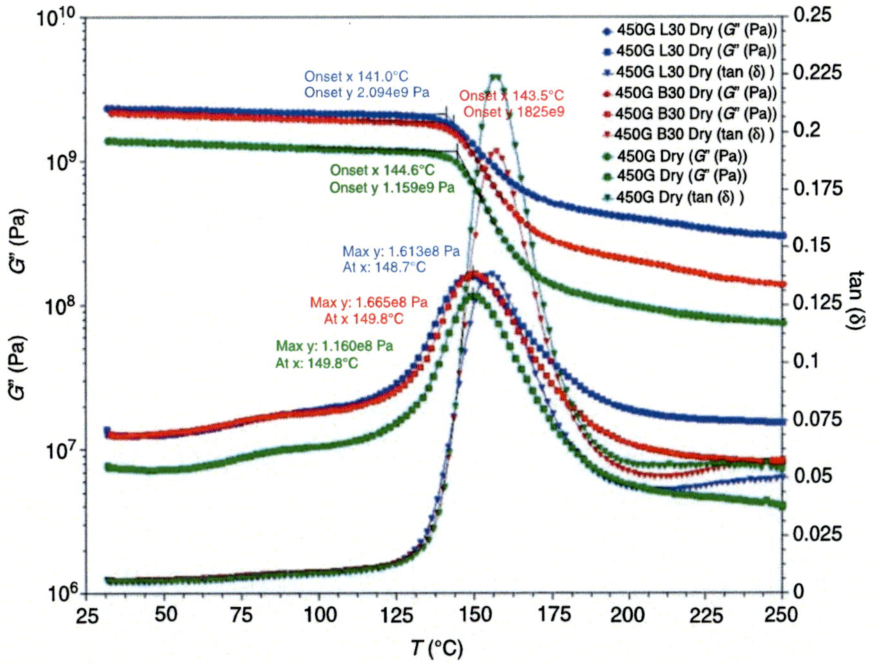

Fig. 18.3 DMA curves of unexposed (dry) 450 G, 450 G B30 and 450 G L30 samples measured in dry condition

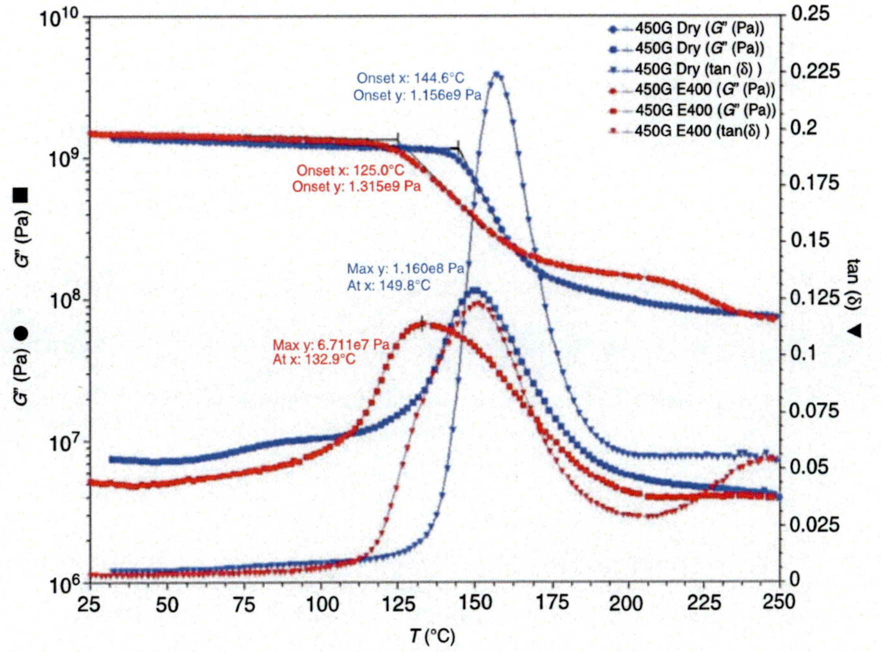

Fig. 18.4 Measured DMA curves of the 450 G samples before (dry) and after the E400 exposure (wet)

three 450 G grades have the same PEEK resin matrix that governs the measured T_g values. We can see also from this chart that (1) the torsional storage modulus curves of 450 G, 450 G B30 and 450 G L30 indicated reasonable differences in their modulus values; (2) after the glass transition, the unfilled PEEK and the PEEK composites still retained certain useful elastic modulus because of its semi-crystalline polymer structure.

The measured DMA curves of the 450 G, 450 G B30 and 450 G L30 samples before (dry) and after exposure in 3% NaCl brine at 204°C, 34.5 MPa for 240 h (E400) are presented in Figs. 18.4, 18.5 and 18.6, respectively. We found from these

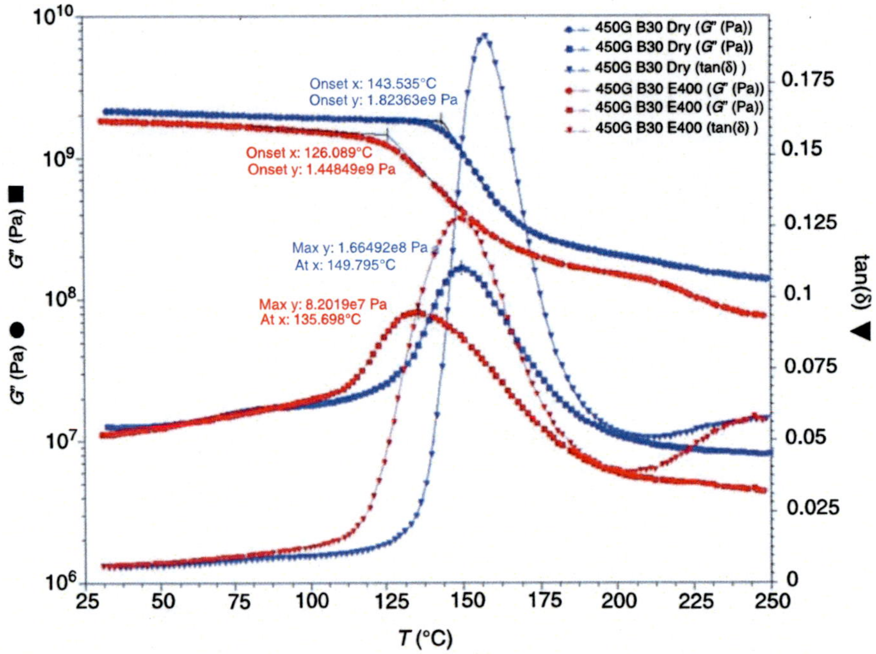

Fig. 18.5 Measured DMA curves of the 450 G B30 samples before (dry) and after the E400 exposure (wet)

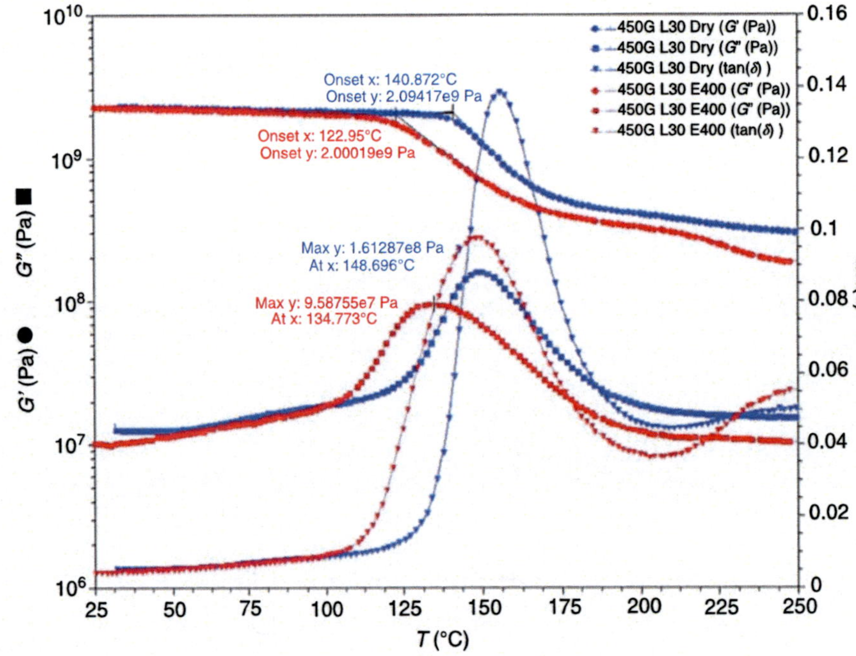

Fig. 18.6 Measured DMA curves of the 450 G L30 samples before (dry) and after the E400 exposure (wet)

curves that the measured wet T_gs of the 450 G, 450 G B30 and 450 G L30 samples after the E400 exposure shifted from their original dry T_g to the lower values by 16.9°C, 14.1°C and 13.9°C, respectively. In general, such as in the case of the unfilled PEEK 450 G, when the temperature is below the wet T_g, the modulus of the exposed wet sample is still close to the one of the unexposed dry sample, but above the wet T_g, the modulus of the exposed wet sample dropped quickly even though the temperature is still below its dry T_g. This phenomenon is known as the moisture-induced thermomechanical degradation that is supported by classical polymer science free volume theories [12] and is considered in general as a reversible process – a drying process to the exposed wet sample can usually recover its original dry T_g [5]. However, the material structural

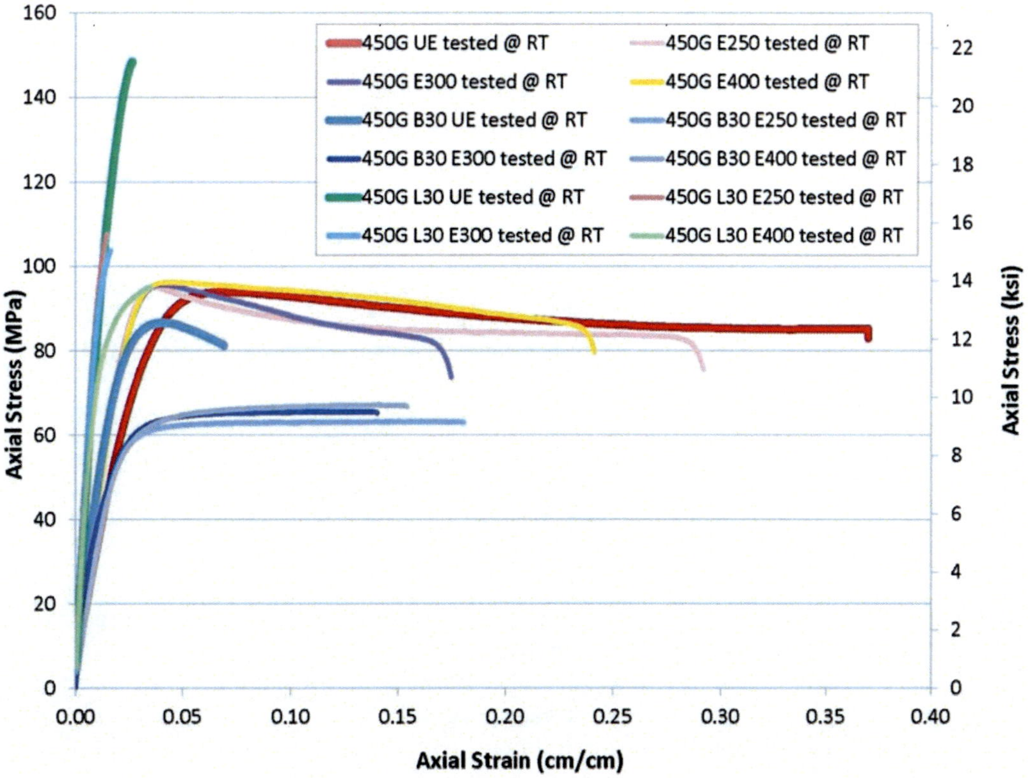

Fig. 18.7 Room-temperature tensile stress–strain curves of 450 G, 450 G B30 and 450 G L30 in unexposed (UE) and exposed (E250. E300 and E400) conditions

damage in the filled 450 G B30 and 450 G L30 samples introduced by the HP/HT hot-wet exposure as shown in the previous section is not reversible. As shown in the Fig. 18.5, a drying process to the exposed 450 G B30 sample may bring its wet T_g value back to its original dry T_g value but can't recover its modulus to its original one.

Although we presented the DMA data for the test grades only after the E400 exposure, we believe that a similar or a little less wet T_g shift will occur with these grades after the E300 and E250 exposure and the wet T_g of each grade after the E250 and E300 exposure should fall in the current ranges since the water absorption values of all the test grades in these exposure conditions has been identified as presented in Table 18.2.

Mechanical Properties: Tensile tests for 450 G, 450 B30 and 450 G L30 specimens were first conducted at room temperature (RT), 121°C (~28°C below their dry T_g), 149°C (near their dry T_g) and 204°C (far above their dry T_g) in unexposed (UE) dry condition. Then, the tensile specimens of each grade were exposed in 3% NaCl brine at 121°C (E250), 149°C (E300) and 204°C (E400), under 34.5 MPa for 240 h, respectively. And then, the exposed tensile specimens of each grade were tested at room temperature and the corresponding exposure temperature, respectively. For convenience in data presentation and analysis, the discussion is grouped first for room-temperature tensile tests and than for elevated-temperature tensile tests.

Room-Temperature Tensile Properties and Failure Mode Discussion: Room-temperature tensile stress–strain curves of the unexposed (UE) and E250, E300 and E400 exposed 450 G, 450 G B30 and 450 G L30 specimens are presented in Fig. 18.7, respectively, where the three stress–strain curves determined from the unexposed dry specimens are plotted with thicker lines as the baselines in this study. From these three curves, we can see that the unfilled PEEK 450 G has decent tensile yield strength around 95 MPa, tensile modulus around 3.5 GPa and good elongation at break around 37% at room temperature in dry condition. The 30% glass fibers in 450 G L30 grade enhanced its mechanical properties significantly. The tensile strength and tensile modulus of 450 G L30 at room temperature in dry condition reaches about 148 MPa and 11.8 GPa, respectively. However, the reinforcement function of the 30% glass beads in 450 G B30 grade is not as good as the glass fibers. Although the tensile modulus of the 450 G B30 increased to 5.6 GPa, its tensile yield strength at room temperature in dry condition is only 86 MPa, about 10 MPa lower than the one of the unfilled 450 G specimens. This reflects the effect of reinforcement geometry on the enhancement efficiency of the mechanical properties in the filled PEEK composite grades.

Fig. 18.8 Failed room-temperature test specimens of 450 G in unexposed (UE) and E250, E300 and E400 exposed conditions (from *top down*), showing necking tensile failure mode

The purpose of testing specimens of each grade at room temperature before and after the hot-wet exposure and plot all the room-temperature test results in one chart is trying to identify effectively the hot-wet-exposure-induced material structural damage and the structural damage-induced mechanical degradation with the exposed test specimens since this comparative scheme excludes any other effects to the mechanical characterization beside the hot-wet-exposure-induced material structural damage, such as glass-transition temperature shifting (room temperature is far below both the dry T_g and the wet T_g of the test material) and preheating damage to the exposed wet specimens in the elevated-temperature testing. Comparing the room-temperature stress–strain curves of the unfilled 450 G specimens after E250, E300 or E400 exposure with the one from the unexposed (UE) 450 G specimen in Fig. 18.7, we can see that the exposed and the unexposed 450 G specimens possess equivalent yield strengths in the range of 94.4–96.4 MPa, and the exposed 450 G specimens have a little higher modulus (~3.95 GPa vs. 3.48 GPa) and lower elongation at break (18 ~ 32% vs. 37%). This indicates that the unfilled 450 G PEEK resin possesses excellent HP/HT hydrolysis resistance with no degradation in mechanical performance at room temperature, and we can conclude further that no material structural degradation is expected in the test specimens after the E250, E300 and E400 HP/HT hot-wet exposure.

For the 450 G B30 and 450 G L30 grades, the situation was quite different. We can see from the stress–strain curves in Fig. 18.7 that the tensile yield strengths of 450 G B30 specimens after E250, E300 or E400 hot-wet exposure dropped from ~87 MPa (UE) to ~55 MPa (−37%) with gradually lower tensile modulus (−8–40%). The tensile failure strengths of the 450 G L30 specimens after E250, E300 and E400 hot-wet exposure dropped from ~148 MPa (UE) to 107–95 MPa (−28–36%) with very slightly decrease in tensile modulus and no ductile deformation until after the E400 exposure. The mechanisms causing these degradations are obvious according to the early discussion on glass-fiber/resin and glass-bead/resin interface debonding. These fiber/resin and bead/resin debonding process with further weakening in fiber mechanical strength during the HP/HT hot-wet environmental exposure resulted in irreversible material structural degradation. The degree of the degradation depends on the exposure temperature and the duration.

The room-temperature failure mode of the unfilled 450 G specimens is typically a yielding with necking no matter their environmental exposure conditions, as shown in Fig. 18.8, characterized by an engineering stress–strain curve with a maximum yield strength followed by stress drop until break, as demonstrated in Fig. 18.7, due to the local necking instability that weakened the test specimens. In this necking process, the necking was not stable and could not propagate because of the limited flow ability of the semi-crystalline polymer structure of PEEK polymer at room temperature.

The room-temperature failure mode of the unexposed (UE) and exposed (E250, E300, E400) glass-bead filled 450 G B30 specimens is believed to be a ductile failure according to their stress–strain curves. However, unlike the unfilled 450 G specimens, they were failed without necking, and the plastic deformation is small (<10%) and uniform through the gauge section of the specimens and could not be seen easily from the failed specimens, as shown in Fig. 18.9. The glass beads distributed in the PEEK resin matrix interfered the necking instability and reduced plastic flow ability of the resin matrix at room temperature and might distribute the plastic deformation in the gauge section of the specimen under the tensile loading.

The room-temperature failure mode of the unexposed (UE) and exposed (E250, E300, E400) glass-fiber filled 450 G L30 specimens is typically brittle fracture, as shown in Fig. 18.10, with strain at break 1.36–3.48% although the E400 exposed specimen showed certain ductility under high tensile stress, where we still could not distinguish the plastic deformation from

Fig. 18.9 Failed room-temperature test specimens of 450 G B30 in unexposed (UE) and E250, E300 and E400 exposed conditions (from *top down*)

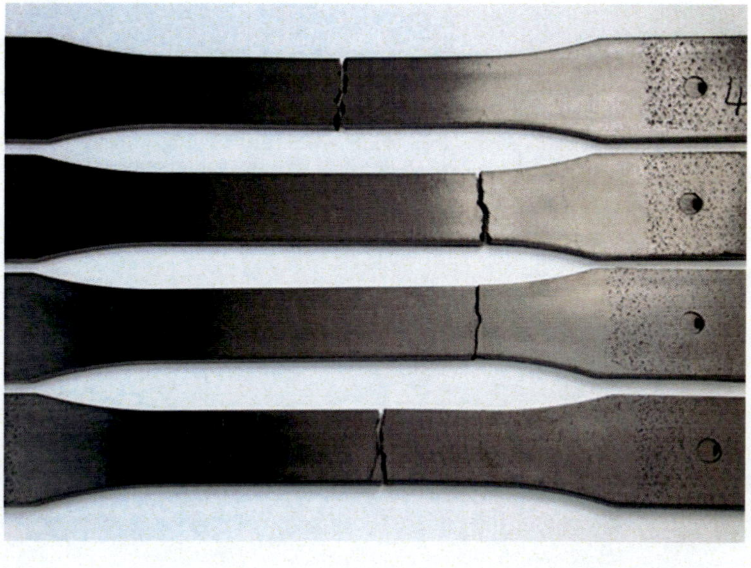

Fig. 18.10 Failed room-temperature test specimens of 450 G L30 in unexposed (UE) and E250, E300 and E400 exposed conditions (from *top down*)

the failed E400 exposed specimen since the deformation was very small (<3.5%). Since the 30% glass fibers are oriented in the longitudinal direction mostly in the gauge section of the specimen, it reduces the ductility of the filled 450 G L30 test specimens greatly.

Elevated-Temperature Tensile Properties and Failure Mode Discussion: Figures 18.11, 18.12, and 18.13 present the typical tensile stress–strain curves of the 450 G, 450 G B30 and 450 G L30 specimens tested in unexposed condition (UE) at room temperature, 121°C, 149°C and 204°C, respectively, and in E250, E300 and E400 exposed conditions at their corresponding exposure temperatures, 121°C, 149°C and 204°C, respectively. To reveal the full range extension of these tests, the tensile stress-extension curves of the unexposed and exposed 450 G and 450 G B30 specimens tested at various temperatures are presented in Figs. 18.14 and 18.15, respectively. From the stress–strain and stress-extension curves presented in these charts we can see all types of tensile deformation and failure behaviors of the three PEEK polymer grades tested at different temperatures, including brittle fracture, ductile failure, necking and cold drawing. These failure modes can be viewed clearly in the photos taken from the failed 450 G, 450 G B30 and 450 G L30 tensile test specimens tested at 121°C, 149°C and 204°C, respectively, in unexposed (UE) and exposed (E250, E300 and E400) conditions, presented in Figs. 18.16, 18.17, and 18.18. For data comparison convenience, measured tensile modulus and tensile strength values of 450 G, 450 G B30

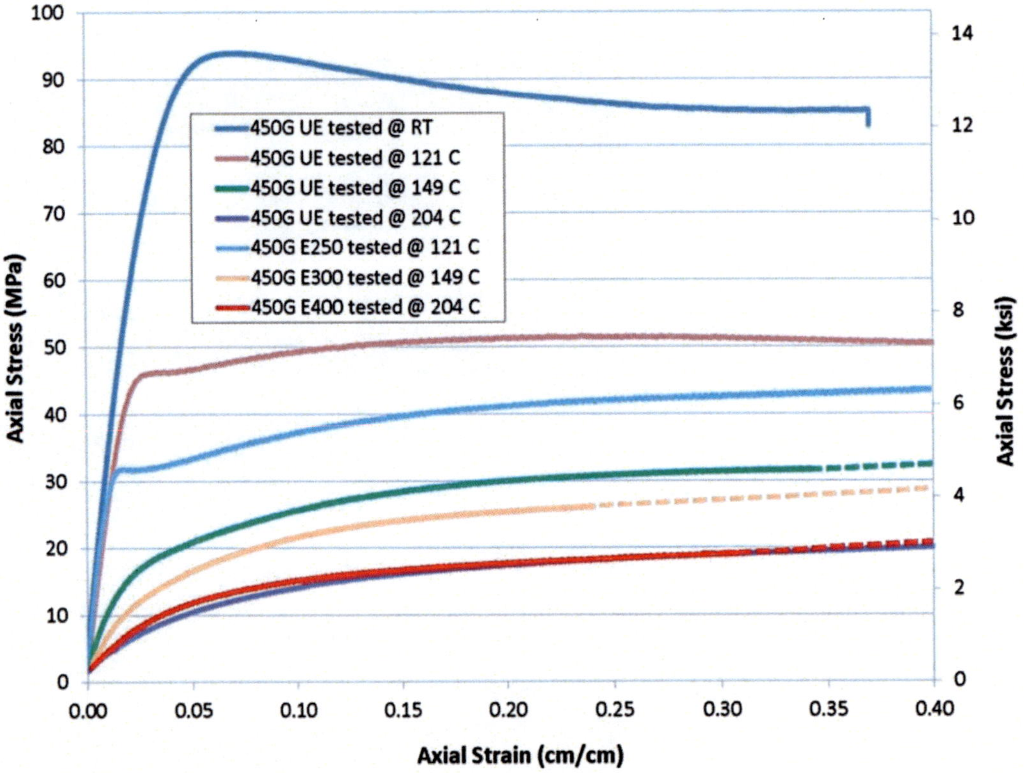

Fig. 18.11 Tensile stress–strain curves of unexposed (UE) 450 G specimens tested at RT, 121°C, 149°C and 204°C, respectively, and E250, E300 and E400 exposed 450 G specimens tested at 121°C, 149°C and 204°C, respectively

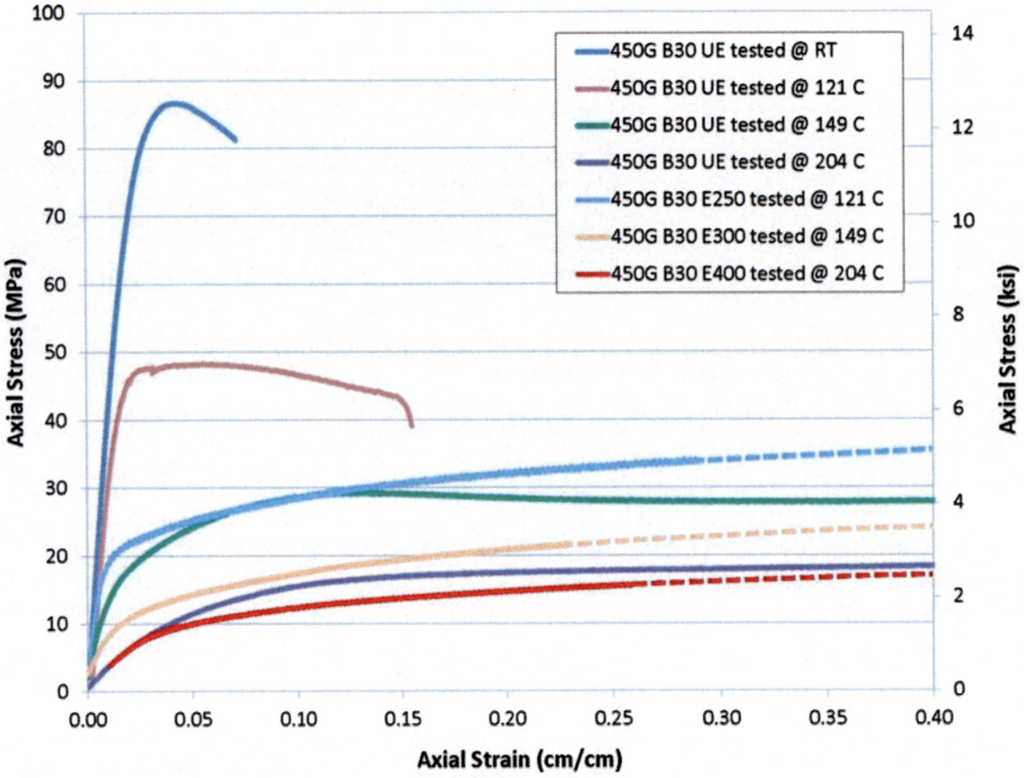

Fig. 18.12 Tensile stress–strain curves of unexposed (UE) 450 G B30 specimens tested at RT, 121°C, 149°C and 204°C, respectively, and E250, E300 and E400 exposed 450 G B30 specimens tested at 121°C, 149°C and 204°C, respectively

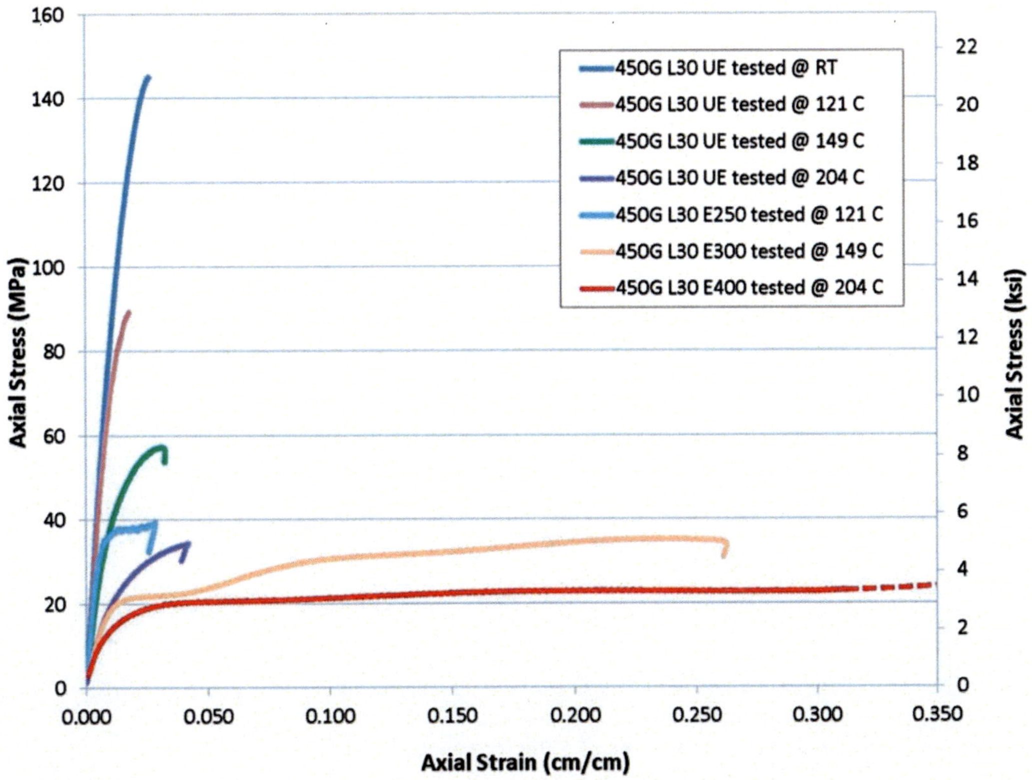

Fig. 18.13 Tensile stress–strain curves of unexposed (UE) 450 G L30 specimens tested at RT, 121°C, 149°C and 204°C, respectively, and E250, E300 and E400 exposed 450 G L30 specimens tested at 121°C, 149°C and 204°C, respectively

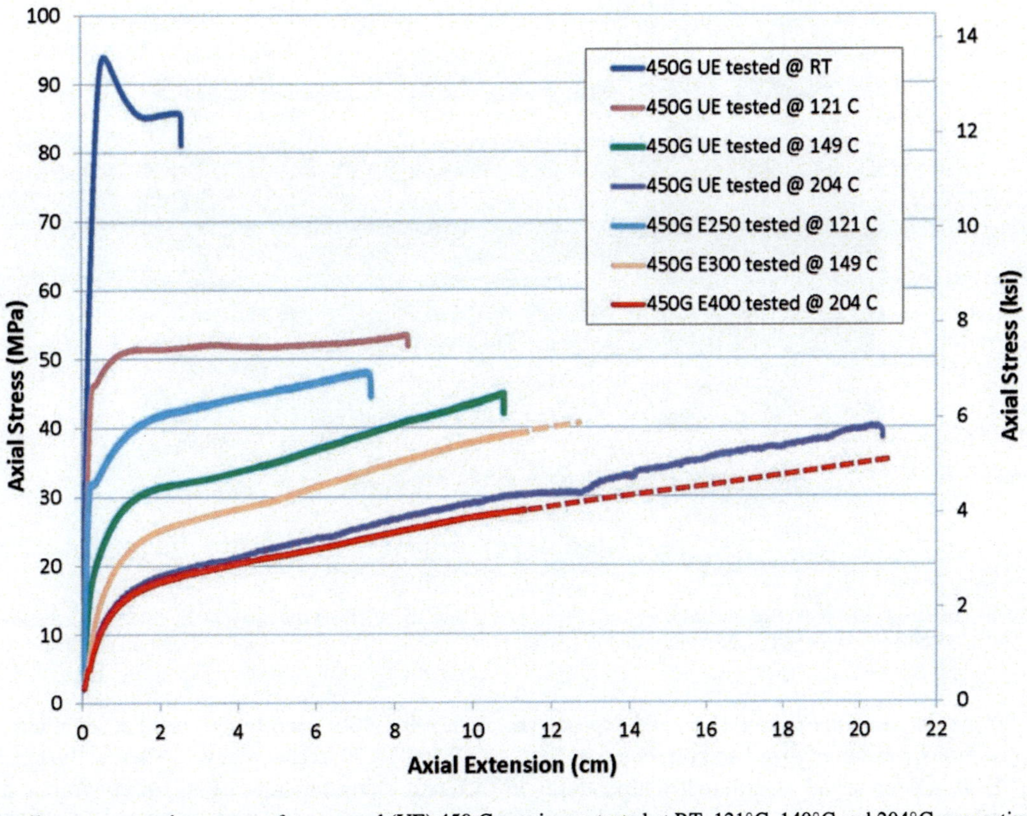

Fig. 18.14 Tensile stress-extension curves of unexposed (UE) 450 G specimens tested at RT, 121°C, 149°C and 204°C, respectively, and E250, E300 and E400 exposed 450 G specimens tested at 121°C, 149°C and 204°C, respectively

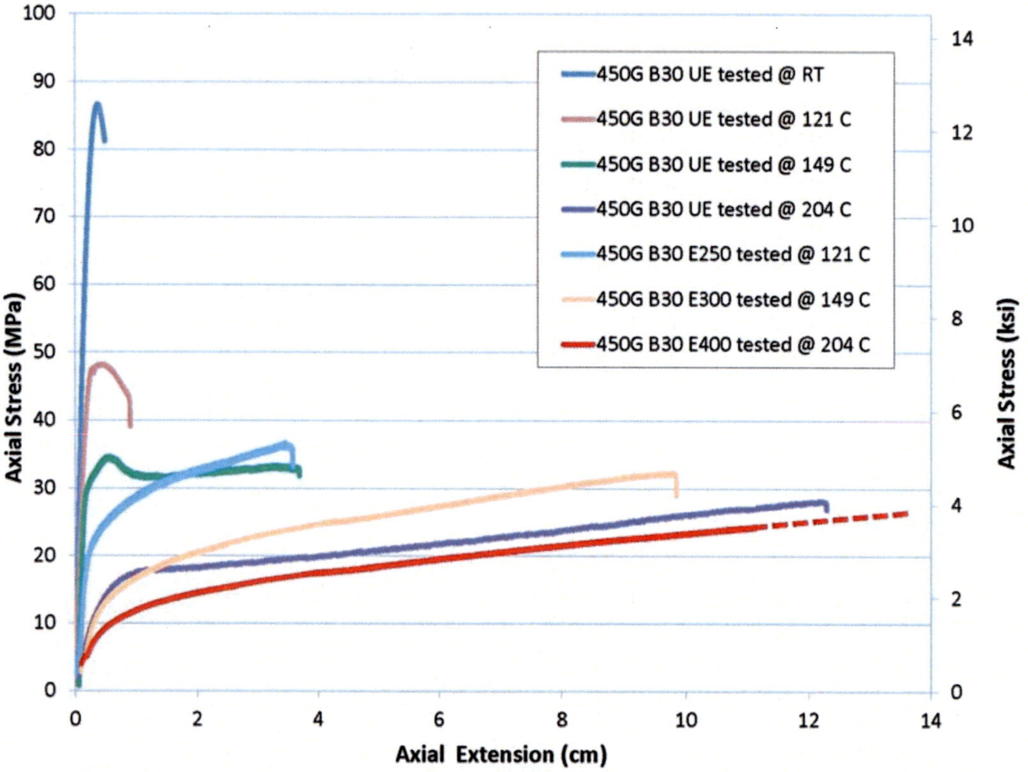

Fig. 18.15 Tensile stress-extension curves of unexposed (UE) 450 G B30 specimens tested at RT, 121°C, 149°C and 204°C, respectively, and E250, E300 and E400 exposed 450 G B30 specimens tested at 121°C, 149°C and 204°C, respectively

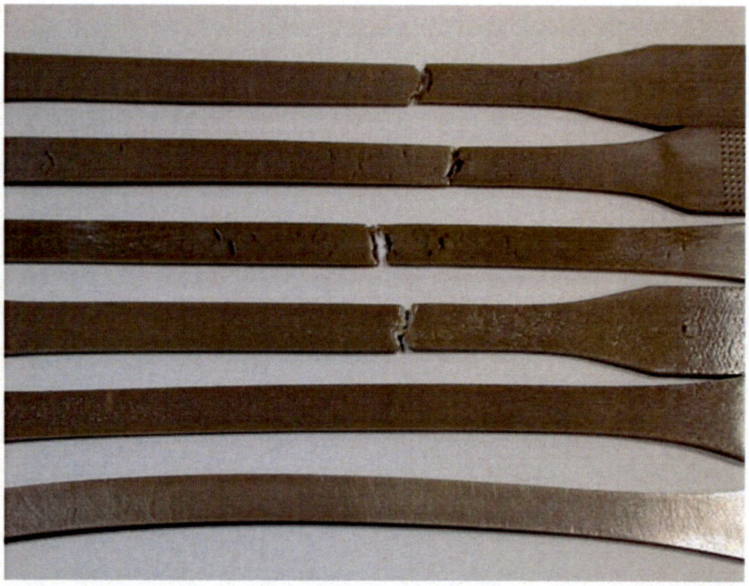

Fig. 18.16 Failed unexposed (UE) 450 G specimens tested at 121°C, 149°C and 204°C (from *top down* in the first three), respectively, and the failed E250, E300 and E400 exposed 450 G specimens tested at 121°C, 149°C and 204°C, (from *top down* in the second three) respectively

and 450 G L30 specimens in unexposed (UE) and exposed (E250, E300, E400) conditions, tested at room temperature and elevated temperatures, are summarized and presented in Figs. 18.19 and 18.20, respectively. To define the tensile strengths of different PEEK grades tested at different temperatures in different environmental exposure conditions properly and consistently, we used the follow guideline in this study: (1) If the test specimen showed substantial plastic deformation and

Fig. 18.17 Failed unexposed (UE) 450 G B30 specimens tested at 121°C, 149°C and 204°C (from *top down* in the first three), respectively, and the failed E250, E300 and E400 exposed 450 G B30 specimens tested at 121°C, 149°C and 204°C, (from *top down* in the second three) respectively

Fig. 18.18 Failed unexposed (UE) 450 G L30 specimens tested at 121°C, 149°C and 204°C (from *top down* in the first three), respectively, and the failed E250, E300 and E400 exposed 450 G L30 specimens tested at 121°C, 149°C and 204°C, (from *top down* in the second three) respectively

its yield strength was determined properly, the yield strength is used to define the tensile strength; (2) if the test specimen showed a brittle fracture failure, the ultimate tensile stress is used to define the tensile strength; (3) if the test specimen showed certain plasticity or nonlinearity and fractured quickly at a low strain (<5%) without any indication of yielding, the ultimate tensile stress is used to define the tensile strength.

Unfilled PEEK 450 G: When the test temperature reached 121°C or higher, the unfilled PEEK 450 G specimen showed even stronger plasticity than that as showed when tested at room temperature. In these elevated-temperature test conditions, the semi-crystalline polymer structures in the test specimens started breakdown and realignment under the tensile loading – the "cold drawing" started [10, 11]. As illustrated in the Fig. 18.16, in the cold-drawing process, the necking in the gauge section is stable in cross-section size and propagated until reaching the widened ends of the test specimens. This was observed from all the 450 G specimens tested at all three elevated temperatures, 121°C, 149°C and 204°C, but with different drawing rates, where a higher test temperature gave a higher drawing rate. Also, we found from the tensile stress-extension curves in Fig. 18.14 that all the drawing extensions are associated with the flow stress increase – the strain hardening. This may be explained by the consideration of molecular orientation and changes in morphology of the semi-crystalline polymer under

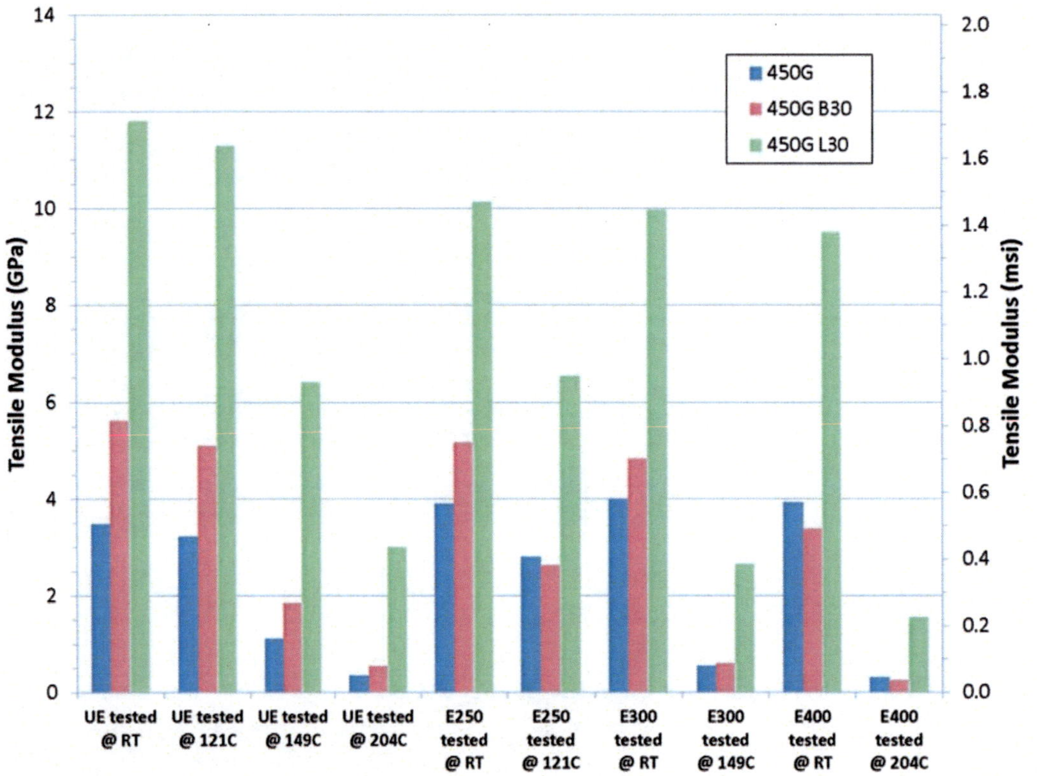

Fig. 18.19 Averaged tensile modulus values of 450 G, 450 G B30 and 450 G L30 specimens tested at various temperatures in unexposed (UE) and E250, E300 and E400 exposed conditions

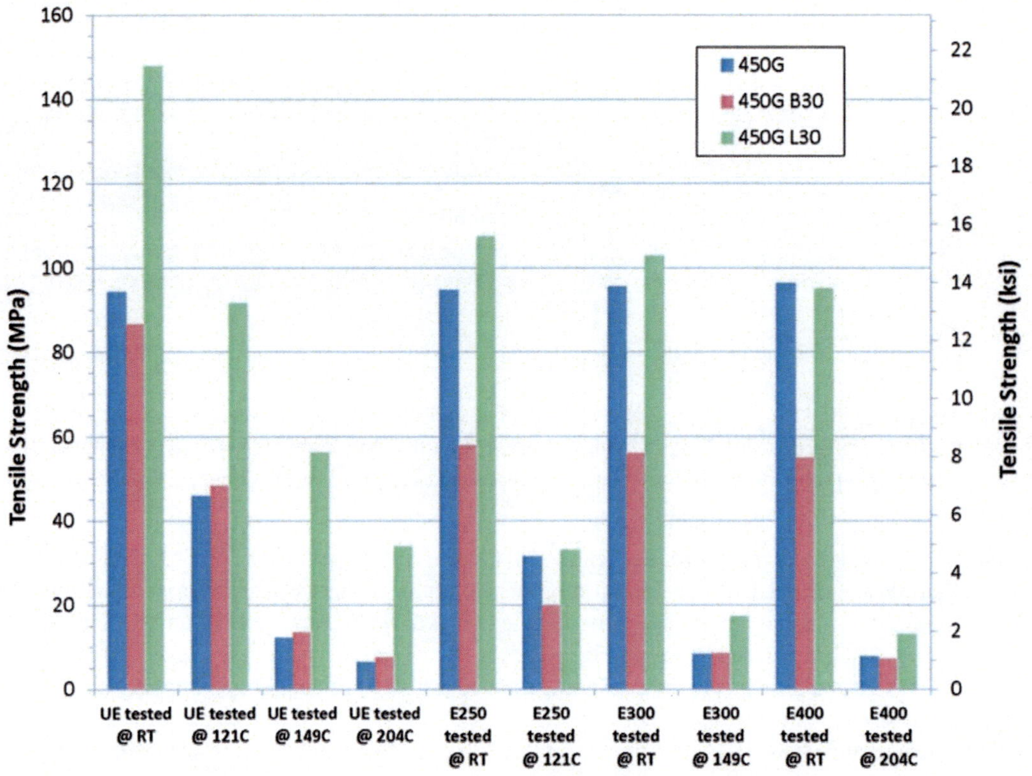

Fig. 18.20 Averaged tensile strength values of 450 G, 450 G B30 and 450 G L30 specimens tested at various temperatures in unexposed (UE) and E250, E300 and E400 exposed conditions

extension when the degree of plastic deformation increases, where the molecular chains are stretched and aligned in the amorphous phases and the crystalline structure changes from a spherulitic to a fibrillar type in the crystalline phases along the drawing direction [10, 11]. This consideration is supported by the observation of the tensile failed 450 G specimen tested at 149°C and 204°C, presented in Fig. 18.16, where the highly drawn specimens with severe transverse rupture are still capable of bearing hundreds pounds of tensile load at the elevated temperatures. This tensile failure mode and failure mechanism can also be applied to the E250, E300 and E400 exposed 450 G specimens tested at 121°C, 149°C or 204°C, respectively, as shown in Fig. 18.16, where the E300 and E400 exposed specimens could not be broken by the test machine at 149°C and 204°C.

From the stress–strain curves and the stress-extension curves presented in Figs. 18.11 and 18.14 we can see that when the tests were conducted at 121°C, a temperature blow the dry T_g of the unexposed (UE) 450 G, the test specimens can retain decent modulus and mechanical strength. When the tests were conducted at 149°C or 204°C, a temperature near or above the dry T_g of the unexposed (UE) 450 G, both the modulus and yield strength of the test specimens dropped quickly and significantly. In comparison of each stress–strain curve determined from each of the E250, E300 and E400 exposed specimens tested at the corresponding exposure temperature (121°C, 149°C or 204°C) with the one determined from the unexposed (UE) specimen tested at the same temperature we can see that (1) when tested at 121°C, the E250 exposed specimens showed a 13% lower elastic modulus (2.8 vs. 3.22 GPa) and about 30% lower yield strength (31.6 vs. 45.9 MPa); (2) when tested at 149°C, the E300 exposed specimens showed about 50% lower elastic modulus and 30% lower yield strength; (3) when tested at 204°C, the E400 exposed specimens and the unexposed (UE) specimens showed very similar and very weak elastic moduli (0.31–0.35 GPa) and yield strengths (6.4–7.8 MPa), as presented in the charts in Figs. 18.19 and 18.20.

The above observation along with the room-temperature test results discussed earlier can be explained further with the viewpoint of the moisture-induced thermomechanical degradation and the consideration of no material structural damage introduced to the 450 G specimens by the hot-wet exposures. Since room temperature is far below both the wet T_g and the dry T_g of the 450 G samples, the room-temperature mechanical properties of the exposed wet samples are retained to equivalent to the room-temperature properties of the unexposed dry samples. Since 121°C is a little lower than the wet T_g but much lower than the dry T_g of the 450 G samples, the E250 exposed samples have a moderately lower modulus and yield strength than the unexposed samples tested at 121°C. Since 149°C is above the wet T_g but near the dry T_g of the 450 G samples, the modulus and yield strength of the exposed wet sample dropped more from those of the unexposed (UE) specimens tested at 149°C. Finally, since the 204°C is far above both the wet T_g and the dry T_g of the 450 G samples, no matter the test specimen is in exposed (E400) or unexposed (UE) condition, their mechanical properties at 204°C are all very week and similar as summarized in the charts presented in Figs.18.19 and 18.20.

Glass-Bead Filled 450 G B30: The filled 450 G B30 specimens showed substantial plasticity to failure when the test temperature reached 121°C or higher in both unexposed (UE) and exposed (E250, E300 and E400) conditions. The unexposed (UE) specimen tested at 121°C, as shown in Figs. 18.12 and 18.17, is considered to have a local yielding with a maximum yield stress (48.4 MPa) followed with stress decrease and fractured at 13.7% strain without local necking observed. With this exception, the deformation modes of all other test cases are basically cold-drawing, as demonstrated in Figs. 18.15 and 18.17. It is also observed that a higher test temperature could facilitate the cold-drawing and gave a higher drawing rate; also the debonding of the glass-bead/PEEK interfaces in the hot-wet exposed specimens could facilitate the drawing process with higher drawing rates.

Measured modulus and tensile strength values of the unexposed (UE) 450 G B30 specimens at 121°C, 149°C and 204°C follow the trends defined by "below", "near" or "far above" the dry T_g of the 450 G B30 samples, as discussed for the unfilled 450 G specimens in the previous section. In evaluation of the mechanical properties of the E250, E300 and E400 exposed 450 G B30 specimens at different temperatures, however, the situation is more complicated because of the additional effect from the hot-wet-exposure induced material structural degradation. Comparing the test results of the E250, E300 and E400 exposed 450 G B30 specimens with the test results of the unexposed (UE) 450 G B30 specimens tested at 121°C, 149°C and 204°C, respectively, quite large differentiations are observed in both modulus and yield strength values at 121°C and 149°C even 121°C is below the wet T_g of the exposed test samples. When tested at 204°C, the mechanical properties of the exposed and unexposed (UE) specimens become quite similar since the test temperature is far above both the dry T_g and wet T_g, where the PEEK resin matrix in the 450 G B30 specimens becomes soft and the glass-bead/resin interface debonding could not give much effect to the mechanical properties of the material. The comparative data of the E250, E300 and E400 exposed and unexposed (UE) 450 G B30 specimens tested at 121°C, 149°C and 204°C are presented in Figs. 18.19 and 18.20, respectively.

Glass-Fiber Filled 450 G L30: Because of the effective reinforcement of the glass fibers in the 450 G L30 specimens, the stiffness, strength and rigidity of the unexposed (UE) 450 G L30 specimens are enhanced significantly, especially when tested at elevated temperatures. As shown in the charts in Figs. 18.19 and 18.20, the measured tensile modulus values of the unexposed (UE) 450 G L30 specimens at 121°C, 149°C and 204°C increased from 3.22, 1.11 and 0.35 GPa (measured from the unfilled 450 G specimens) to 11.3, 6.41 and 3.01 GPa, respectively, and the measured tensile strengths of the unexposed 450 G L30 specimens at 121°C, 149°C and 204°C increased from 45.9, 12.2 and 6.4 MPa (measured from the unfilled 450 G specimens) to 91.6, 56.2 and 34.0 MPa, respectively. These unexposed 450 G L30 specimens showed a brittle fracture failure at 121°C since the test temperature is below the dry T_g of the PEEK matrix resin and the fiber reinforcement resisted its plastic flow and deformation. When tested at 149°C and 204°C, the test specimens showed a slight ductile deformation but fractured at a very small strain (<4.5%) as shown in Fig. 18.13, and we can see only very localized plastic deformation at the fracture surfaces as shown in Fig. 18.18. In these tests, although the test temperature is near or much higher than the dry T_g of the PEEK matrix resin, the effective reinforcement of the glass fibers in the 450 G L30 composite specimens in unexposed condition enhanced significantly their rigidity.

Because the HP/HT hot-wet exposure introduced the glass-fiber/resin interface debonding and the glass-fiber mechanical degradation in the hot-wet exposed 450 G L30 specimens, as identified in the early section and in the reference study [9], tensile properties of the E250, E300 and E400 exposed 450 G L30 specimens tested at various elevated temperatures degraded substantially. As presented in Figs. 18.19 and 18.20, the measured tensile modulus values of the E250, E300 and E400 exposed 450 G L30 specimens at 121°C, 149°C and 204°C decreased from their original unexposed values 11.3, 6.41 and 3.01 GPa to 6.53, 2.66 and 1.56 GPa, respectively, and the measured tensile strengths of the E250, E300 and E400 exposed 450 G L30 specimens at 121°C, 149°C and 204°C decreased from their original unexposed values 91.6, 56.2 and 34.0 MPa to 33.2, 17.3 and 13.2 MPa, respectively. Comparing these elevated-temperature test data of the exposed glass-fiber filled 450 G L30 specimens to the test data measured from the exposed unfilled 450 G specimens at the same temperatures and with the same exposure conditions, respectively, we found that (1) tensile modulus values of the E250, E300 and E400 exposed 450 G L30 specimens at 121°C, 149°C and 204°C are about 2.3–5 times of those measured from the exposed 450 G specimens, but the measured values are only around 1.5–2.7 GPa at 204°C–149°C that is quite low; (2) tensile strength of the E250 exposed 450 L30 specimens at 121°C is about the same as the one measured from the E250 exposed 450 G specimens (~32 MPa); tensile strengths of the E300 and E400 exposed 450 G L30 specimens at 149°C and 204°C are only about twice or less than twice of the ones measured from the exposed 450 G specimens, and the measured values are only around 13–17 ksi at 204°C and 149°C, respectively, that is very weak. These test results from the HP/HT hot-wet exposed 450 G L30 specimens can be further explained by their tensile failure modes. For the E250 exposed specimen tested at 121°C, it showed certain plasticity but failed at a very small strain about 3% as shown in Fig. 18.13, and the failed test specimen showed no major plastic deformation but some signs of local transverse rupture indicating a potential initiation of cold-drawing deformation as shown in Fig. 18.18, where although the test temperature was below the wet T_g of the E250 exposed specimens, the glass-fiber/resin interface debonding released the restriction of the glass fibers to the plastic flow of the PEEK matrix resin to some extent. When tested at 149°C and 204°C, as shown in Fig. 18.18 that the failure modes of the E300 and E400 exposed specimens became cold-drawing with major plastic deformation and transverse ruptures; in these cases, the glass-fiber/resin interface debonding in the exposed test specimens released the restriction of the glass fibers to the plastic flow of the PEEK matrix resin in the 450 G L30 specimens and the test temperatures were above or far above the wet T_g of the E300 and E400 exposed specimens that facilitated the cold-drawing deformation of the PEEK matrix resin in the HP/HT hot-wet exposed 450 G L30 test specimens.

Based on above comparative analysis for the glass-fiber filled 450 G L30 PEEK composite, we can see that although the glass-fiber reinforcement enhanced the mechanical properties of the PEEK composite at room temperature and elevated temperatures very efficiently and significantly to a high level in unexposed (UE) condition (with basically brittle fracture failure mode) compared with the mechanical properties of the unfilled PEEK 450 G tested at the same temperatures in unexposed (UE) condition (with basically plastic failure modes). However, after exposure in 3% NaCl brine at 121°C, 149°C or 204°C under 5,000 psi for 240 h, respectively, mechanical properties of the filled 450 G L30 composite specimens degraded severely to a much lower level and became compatible to the mechanical properties of the unfilled virgin PEEK 450 G tested at the same temperatures and with the same exposure conditions, especially in mechanical strength (returned to plastic failure modes). This indicates that after the HP/HT hot-wet exposure the glass-fibers in the 450 G L30 composite specimens lost their reinforcement function partially or completely because of the fiber/resin interface debonding. Comparing also the mechanical test results of the E250, E300 and E400 exposed 450 G L30 composite specimens tested at 121°C, 149°C or 204°C with the room-temperature test results and with the test results from the unexposed 450 G L30 specimens tested at 121°C, 149°C or 204°C, respectively, we can see that hot-wet exposure or wet condition only (tested at room temperature) or temperature only (tested in unexposed dry condition) can only partially degrade their mechanical properties, as shown in Figs. 18.19 and 18.20; a combined temperature (tested at an elevated temperature) and hot-wet exposure condition will give the exposed composite specimens a full range of degradations involving temperature degradation, moisture-induced thermomechanical degradation and the HP/HT hot-wet-exposure induced material structural degradation. In these cases, a HP/HT in-situ mechanical test method is desirable [13, 14].

18.4 Conclusions

Based on the investigation and the experimental and analysis results in HP/HT hot-wet exposure, dynamic mechanical thermal analysis and mechanical testing of the Victrex unfilled PEEK 450 G and the 30% glass-bead and the 30% glass-fiber filled PEEK composites 450 G B30 and 450 G L30, the following conclusions may be reached:

1. Unfilled PEEK resin 450 G possesses good mechanical retention at 121°C and relatively large mechanical degradation at 149°C and 204°C, a temperature near or above its T_g, in unexposed conditions. It possesses excellent HP/HT hot-wet resistance with a small water absorption rate, small wet T_g shift, excellent geometry stability, stable material structural integrity and good mechanical retention up to 204°C and 5,000 psi.
2. 30% glass-fibers in the filled PEEK 450 G L30 composite enhanced its mechanical properties at room temperature and elevated temperatures efficiently and significantly in unexposed dry conditions. However, after exposure in 3% NaCl brine at 121°C, 149°C or 204°C under 5,000 psi for 240 h, mechanical properties of the 450 G L30 composite degraded severely to a much lower level and became compatible to the mechanical performance of the unfilled PEEK resin 450 G tested at the same temperatures with the same exposure conditions. This indicated the glass-fibers in the 450 G L30 composite lost their reinforcement function because of the fiber/resin interface debonding. Temperature only or wet condition only can only partially degrade the composite mechanical properties; a combined temperature and hot-wet exposure condition will give the composites a full range of degradations involving temperature degradation, moisture-induced thermomechanical degradation and the HP/HT hot-wet-exposure induced material structural degradation.
3. Although the glass beads in filled PEEK 450 G B30 composite can help to reduce component geometry distortion in the molding process, it is not an efficient reinforcement technique for mechanical strength of 450 G B30 composite at room temperature and elevated temperatures in both unexposed dry and exposed wet conditions.

Acknowledgement The authors would like to thank Baker Hughes Completions and Production Technology and Research for their permission to publish this paper. The authors would also like to express sincere appreciation to Bennett Richard, director, research and technology, for his constant support for the research projects. Special thanks are due to John Grasmeder, technical director, Victrex, for his kind support and cooperation to this research project by providing all the Victrex PEEK tensile test coupons.

References

1. Rose JB, Staniland PA (1982) Thermoplastic aromatic polyetherketone. US Patent 4,320,224, 16 Mar 1982
2. Victrex PEEK Polymer Chemical Resistance Guide (2012) Victrex Polymer Solutions
3. Yuan Y, Goodson J (2007) Hot-wet downhole conditions affect composite selection. Oil & Gas J 105(34):52–63
4. Ren J, Gerrard D, Goodson J (2010) Properties of PAEK materials and their potential applications in HP/HT sealing environment. In: Proceedings of the 7th MERL international conference – oilfield engineering with polymers 2010, London, 23–25th Oct 2010
5. Yuan Y, Goodson J (2005) HT/HP hot-wet thermomechanical behavior of fiber-reinforced high-temperature polymer composites. In: Proceedings of the fourth international conference on composite materials for offshore operations, Houston, 4–6 Oct 2005
6. Browning CE (1976) The mechanisms of elevated temperature property losses in high performance structural epoxy resin matrix materials after exposure to high humidity environments. Ph.D. dissertation, University of Dayton
7. Morgan RJ, Shin EE, Lincoln JE, Zhou J, Drzal LT, Wilenski MS, Lee A, Curliss D (1998) Durability characterization of bismaleimide and polyimide-carbon fiber composites. In: Proceedings of the 43 rd international SAMPE symposium, Anaheim, May 31–June 4 1998, pp. 9 106–119
8. Agarwal BD, Broutman LJ (1990) Analysis and performance of fiber composites, 2nd edn. Wiley, New York
9. Yuan Y (2003) BHI internal report on hot-wet resistance of reinforcement fibers
10. Ward IM, Sweeney J (2004) The mechanical properties of solid polymers, 2nd edn. Wiley, New York
11. Hertzberg RW (1996) Deformation and fracture mechanics of engineering materials, 4th edn. Wiley, New York
12. Bueche F (1962) Physical properties of polymers. Interscience, New York
13. Yuan Y, Goodson J (2007) HT/HP hot-wet thermomechanical properties and HT/HP in-situ mechanical test method of high-temperature polymer composites. In: Proceedings of the 52nd international SAMPE symposium and exhibition, Baltimore, 3–7 June 2007
14. Yuan Y, Goodson J (2008) HTHP in-situ mechanical test rig and test method for high-temperature polymers and composites, SPE 113516. In: Proceedings of the 2008 SPE Europec/EAGE annual conference and exhibition, Rome, 9–12 June 2008

Chapter 19
Reinforcement of Epoxy Resins with POSS for Enhancing Fracture Toughness at Cryogenic Temperature

Kunal Mishra and Raman P. Singh

Abstract Although epoxy resins have been widely used for engineering applications, because of their good mechanical properties, they are usually brittle and vulnerable to cracking.

This problem becomes more severe when thermosetting epoxy resins are employed in extremely low temperature environments. Hence improvement in fracture toughness is desired. Therefore, fillers such as clay, carbon nanotube, etc. have been used to enhance mechanical properties. Though these reinforcements showed promising results at room temperature, at cryogenic temperature they deteriorate the mechanical properties of epoxy resins. This is due to free space, free volume and thermal contraction effect. This study reports on the mechanical characterization of POSS–epoxy nanocomposites at cryogenic temperature (77 K). POSS (Polyhedral oligomeric silsesquioxne) is a hybrid organic–inorganic nanoadditive that directly interact with epoxy resin. These nanocomposites were prepared using DGEBF (diglycidyl ether of bisphenol F) based epoxy resin and three different functionalities of POSS, namely, glycidyl, trisilanol phenyl, and methacryl. POSS was added at various weight fractions of 0.5%, 1%, 3%, 5% and 8% and an amine-based hardener was used to cure epoxy resin. The test results show that the addition of POSS leads to significant improvement in fracture toughness at 77 K. There is no change in flexural modulus observed. Differential scanning calorimetry is used to measure the glass transition temperature.

Keywords Polyhedral oligomeric silsesquioxane (POSS) • POSS–epoxy nanocomposites • Cryogenic temperature • Fracture toughness • Flexural modulus and strength

19.1 Introduction

Epoxy resins are used in various engineering applications due to their superior mechanical properties that include high glass transition temperature, high modulus, high creep resistance, and good resistance against chemicals. With the rapid developments in spacecraft and superconducting cable technologies, etc., epoxy resins have been increasingly employed in cryogenic engineering application as an adhesive, in coating or as matrix in fiber–matrix composited. The properties of epoxy resins are a result of the high degree of chemical cross-linking introduced by diglycidyl ether of bisphenols. Nonetheless this high cross-linked density also tends to make these resins brittle and vulnerable to fracture [1, 2].

At cryogenic temperature epoxy resins are more brittle, due to the thermal contraction, which make them unsuitable for cryogenic applications [3, 4]. This demands the improved mechanical properties of epoxy resins at cryogenic temperature so that they can be readily used in cryogenic applications. In order to develop high performance epoxy resins for cryogenic applications researchers have worked on strengthening and toughening of epoxy resin using hyperbranched polymer [5], flexible diamines [6], exfoliated MMT [7], silica nanoparticles [8], organoclay [9], polyurethane etc. Some of these fillers showed promising results in tensile properties and impact properties but the limited research have been done on measuring the fracture toughness of filler–epoxy composites, some research showed decreasing of the fracture toughness with the inclusion of filler.

K. Mishra (✉) • R.P. Singh
Mechanics of Advanced Material Laboratory, Department of Mechanical and Aerospace Engineering, Helmerich Research Center, Oklahoma State University, 526 N. Elgin Ave, Tulsa, OK 74106, USA
e-mail: kunalm@okstate.edu

E. Patterson et al. (eds.), *Composite Materials and Joining Technologies for Composites, Volume 7: Proceedings of the 2012 Annual Conference on Experimental and Applied Mechanics*, Conference Proceedings of the Society for Experimental Mechanics Series 44, DOI 10.1007/978-1-4614-4553-1_19, © The Society for Experimental Mechanics, Inc. 2013

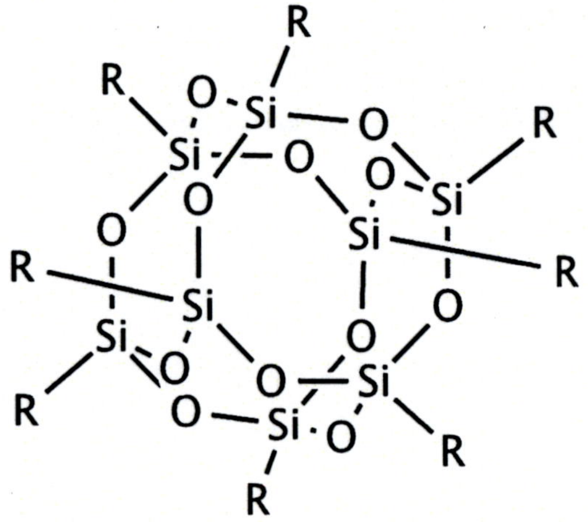

Fig. 19.1 General structure of POSS

Researcher reported that this decrease in the value of the fracture toughness of filler incorporated resin is due to the interference in shrinkage of polymer chain by this filler. This leads to the low intermolecular force between the polymer chains.

In the past decade, due to the development of organic–inorganic hybrid nanoparticles, researchers have showed great interest in the development of hybrid nanocomposites. Such hybrid nanocomposites combine the advantages of inorganic materials (rigidity, high stability) and organic polymers (flexibility, reactivity and processability). Polyhedral oligomeric silsesquioxane (POSS) is one such kind of hybrid material that possess both organic and inorganic properties and has a size between 1 and 3 nm and it is made up of organic–inorganic monomer silsesquioxane ($RSiO_{1.5}$), as shown in Fig. 19.1. It possesses an inorganic rigid cage type structure containing silicon and oxygen, and selectable organic groups (R) attached to the silicon atoms. Based on the selection of the organic group (R), POSS can be made to be either reactive or non reactive. POSS can interact with epoxy resin in two ways: First, reactive organic groups of POSS react with the epoxy polymer chain forming covalent bonds. Second, POSS can show compatibility either by similarities in chemical structure or by specific polar interaction between non-reactive organic groups on the POSS molecule and the epoxy polymer chain.

Dispersion of POSS in epoxy resin is dependent on interaction of organic group (R) of POSS with the polymer. This dispersion can be at a molecular level (reaction between POSS and epoxy resin) or at micro level (non-reactive compatibility). Usually, positive reinforcement is achieved when there is a good interaction between the polymer and POSS. This dual nature of POSS along with the fact that it is nanosized filler motivates its selection as filler for epoxy resins [10, 11]. Various researches have been done on the characterization of POSS reinforced materials. These have included the study of the formation of nanoscale structure, the synthesis of novel POSS and enhancement of physical properties including thermal degradation and glass transition temperature [12–19].

In previous research of our group we synthesize the POSS–epoxy nanocomposites and characterized these nanocomposites at room temperature. The focus of this study is to synthesize POSS–epoxy nanocomposites to characterize the mechanical properties of these nanocomposites at cryogenic temperature.

19.2 Materials and Methods

19.2.1 Materials

The polymer resin system used in this work consists of EPON 862 (Hexion Specialty Chemicals, Columbus, OH), a diglycidyl ether of bisphenol-F based resin, that is cured with curing agent EPIKURE 3274 (Hexion Specialty Chemicals, Columbus, OH), a low viscous aliphatic amine. Both the resin and the curing agent were purchased from Miller-Stephenson Chemical Company, (Tulsa, OK). The reason behind selecting this is its superior mechanical properties, chemical resistance, and ease of fabrication. Also this system has been very well studied in literature. Figure 19.2 shows the molecular structure of diglycidyl ether of bisphenol-F.

Three different functionalities of POSS were used in this work, namely, trisilanol phenyl, methacryl, and glycidyl POSS. These POSS were purchased from Hybrid Plastics (Hattiesburg, MS). The specific variant of these POSS were selected based on their interaction with the epoxy resin. Trisilanol phenyl POSS is compatible with epoxy resin due to hydrogen bonding and $\pi - \pi$ interaction. Methacryl POSS reacts with epoxy resin forming covalent bonds. Glycidyl POSS can potentially exhibit both compatibility and reactivity with the epoxy resin. Figure 19.3 shows molecular structures of these POSS.

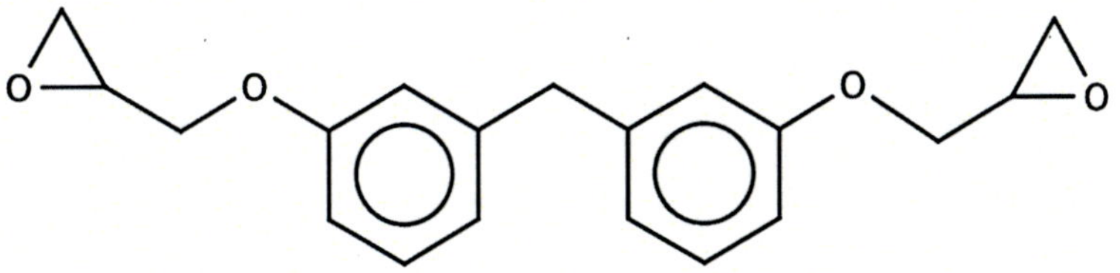

Fig. 19.2 Molecular structure of diglycidyl ether of bisphenol–F

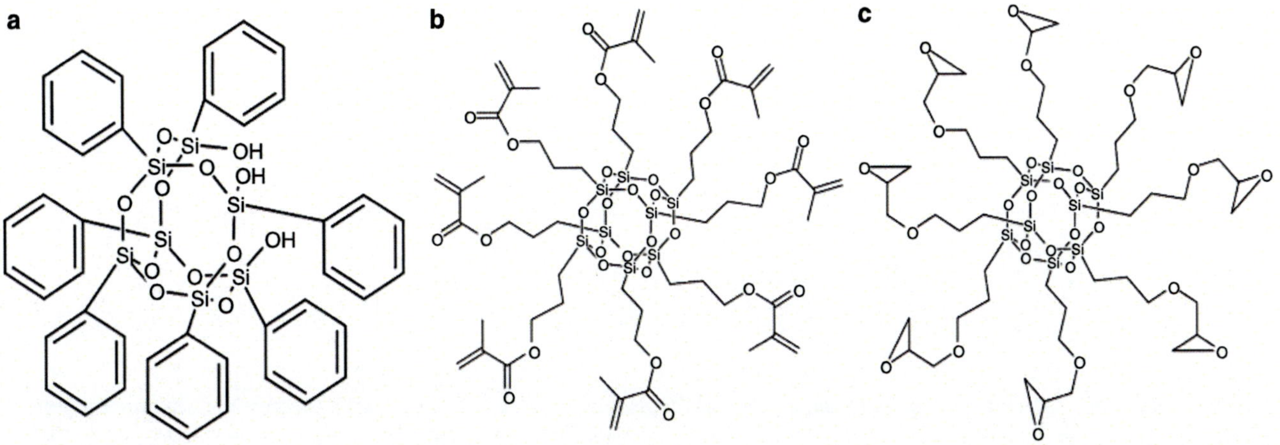

Fig. 19.3 Molecular structure of (a) trisilanol phenyl, (b) methacryl and (c) glycidyl POSS, these POSS selected according to the type of interaction with DGEB–F

19.2.2 Sample Preparation

The liquid epoxy resin (EPON 862) was mechanically mixed with POSS overnight at 50 °C. The mixture was then cooled to room temperature (25 °C) under ambient conditions. The curing agent (EPIKURE 3274) was then added to this mixture in the ratio of 100 parts of epoxy resin to 40 parts by wt% of curing agent. Finally, everything mechanically mixed at room temperature for 10 min at 400 rpm. The mixture was then placed in a vacuum chamber for 30 min to remove gas bubbles that were introduced during mixing. The degassed mixture was poured into a pre-prepared mold, and cured at room temperature for 24 h. The cast plate was then taken out from the mold and put into curing oven for post curing at 121 °C for 6 h. The same procedure was used for preparing neat resin as the baseline material.

19.2.3 Characterization

Various techniques were used to characterize the mechanical, and physical properties of nanocomposites as discussed below [20].

19.2.3.1 Mechanical Properties

Fracture toughness was determined using single edge notch bend testing as per ASTM D–5045. Samples were machined from the cast plate with nominal dimensions of 54.0 × 12.7 × 6.3 mm. A 4.5 mm deep notch was cut using a diamond precision saw, and then the tip of the notch was tapped lightly with a fresh razor blade using a hammer, to initiate a natural pre-crack. The length of notch was kept between 11.4 and 14 mm according to the requirement of the ASTM standard. The pre-cracked single edge notch specimens were loaded under three-point bending using universal testing machine (Instron 5567, Norwood, MA). Tests were performed in a displacement-controlled mode at a fixed crosshead speed of 0.5 mm/min

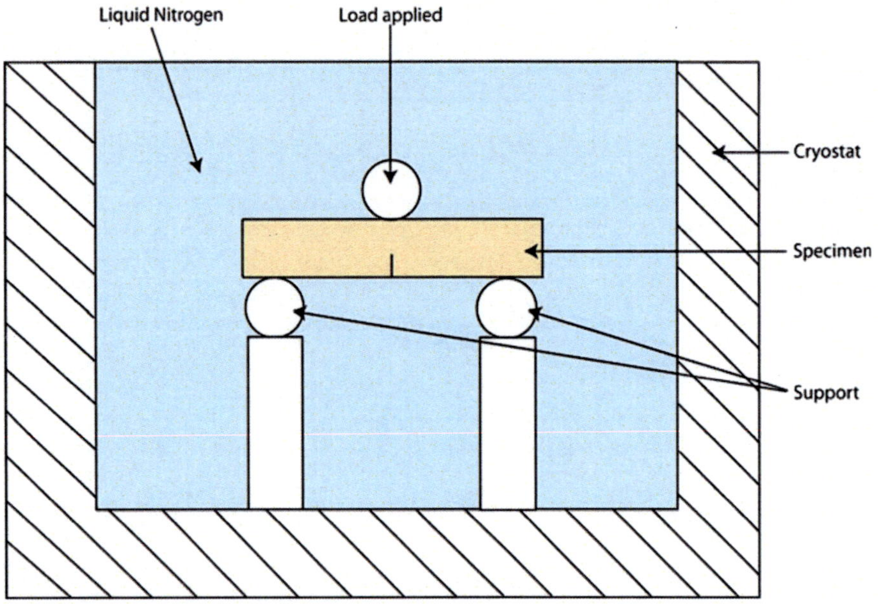

Fig. 19.4 Schematic of single edge notch bend testing under cryogenic condition

till the point of specimen failure. The load and crosshead displacement were recorded during testing. The fracture toughness of nanocomposites was measured in terms of critical stress intensity factor (K_{Ic}) calculated from the peak load, as per ASTM D–5045 [21].

Flexural modulus and strength was determined using three point bend testing according to ASTM D–790 [22]. Samples were machined from cast plate with nominal dimensions of 110 × 25.4 × 6.3 mm. The flexural strength and modulus was calculated as per ASTM standard.

The cryogenic temperature condition was achieved by samples in liquid nitrogen filled cryogenic setup designed in our laboratory. The entire test was conducted while the specimen and its loading fixture were submerged in liquid nitrogen (Fig. 19.4).

19.2.3.2 Differential Scanning Calorimetry

Differential scanning calorimetry for thermal analysis of samples was carried out using a Q 2000 DSC (TA instruments, Inc.). After calibration with high purity Indium, samples weighed around 5–10 mg was placed on the DSC cell and then heated at scan rate of 10 °C/min within the range −10 °C to 210 °C. Then, the specimen was instantaneously cooled to −10 °C using liquid nitrogen after the first scan. The second scan was then performed in similar fashion. The obtained data is extracted from second heating cycle.

19.3 Results and Discussions

19.3.1 Fracture Toughness

The load–displacement curve of neat resin and 5 wt% POSS–epoxy nanocomposites at both room temperature and cryogenic temperature are shown in Figs. 19.5 and 19.6. It is clear from the figures that epoxy resin and POSS–epoxy nanocomposites fail at the higher load at cryogenic temperature compared to failure at room temperature. At cryogenic temperature all POSS–epoxy nanocomposites shows brittle failure but at room temperature glycidyl POSS–epoxy nanocomposites shows that ductile failure. Figure 19.7 and 19.8 shows the fracture toughness of neat resin and POSS–epoxy nanocomposites at room temperature as well as at cryogenic temperature. It has been apparent from figures that the fracture toughness of the neat resin increases with the lowering of temperature, which may be due to the shrinkage of the bond causing high

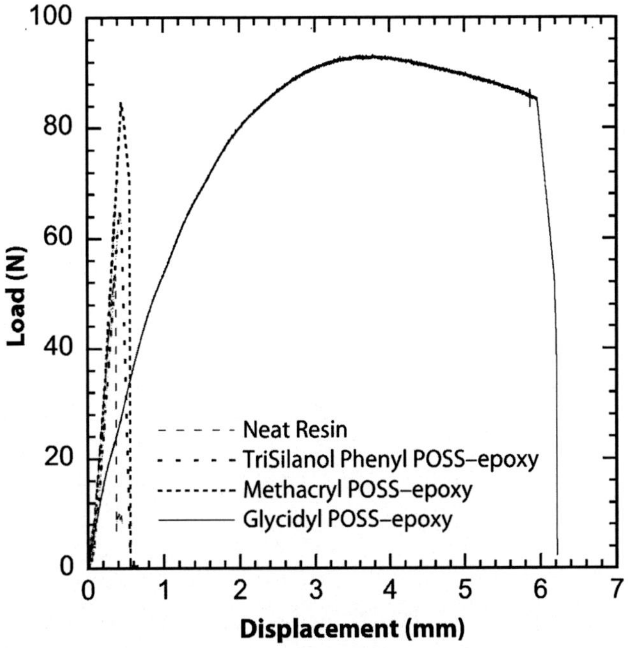

Fig. 19.5 Load displacement figure of neat resin and 5 wt% POSS incorporated nanocomposites, at room temperature

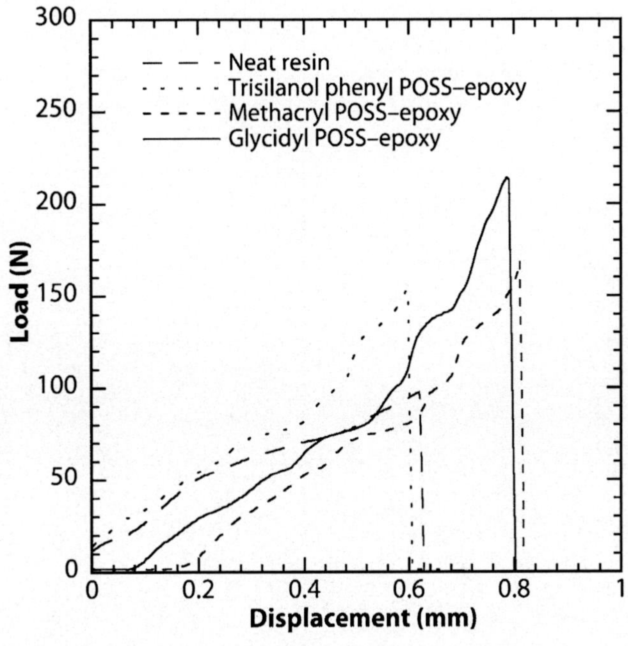

Fig. 19.6 Load displacement figure of neat resin and 5 wt% POSS incorporated nanocomposites, at cryogenic condition

intermolecular force. At room temperature the fracture toughness of all POSS reinforced nanocomposites reaches maximum at 5 wt% of POSS content. For glycidyl POSS–epoxy nanocomposites the fracture toughness increases up to 2.4 times compared to neat resin.

At cryogenic temperature the fracture toughness of neat epoxy increased by 120% compare to the fracture toughness at room temperature because of the shrinkage of polymer chain in network hence increasing the intermolecular force.

Similar to the room temperature the fracture toughness increases with the incorporation of POSS attaining the maximum then decreases. Trisilanol phenyl POSS incorporated composite shows maximum fracture toughness at 3 wt% but after that it starts decreasing while in case of methacryl and glycidyl POSS incorporated composite it shows maximum fracture toughness at 5 wt%. Especially among all the three POSS used, glycidyl POSS incorporated composite shows maximum value. In case of glycidyl POSS–epoxy composite the fracture toughness increases up to 1.5 times than that of neat resin.

Fig. 19.7 Fracture toughness of neat resin and POSS incorporated nanocomposites of different wt%, at room temperature condition

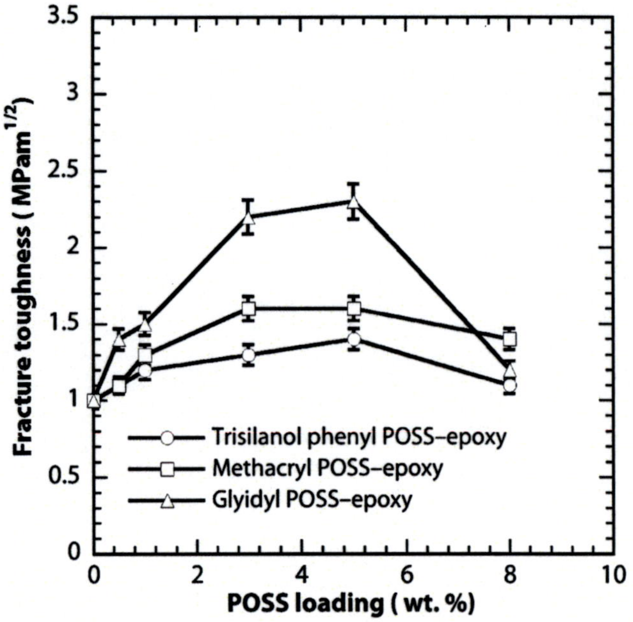

Fig. 19.8 Fracture toughness of neat resin and POSS incorporated nanocomposites of different wt%, at cryogenic condition

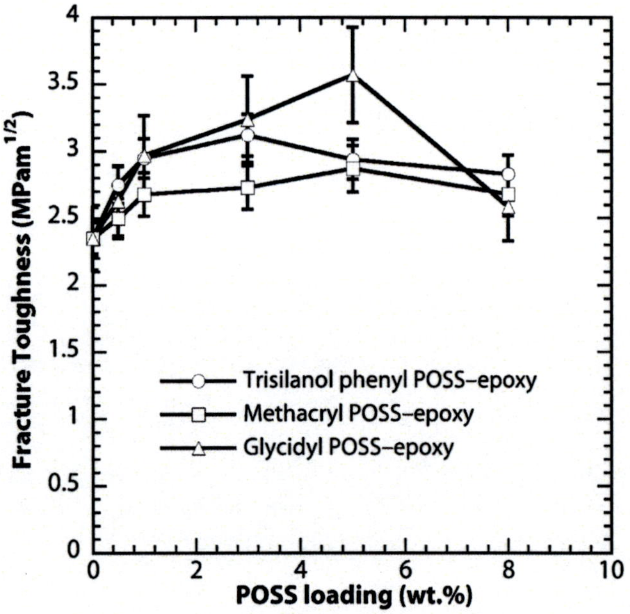

This increase of fracture toughness at cryogenic temperature due to incorporation of POSS can be explained in terms of free volume. Free volume is the unoccupied space between molecules. When POSS molecule incorporated in epoxy resin the free volume in the polymer network increases because of the steric effects associated with POSS structure. In order to understand the dependence of mechanical properties of epoxy resin on POSS content, schematic illustration of cured network in the pure epoxy and hydrogen bonds, free volume in the POSS–epoxy modified systems is presented in Fig. 19.9.

These free volumes due to introduction of POSS still existed in the cryogenic condition contributing high intermolecular force hence high fracture toughness. These free spaces and volumes also decrease the high internal stress generated in the polymer network because of very low temperature.

As shown in Fig. 19.10 room temperature where flexural modulus decreases with the incorporation of glycidyl POSS because of the introduction of plasticity in the nanocomposite with increasing glycidyl POSS content. This plasticization in nanocomposite is also may be due to the free spaces and free volume in the polymer network.

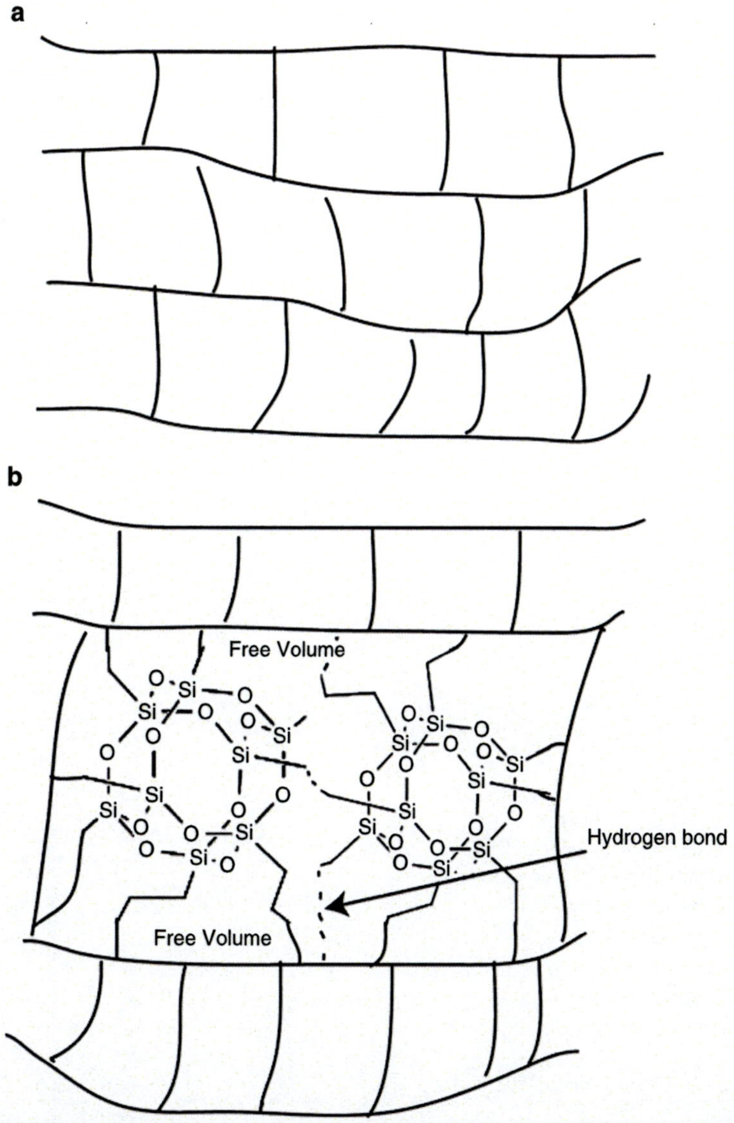

Fig. 19.9 Schematic of illustration of (**a**) pure epoxy and (**b**) free volume in POSS modified epoxy chain

While at cryogenic condition there is no variation in the value of flexural modulus with incorporation of either POSS. This may be because at very low temperature there is shrinkage of bonds hence the flexibility of bonds gets restricted. Due to the shrinkage of bonds and free space/volume the flexural modulus of neat resin and POSS–epoxy nanocomposites at cryogenic condition is more than that at room temperature as shown in Figs. 19.10 and 19.11.

19.3.2 Differential Scanning Calorimetry

To support the increase in the free volume we have done some DSC test. The glass transition temperature (T_g) for 5 wt% of POSS–epoxy nanocomposites has been investigated. T_g depends primarily on chain flexibility, crosslinking of composite, intermolecular attraction, and so forth. Table 19.1 shows the variation of T_g with the incorporation of POSS.

Gycidyl POSS–epoxy and methacryl POSS–epoxy nanocomposite displayed lower T_g compared to neat resin. This is because of increase in mobility of the polymer chain by addition of these POSS. Glycidyl and methacryl POSS reacts with the epoxy resin and merge into polymer network, hence increased the free volume in nanocomposites. T_g for trisilanol phenyl POSS–epoxy increased with the incorporation of POSS. This might be due to the restriction of the chain mobility because of steric hindrance due of big phenyl group attached to trisilanol phenyl POSS.

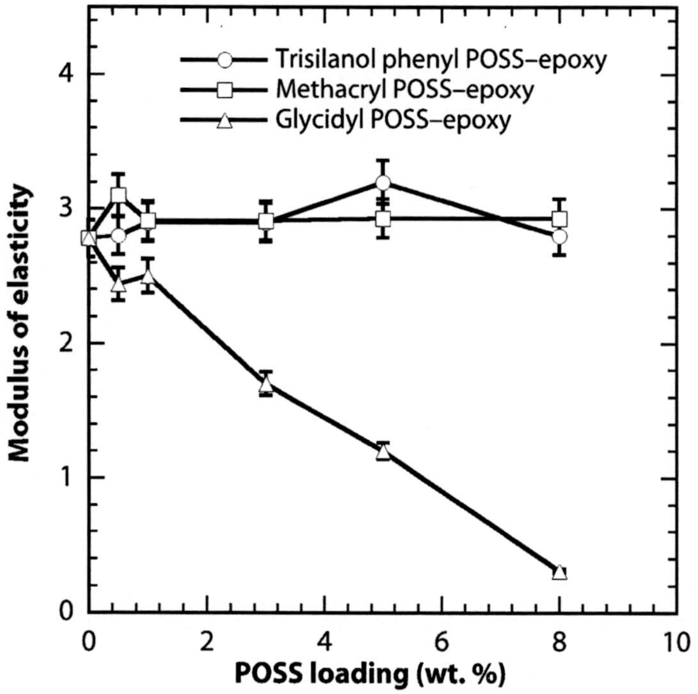

Fig. 19.10 Flexural modulus of neat resin and POSS incorporated nanocomposites for different wt%, at room temperature condition

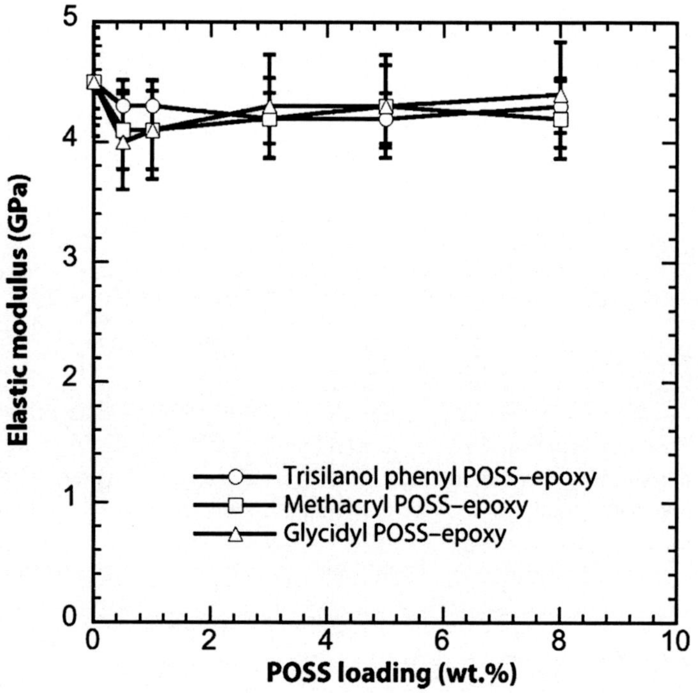

Fig. 19.11 Flexural modulus of neat resin and POSS incorporated nanocomposites for different wt%, at cryogenic condition

Table 19.1 Glass transition temperature for neat resin and 5 wt% of POSS incorporated nanocomposite

	Glass transition temperature (°C)
Neat resin	41.1
Trisilanol phenyl POSS–epoxy	45.9
Methacryl POSS–epoxy	39.2
Glycidyl POSS–epoxy	33.5

19.4 Conclusions

POSS–epoxy nanocomposites were synthesized using simple mechanical mixing. It was confirmed from the result that with the increasing POSS content the fracture toughness increases in both room temperature and cryogenic condition. At room temperature 5 wt% of POSS shows the maximum fracture toughness, especially in the case of glycidyl POSS–epoxy nanocomposites it shows almost 2.4 times increase in the fracture toughness. At cryogenic temperature methacryl and glycidyl POSS shows maximum at 5 wt% while trisilanol phenyl POSS shows maximum at 3 wt%. Similar to room temperature in cryogenic condition also glycidyl POSS–epoxy nanocomposite shows maximum fracture toughness among all three POSS, it shows almost 1.5 times increase in the fracture toughness compare to neat resin. At the cryogenic temperature, the fracture toughness of the epoxy was 2.2 times higher than that at the room temperature. Flexural modulus values at room temperature shows no variation in the value with the incorporation of POSS for trisilanol and phenyl POSS but for glycidyl POSS it starts decreasing because plasticization starts to act. While at cryogenic temperature the value of flexural modulus of POSS–epoxy nanocomposites was almost similar as of neat resin because of the shrinkage of bonds. Change in the glass transition temperature with the incorporation of POSS supports the increase in the free volume with POSS addition.

Acknowledgements We gratefully acknowledge that this work is funded in part or fully by a grant through the Oklahoma Nanotechnology Applications Project (ONAP) (Grant no. O9-20) and NASA Experimental Program to Stimulate Competitive Research (EPSCOR) (Grant no. NNXO).

References

1. Ellis B (1993) Chemistry and technology of epoxy resin. Blackie academic professional, New York
2. Dusek K (ed.) (1986) Epoxy resins and composites II. Advances in polymer science, Springer-Verlag, Berlin, Heidelberg. 75: xiii+180, Price DM 118.00. ISBN 3-540-15825-1
3. Ueki T, Nishijima S, Izumi Y (2005) Designing of epoxy resin systems for cryogenic use. Cryogenics 45:141–148
4. Sawa F, Nishijima S, Okada T (1995) Molecular design of an epoxy for cryogenic temperature. Cryogenics 35:767–769
5. Yang JP, Chen Z-K, Yang G, Fu S-Y, Ye L (2008) Simultaneous improvements in the cryogenic tensile strength, ductility and impact strength of epoxy resins by a hyperbranched polymer. Polymer 49:3168–3175
6. Yang G, Fu SY, Yang JP (2007) Cryogenic mechanical behaviors of epoxy resins modified by flexible diamines. Polymer 48:302–310
7. Huang CJ, Fu SY, Zhang YH, Lauke B, Li LF, Ye L (2005) Cryogenic properties of SiO_2/epoxy nanocomposites. Cryogenics 45:450–454
8. Yang J-P, Yang G, Xu G, Fu S-Y (2007) Cryogenic mechanical behaviors of MMT/epoxy nanocomposites. Compos Sci Technol 67:2934–2940
9. Kim BC, Park SW, Lee DG (2008) Fracture toughness of the nano-particle reinforced epoxy composite. Compos Struct 86:69–77
10. Ni CH, Ni GF, Zhang SW, Liu XY, Chen MQ, Liu LH (2010) The preparation of inorganic/organic hybrid nanomaterials containing silsesquioxane and its reinforcement for an epoxy resin network. Colloid Polym Sci 288(4):469–477
11. Matejka L, Dukh O, Meissner B, Hlavata D, Brus J, Strachota A (2003) Block copolymer organic–inorganic networks. Formation and structure ordering. Macromolecules 36(21):7977–7985
12. Ramirez C, Rico M, Torres A, Barral L, Lopez J, Montero B (2008) Epoxy/poss organic–inorganic hybrids: Atr-ftir and dsc studies. Eur Polymer J 44:3035–3045
13. Laine RM, Choi JW, Lee I (2001) Organic–inorganic nanocomposites with completely defined interfacial interactions. Adv Mater 13 (11):800–803
14. Sanchez-Soto M, Schiraldi DA, Illescas S (2009) Study of the morphology and properties of melt-mixed polycarbonate-poss nanocomposites. Eur Polymer J 45:341–352
15. Hsiao BS, White H, Rafailovich M, Mather PT, Jeon HG, Phillips S, Lichtenhan J, Schwab J (2000) Nanoscale reinforcement of polyhedral oligomeric silsesquioxane (poss) in polyurethane elastomer. Polymer 49:437–440
16. Mather PT, Jeon HG, Romo-Uribe A, Haddad TS, Lichtenhan JD (1999) Mechanical relaxation and microstructure of poly (norbornyl–poss) copolymers. Macromolecules 32:1194–1203
17. Schwab JJ, Lichtenhan JD (1998) Polyhedral oligomeric silsesquioxane (poss)-based polymers. Appl Organomet Chem 12:707–713
18. Lichtenhan JD (1995) Polyhedral oligomeric silsesquioxanes building blocks for silsesquioxane-based polymers and hybrid materials. Comments Inorg Chem 17:115–130
19. Song L, He QL, Hu Y, Chen H, Liu L (2008) Study on thermal degradation and combustion behaviors of pc/poss hybrids. Polyme Degrad Stab 93:627–639
20. Anderson TL (1991) Fracture mechanics: fundamentals and applications. CRC Press, Boca Rotan
21. American society for testing and materials, Philadelphia, PA, ASTM D5045-99: Standard test methods for plain-strain fracture toughness and strain energy release rate of plastic materials
22. American society for testing and materials, Philadelphia, PA, ASTM D790-07: standard test methods for flexural properties of unreinforced and reinforced plastics and electrical insulating materials

Chapter 20
Ballistic Impact Behaviors of GLARE 5 Fiber-Metal Laminated Plates

A. Seyed Yaghoubi and B. Liaw

Abstract In this study, GLARE 5 fiber-metal laminated (FML) plates of dimensions: 152.4 × 101.6 mm (6″ × 4″) with various stacking sequences and thicknesses were impacted by a 0.22 caliber bullet shaped projectile using a high-speed gas gun. The effect of stacking sequence was considered for the GLARE 5 (3/2) panels with cross-ply $[0°/90°]_s$, unidirectional $[0_4°]$, angle-ply $[\pm 45°]_s$ and quasi-isotropic [0°/±45°/90°] lay-up orientations; whereas, the thickness effect was studied for the GLARE 5 cross-ply panels. The experimentally obtained incident projectile impact velocities versus the residual velocities were plotted for each type of specimen. The data were then fitted numerically according to the classical Lambert-Jonas equation for the determination of the ballistic limit velocity, V_{50}. The 3D dynamic nonlinear finite element (FE) code, LS-DYNA, was used to validate the experimental results. Good agreement between experimental and FE results was obtained.

Keywords Ballistic impact • GLARE 5 • Fiber-metal laminates • LS-DYNA

20.1 Introduction

GLARE 5 is a fiber-metal laminate (FML) made of alternating layers of 2024-T3 aluminum alloy sheets and unidirectional S2-glass/epoxy laminates. Seyed Yaghoubi et al. [1] studied velocity measuring approaches for the determination of ballistic limits of GLARE 5 FMLs plates. It was experimentally shown that the projectile was further accelerated along the ballistic trajectory after emerging from the gun barrel. They concluded that the ballistic limit velocity increased parabolically as the panel thickness increased and also increased as the metal volume fraction decreased. Vaidya et al. [2] did research on ballistic performance of graphite/epoxy and S2-glass/epoxy composites with polycarbonate facing. They concluded that woven S2-glass/epoxy were more damage resistant as compared to the prepreg graphite/epoxy laminates. They also reported that the ballistic limit and absorbed energy increased with increase in thickness of the laminate for a given polycarbonate thickness. Naik et al. [3] considered ballistic impact behavior of typical woven fabric E-glass/epoxy thick composites. They observed that for the same mass and diameter of the projectile, as the thickness of the target increased, ballistic limit velocity increased. Gellert et al. [4] studied the effect of target thickness on the ballistic perforation of glass fiber reinforced plastic composites. They concluded that thicker targets are more ballistically efficient, especially against blunt projectiles.

The main objective of this study was to investigate incident versus residual velocity of GLARE 5 FML plates with various stacking sequences and thicknesses. Experimentally obtained velocity measurements were fitted numerically according to the classical Lambert-Jonas equation for the determination of the ballistic limit velocity, V_{50}. Finally, the 3D dynamic nonlinear FE code, LS-DYNA, was used to validate the experimental results.

A. Seyed Yaghoubi (✉) • B. Liaw
Department of Mechanical Engineering, The City College of New York, Convent Avenue and 138th Street, New York, NY 10031, USA
e-mail: ASeyed_Yaghoubi@gc.cuny.edu

E. Patterson et al. (eds.), *Composite Materials and Joining Technologies for Composites, Volume 7: Proceedings of the 2012 Annual Conference on Experimental and Applied Mechanics*, Conference Proceedings of the Society for Experimental Mechanics Series 44,
DOI 10.1007/978-1-4614-4553-1_20,

20.2 Experimental Procedures

All GLARE panels considered in this study consisted of 2024-T3 aluminum alloy with a thickness of 0.305 mm (0.012″) per layer and unidirectional S2-glass/epoxy laminated layers, each with a thickness of 0.508 mm (0.020″). Each S2-glass/epoxy layer has a lay-up orientation according to Table 20.1. GLARE 5 panels with various stacking sequences and thicknesses were cut into rectangular plate specimens with dimensions of 152.4 × 101.6 mm (6″ × 4″). The configuration of GLARE 5 panels with different stacking sequences, thicknesses, MVF and typical mechanical properties of constituents are described in Tables 20.1 and 20.2, respectively [5–7]. In Table 20.1, the term MVF represents metal volume fraction and is defined as the ratio of the sum of the thicknesses of all aluminum layers over the total thickness of the FML [8]:

$$\mathrm{MVF} = \frac{\sum t_{\mathrm{aluminum}}}{t_{\mathrm{FML}}} \tag{20.1}$$

The ballistic impact tests were performed using an in-house-designed gas gun. The gun consists of a pressurized tank, a gas storage vessel, a solenoid valve and a stainless-steel barrel. The pictures of the gas gun setup and specimen container are shown in Fig. 20.1. Compressed helium/air was used as the propellant. A 0.22 caliber copper bullet was chosen as the projectile.

Table 20.1 GLARE 5 panels tested in this study with different stacking sequences and thicknesses

Lay-up configuration (m/n)	Prepreg plies & orientation	Total thickness	MVF
2/1	$[0°/90°]_s$ Cross-ply	1.117 mm (0.044″)	0.546
3/2	[0°4] Unidirectional	1.930 mm (0.076″)	0.474
3/2	$[0°/90°]_s$ Cross-ply	1.930 mm (0.076″)	0.474
3/2	$[\pm 45°]_s$ Angle-ply	1.930 mm (0.076″)	0.474
3/2	$[0°/\pm 45°/90°]$ Quasi-isotropic	1.930 mm (0.076″)	0.474
4/3	$[0°/90°]_s$ Cross-ply	2.743 mm (0.108″)	0.445
5/4	$[0°/90°]_s$ Cross-ply	3.556 mm (0.140″)	0.429
6/5	$[0°/90°]_s$ Cross-ply	4.368 mm (0.172″)	0.419

The configuration notation *m*/*n* means the panel is composed of m aluminum-alloy layers interlaced with *n* fiber-reinforced epoxy layers

Table 20.2 Typical values of mechanical properties [5–7]

Mechanical property	Aluminum 2024-T3	UD S2 Glass/FM 94 Epoxy Prepreg
Tensile ultimate strength (MPa)	L 455	1900
	T 448	57
Tensile yield strength (MPa)	L 359	–
	T 324	–
Tensile modulus (GPa)	L 72	54
	T –	9.4
Ultimate strain (%)	L 19	3.5
	T –	0.6
Compressive yield strength (MPa)	L 303	–
	T 345	–
Compressive modulus (GPa)	L 74	54.4
	T –	11
Density (g/cm^3)	2.79	2.00

The symbols, L and T, stand for longitudinal (the rolling direction for the metal) and transverse directions, respectively

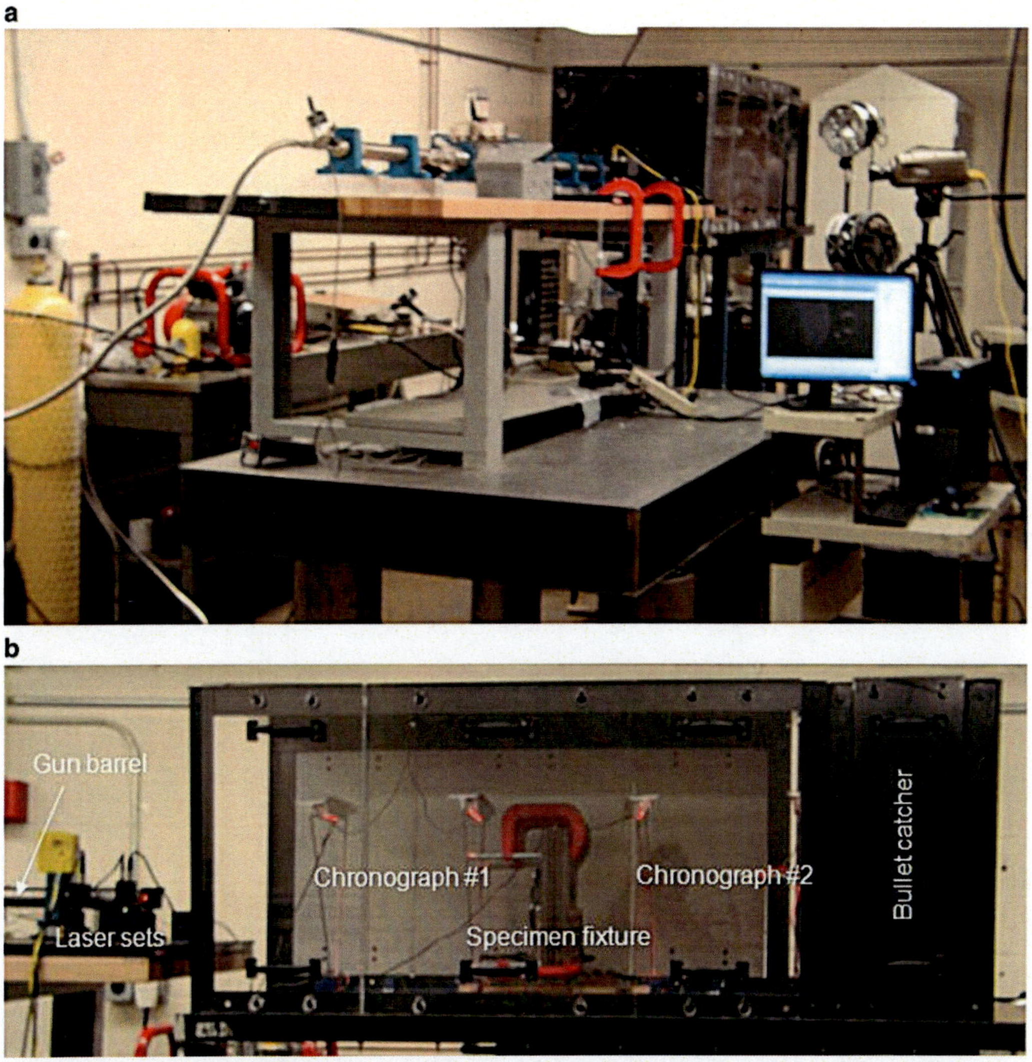

Fig. 20.1 **(a)** The high-speed gas gun setup for ballistic impact tests and **(b)** specimen container.

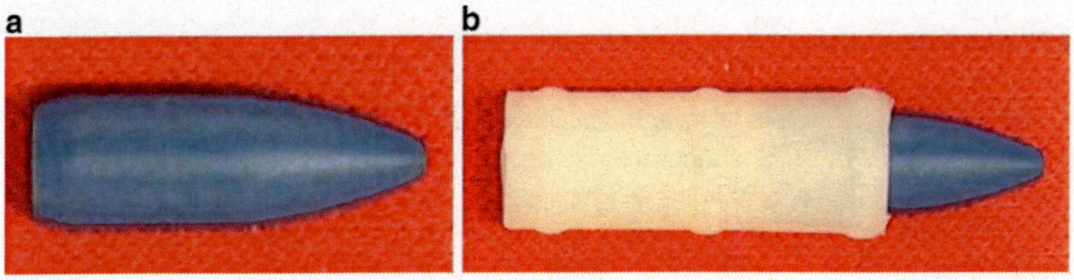

Fig. 20.2 **(a)** 0.22 caliber copper bullet and **(b)** bullet mounted in a pair of plastic sabots

The bullet mounted inside a pair of plastic sabots and then inserted into the gun barrel. Figure 20.2 illustrates the 0.22 caliber copper bullet and the sabot. For the same type of sabots, the projectile's initial velocity can be controlled by changing its position inside the barrel and/or by varying the input pressure of the gun. The plate specimen was clamped over a length of 25.4 mm (1″) from each side (Fig. 20.3). The specimen fixture was extra supported by four C-clamps inside the specimen container (Fig. 20.1b). The composite plate specimen was then impacted by the projectile at the center.

The speed of the projectile was measured in three distinct ways [1]. A pair of diode-lasers/amplified-photodiodes, separated by 101.6 mm (4″), was located near the exit of the gun barrel, forming two optical gates to measure the speed of the projectile at the gun muzzle. For the second measurement, two pairs of chronographs were used to measure the projectile speed before and after impact; one pair was located in front of the target, while the second was located behind the target.

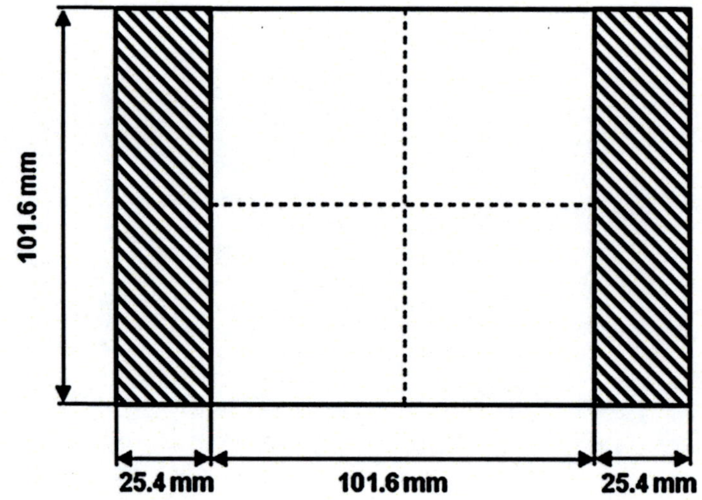

Fig. 20.3 A composite plate specimen clamped at the two ends

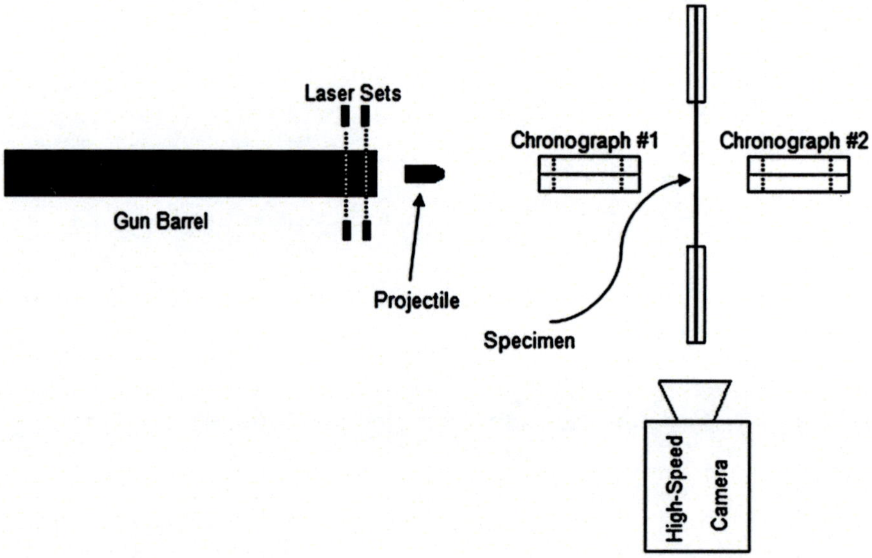

Fig. 20.4 Schematic of the experimental setup

Thirdly, a high-speed camera was used to monitor the bullet motion during the test. In this study, the high-speed camera was set perpendicular to the projectile's ballistic trajectory. Using the captured high-speed video, the bullet speed was then determined before and after the impact. A schematic of the setup is shown in Fig. 20.4.

20.3 Results and Discussion

After determining the speeds of the bullet before and after impact, the incident projectile impact velocity was plotted versus the residual velocity for each type of the specimens listed in Table 20.1. The experimental data was then fitted by least-square regression according to the classical Lambert-Jonas equation [9] for the positive residual velocity values:

$$V_R^P = A\left(V_I^P - V_{50}^P\right) = A \cdot V_I^P - B \tag{20.2}$$

where A and B are two regression coefficients and P is a power. V_R and V_I are the residual and incident velocities of the projectile, respectively, while V_{50} is the ballistic limit velocity, which is defined as the velocity required for a projectle to perforate the target 50% of the time.

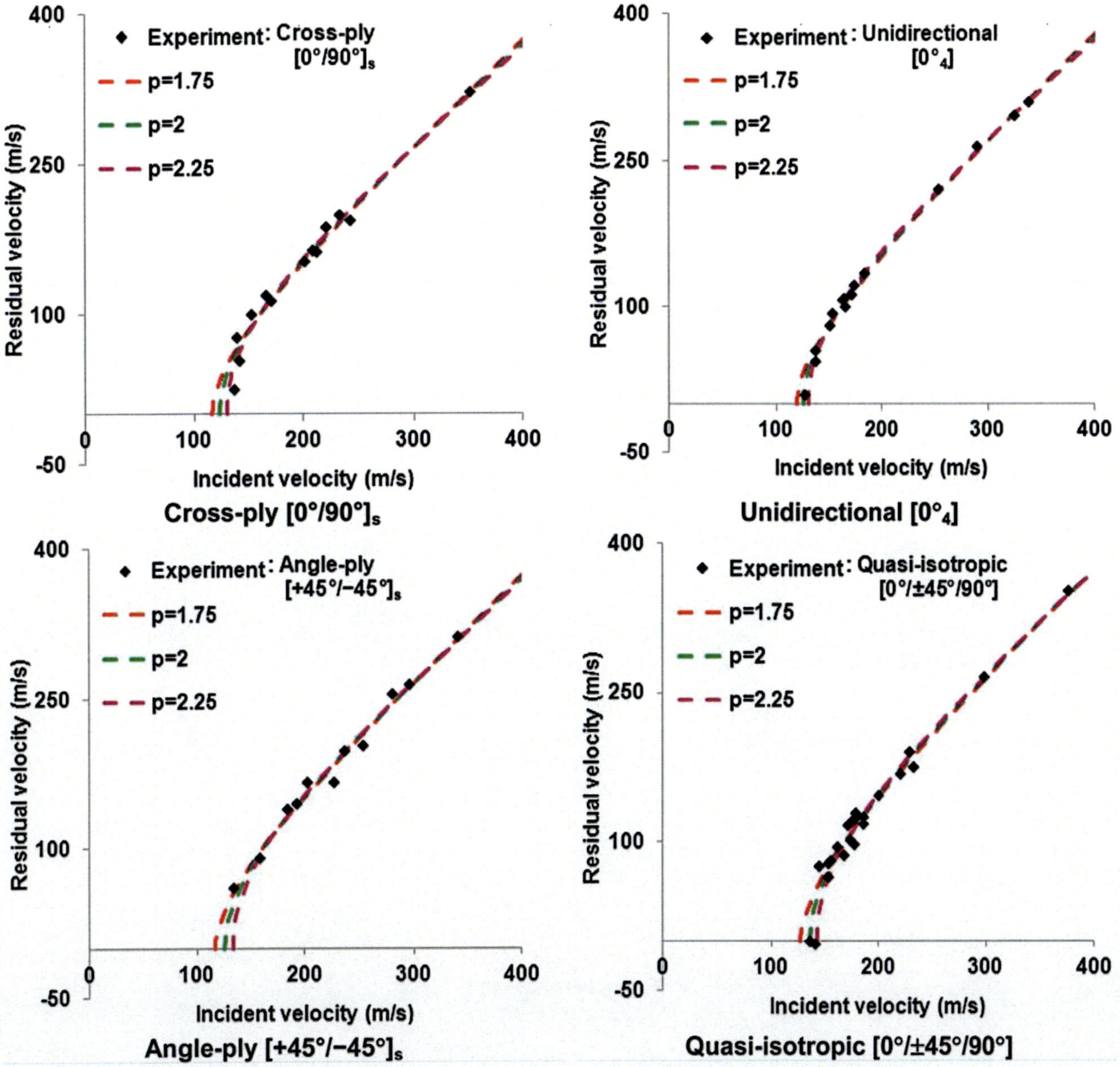

Fig. 20.5 Experimental residual velocity *vs.* incident velocity fitted with the Lambert-Jonas equation using different *P* values for the GLARE 5 (3/2) FML plates with various stacking sequences

Figures 20.5 and 20.6 illustrate experimental residual velocity versus incident velocity data fitted with the Lambert-Jonas equation using different P values for the GLARE 5 FML plates with various stacking sequences and thicknesses, respectively. For each specimen type, several P values were tried according to its V_R versus V_I experimental trend. In the figures, the experimental data are specified by a solid diamond symbol and the Lambert-Jonas curves with different P values are shown by dashed lines. The intersection of the Lambert-Jonas curve with the incident-velocity axis was considered as the V_{50} velocity for a given P value.

As is apparent from Fig. 20.5, cross-ply, unidirectional and angle-ply specimens offered almost the same resistance to the projectile perforation. In addition, the quasi-isotropic specimen showed relatively higher ballistic limit velocity compared to other stacking sequences. It can be seen from Fig. 20.6 that as the panel became thicker, the ballistic limit also increased. It should be mentioned that in the figures the symbols below the incident-velocity axes represent rebound velocities of the projectiles. With current gas gun facility, achieving low velocity range, i.e. lower than 140 m/s, was very diffecult, if not imposible. Therefore, the Lambert–Jonas equation becomes very useful for determining the V_{50}.

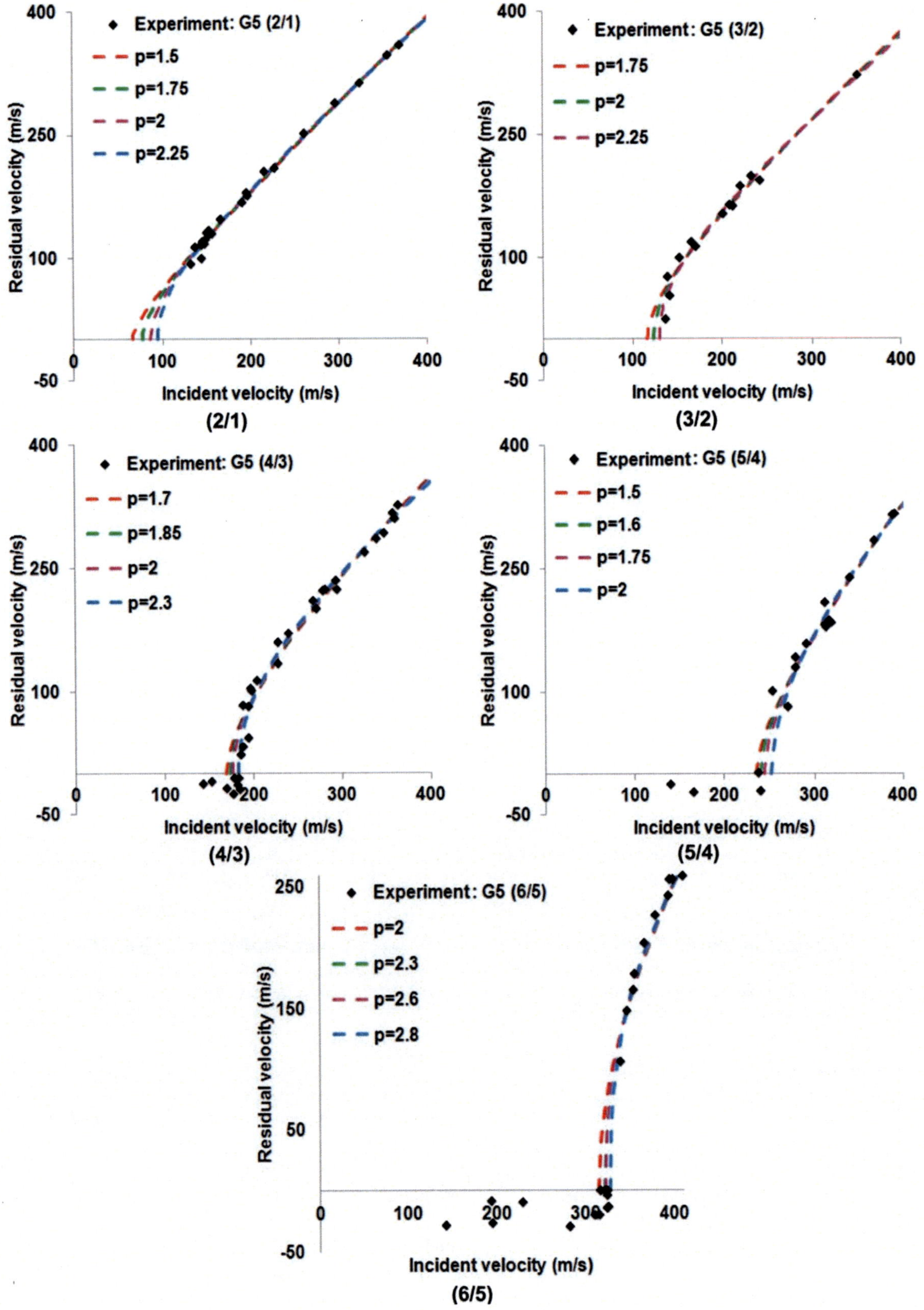

Fig. 20.6 Experimental residual velocity *vs.* incident velocity fitted with the Lambert-Jonas equation using different *P* values for the cross-ply GLARE 5 FML plates with various thicknesses

20.4 Finite Element Simulation

In this study, ballistic impact on GLARE 5 FML plates with various stacking sequences and thicknesses were modeled and analyzed using the commercially available finite element software, LS-DYNA. Figure 20.7 exhibits perspective views of the composite plate. To simulate the experimental results, the aluminum layers as well as the 0.22 copper bullet were modeled using the simplified Johnson-Cook material model. Table 20.3 represents the material parameters used for this model [10, 11]. In the simplified model, thermal effects and damage are ignored [12].

The composite model with Chang-Chang damage criteria was employed for the prepreg layers. This model is an orthotropic material model with optional brittle failure for composites [12]. For this model three possible failure criteria are possible, namely: matrix cracking, compressive and fiber breakage failures. The mechanical properties of the unidirectional S2-glass/epoxy composite are described in Table 20.4 (some data were provided by the Structural Laminates Company). Many of the constitutive models in LS-DYNA do not allow failure and erosion and hence MAT_ADD_EROSION failure criteria were adapted to all the materials used here. In this study a strain-based failure criterion was used for element erosion; that is, when $\varepsilon \geq \varepsilon_{erosion}$, the element was eroded and removed from the calculation.

The interfaces between adjacent layers were modeled with tiebreak interface option to simulate the delamination and debonding. The contact between layers in the FML was defined using the option, CONTACT_AUTOMATIC_SURFACE_TO_SURFACE_TIEBREAK. ERODING_SURFACE_TO_SURFACE contact model was also used between the projectile and FML. This model allows elements to be eroded when certain failure criteria are met.

Figures 20.8 compares FE predictions with experimental residual velocity versus incident velocity data for the GLARE 5 (3/2) FML plates with cross-ply $[0°/90°]_s$, unidirectional $[0°_4]$, angle-ply $[\pm 45°]_s$, and quasi-isotropic $[0°/\pm 45°/90°]$ stacking sequences, respectively. The FE results are shown by a circular symbol. As is apparent from the figure, FE predictions are in excellent agreement with the experimental results. Moreover, excellent agreements between FE predictions and experimental results could be seen for the GLARE 5 FML plates with (2/1), (3/2), (4/3), (5/4) and (6/5) configurations (Fig. 20.9).

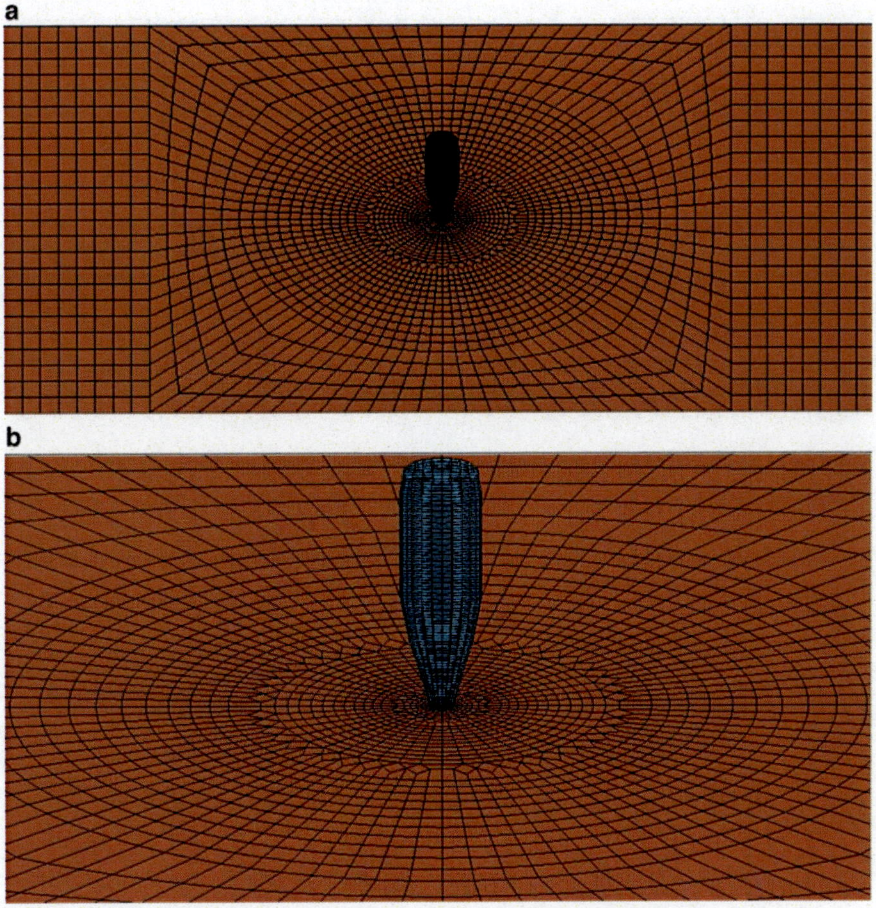

Fig. 20.7 Perspective views of the finite element simulation model: (**a**) far-off view, (**b**) close-up view

Table 20.3 Material model parameters used for the 2024-T3 aluminum-alloy layers and 0.22 copper bullet [10, 11]

Material	Density (kg/m^3)	Young's modulus [GPa]	Static yield limit [MPa]	Strain hardening modulus [MPa]	Strain hardening exponent	Strain rate coefficient	Poisson's ratio
Aluminum	ρ = 2,770	E = 73.084	A = 369	B = 684	n = 0.73	C = 0.0083	$\nu = 0.33$
Bullet	–	E = 117.21	A = 4.36	B = 457.81	n = 0.37575	C = 0	$\nu = 0.33$

Table 20.4 Ply properties of the unidirectional S2-glass/epoxy prepreg

Density (kg/m^3)	Young's modulus [GPa]	Shear modulus [GPa]	Poisson's ratio	Strength [MPa]
ρ = 2,000	$E_1 = 54$	$G_{12} = G_{23} = G_{31} = 5.6$	$\nu_{20} = \nu_{31} = 0.0575$	S_1 =1,900
				S_2 =57
	$E_2 = E_3 = 9.4$		$\nu_{32} = 0.33$	S_{12} =76
				C_2 = 285

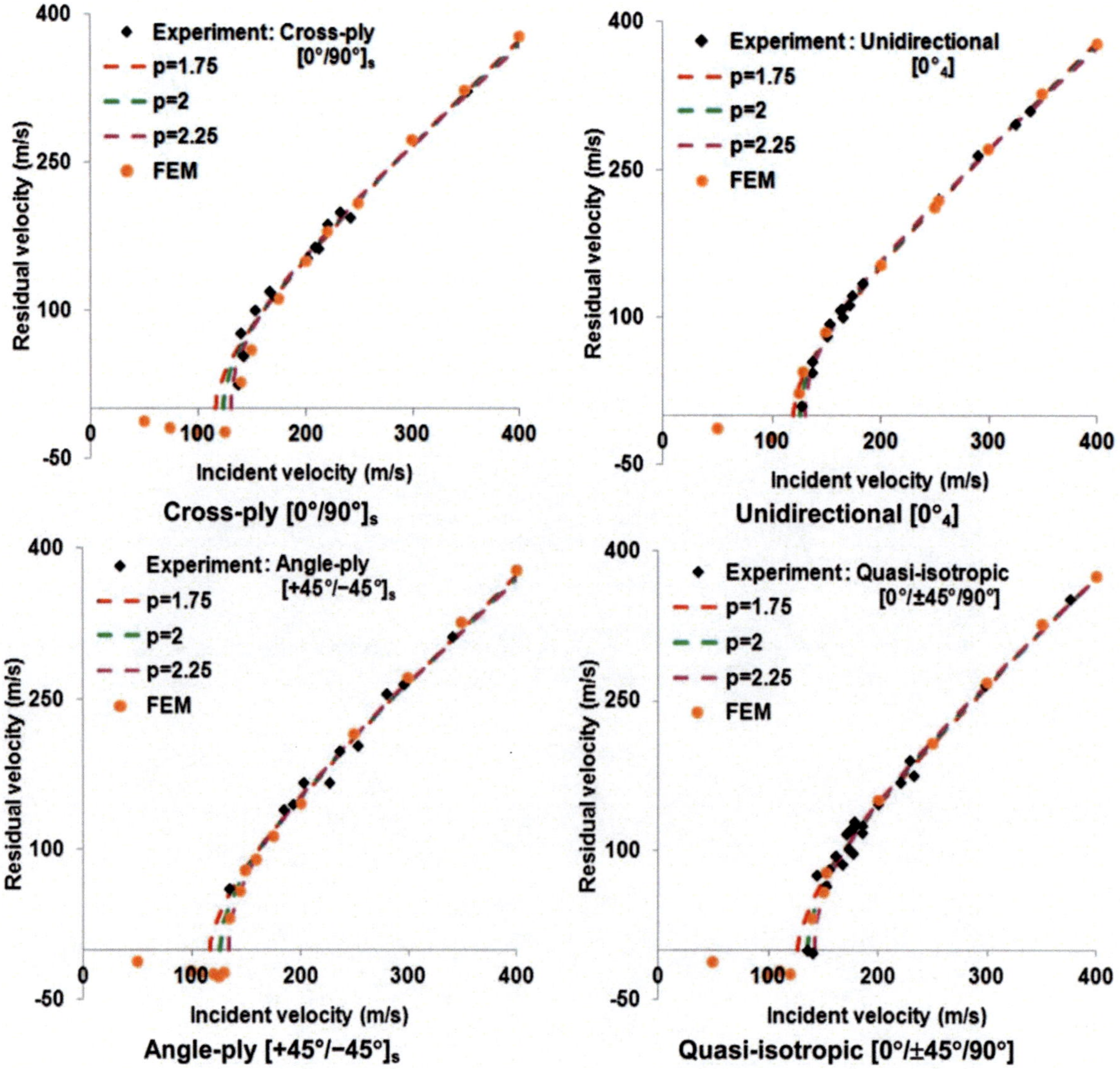

Fig. 20.8 Comparison between FE predictions and experimental residual-incident velocity data for the GLARE 5 (3/2) FML plates with various stacking sequences

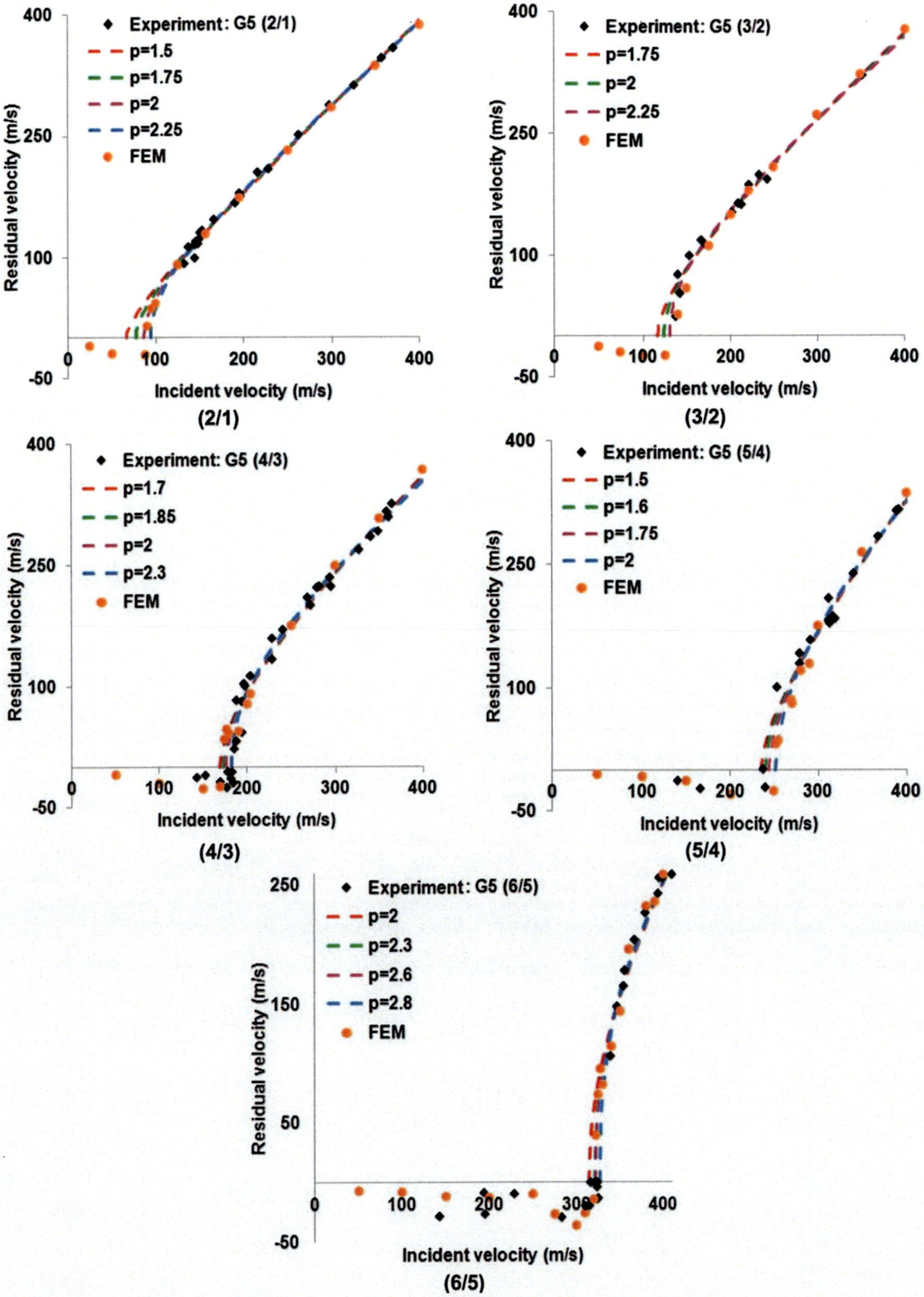

Fig. 20.9 Comparison between FE predictions and experimental residual-incident velocity data for the cross-ply GLARE 5 FML plates with various thicknesses

20.5 Conclusions

This study presents experimental and numerical investigations on the incident versus residual velocity behaviors of GLARE 5 FML plates with various stacking sequences and thicknesses. The following remarks can be concluded from this study.

- Cross-ply, unidirectional and angle-ply specimens offered almost the same resistance to the projectile perforation. In addition, the quasi-isotropic specimen showed relatively higher ballistic limit velocity compared to other stacking sequences.
- By changing the specimen configuration from (2/1) to (6/5), i.e. by increasing the specimen thickness, the ballistic limit was also raised.
- The classical Lambert-Jonas equation was very useful in obtaining the ballistic limit velocities, especially for the cases where obtaining lower incident velocities was experimentally difficult.
- FE obtained incident-residual velocity data were in excellent agreement with their experimental counterparts.

Acknowledgments The authors would like to thank the supports from NASA Faculty Award for Research (FAR) under Grant No. NAG3-2259 and from PSC-CUNY under Grants 63168–00 41 and 64543–00 42. Dr. Kenneth J. Bowles and Dr. John P. Gyekenyesi were the Technical Monitors of the NASA grant. Part of the equipment was procured through ARO Grant No. DAAD19-03-1-0086, of which Dr. Bruce LaMattina was the Program Manager.

References

1. Seyed Yaghoubi A, Chow MF, Liaw BM (2011) Velocity measuring approaches for the determination of ballistic limits of GLARE 5 fiber-metal laminate plates. Experimental mechanics in the age of modeling and simulation. In: SEM annual conference & exposition on experimental and applied mechanics, Mohegan Sun 13–16 June 2011
2. Vaidya UK, Kulkarni M, Haque A, Hosur MV, Kulkarni R (2000) Ballistic performance of graphite/epoxy and S2-glass/epoxy composites with polycarbonate facing. Mat Tech and Adv Perf Mat 15(3):202–214
3. Naik NK, Doshi AV (2008) Ballistic impact behavior of thick composites: parametric studies. Compos Struct 82:447–464
4. Gellert EP, Cimpoeru SJ, Woodward RL (2000) A study of the effect of target thickness on the ballistic perforation of glass-fiber-reinforced plastic composites. Int J Impact Eng 24:445–456
5. Data sheets (1998) Aviation equipment structures, Inc., Costa Mesa, California
6. QA Reports (1994) B0319B-2, B1008B-1, B0904A-3. Structural Laminates Company, New Kensington
7. Alloy 7475 Plate and Sheet, ACRP-053-B. Alcoa Mill Products, Bettendorf http://www.alcoa.com/mill_products/catalog/pdf/alloy7475tech-platesheet.pdf
8. Vlot A, Gunnink JW (eds) (2001) Fiber metal laminates, an introduction. Kluwer, Dordrecht
9. Lambert JP, Jonas GH (1976) Towards standardization of in terminal ballistics testing: velocity representation. Ballistic Research Laboratories, Aberdeen Proving Ground, Maryland, Report No. BRL-R-1852
10. Kay G, Goto D, Couch R (2007) Statistical testing of aluminum, titanium, lexan and composites for transport airplane rotor burst fragment shielding. FAA Rep. No. DOT/FAA/AR-07/26, Federal Aviation Administration, Washington, D.C.
11. http://www.varmintal.com/aengr.htm
12. LS-DYNA Keyword User's manual (2007) Livermore Software Technology Corporation (LSTC). Volume 1, version 971

Chapter 21
Impact of Aluminum, CFRP Laminates, Fibre-Metal Laminates and Sandwich Panels

Shengqing Zhu and Gin Boay Chai

Abstract In the present paper, a thorough experimental investigation for the impact behavior has been conducted for panels of four different materials, namely monolith aluminum alloy (AL), Carbon Fiber-Reinforced Laminates (CFRP), Fiber-Metal Laminates (FML) and Composite Sandwich Panel of CFRP with nomex honeycomb core (CSP). The details of impact response, energy-absorption, stiffness, and failure modes are discussed and presented. The results of the experimental investigations, the finite element analysis and an analytical model are included in this paper. The ductility and impact resistance of AL, CFRP, FML and CSP plates are discussed and compared. CSP and FML panels are found to have good impact resistance and ductility. Failure mechanism of FML under impact loading is given in some details.

Keywords Impact dynamics • Composites • Failure mode • Fibre-metal laminates • Sandwich panel

21.1 Introduction

The components of the structures in aircraft and aerospace components are preferred to be made of light-weight materials such as fiber-reinforced plastic (FRP), sandwich panel and lightweight metals due to their weight efficiency. Fiber-metal laminate (FML) is a new structural material consisting of metal sheets with FRP layers. FML combines the advantage of lightweight of FRP with the advantage of high impact resistance of aluminum alloy and thus is exploited in the aerospace and aircraft industry [1].

During service, structures may be subjected to impact by foreign objects such as runaway debris and dropping of tools. In the category of low-velocity impact, the dynamic response of the structure is of great importance. The key reason is that the contact duration is sometimes too short for the entire structure to respond to the impact and thus more elastic strain energy can be absorbed. Another reason is that the material, being ductile or brittle, gets stiffer when subjected to loading of high strain rate [2, 3]. The impact damage of metals is easy to detect since the damage is initiated with yielding at the impacted surface. In contrast, damage in composites often begins below the impacted surface and usually in the form of an internal delamination [4, 5]. As a result, the impact response and damage in composites still need to be resolved.

21.2 Experimental Set-up

Four types of panels were fabricated as shown in Table 21.1 and used to conduct the impact tests on Instron Dynatup 8250. All specimens have the same size (100 × 100mm) but of different thicknesses as shown in Table 21.1. GLARE is one type of FML and GRP in the table means glass reinforced plastics.

S. Zhu • G.B. Chai (✉)
School of Mechanical and Aerospace Engineering, Nanyang Technological University, 50 Nanyang Avenue, Singapore 639798, Singapore
e-mail: mgbchai@pmail.ntu.edu.sg

E. Patterson et al. (eds.), *Composite Materials and Joining Technologies for Composites, Volume 7: Proceedings of the 2012 Annual Conference on Experimental and Applied Mechanics*, Conference Proceedings of the Society for Experimental Mechanics Series 44, DOI 10.1007/978-1-4614-4553-1_21,

Table 21.1 Configuration of tested specimens and impact velocity

Specimen	Thickness (mm)	Areal density (kg/m^2)	Lay-up	Velocity (m/s)
Al 1100-H14	2.00	5.40	–	0.802,2.31
CFRP	1.75	1.93	[0/0/90/0/0/45/-45]s	2.31
CFRP	1.75	1.93	[45/0/0/90/0/0/-45]s	2.31
CFRP	1.75	1.93	[45/0/0/90/0/-45/0]s	2.31
CSP	17.5	3.88	[0/90/0/90/0]$_s$ 13:15*	1.14, 2.11, 2.20, 2.27, 2.31
CSP	22.5	4.05	[+45/-45/0/90/0]s 3:20	1.65
GLARE1	0.970	1.95	Al/G.E./Al*	3.08
GLARE2	0.973	1.95	Al/G.E./Al	3.14
GLARE3	0.964	1.95	Al/G.E./Al	3.14

Note: 13:15 means a honeycomb core with cell size 13 mm and core thickness 15 mm

21.3 Experimental and FE Results

All the tests were simulated using a commercial finite element software ABAQUS/Explicit. The finite element results were compared with experimental results in this section. For the four types of panels, the variation of contact force with displacement of the impactor was divided into three stages (Fig. 21.1):

1. *Elastic stage*: the response in this stage is nearly linear elastic in which contact force is proportional to displacement with transverse stiffness K1. This stage ends with initial threshold force P_1 at u_1.
2. *Plasticity and/or facesheet damage and/or core buckling stage*: the response the second stage is non-linear. For Al panels, laminate panels and FML, this stage begins with localization around the contact area. For sandwich panel, indentation happens in this stage due to core bucking. This stage ends with initial threshold force P_m at u_f. In this paper, the slope of the first stage and an approximate slope of the second stage K_1 ,K_2 is adopted

$$K_1 = \frac{P_1}{u_1} \quad ,K_2 = \frac{P_m - P_1}{u_f - u_1} \tag{21.1}$$

3. *Unloading stage or fracture stage*: the impactor is sprung back by panels and the contact force drop dramatically to zero provided that there is no penetration over the target. If the impact energy is large enough to perforate the panels, the impactor may not be rebounded and the contact force will decrease sharply at the point of maximum load.

The impact response and damage of the panels will be presented sequentially in the following sections for AL, CFRP, CSP and FML panels.

21.3.1 AL Panels

The results of both the experiment and FEA were plotted in Fig. 21.2. The numerical results agree very well with the experimental results. The curve of load-displacement is consistent with the three-stage diagram in Fig. 21.1 and the stiffnesses in the first two stages are summed in Table 21.3. Aluminum is well-known for its super ductility during deformation [6, 7]. In present tests, the AL panels undergo large deflection up to twice of its thickness without fracture. As shown in Fig. 21.2b that most kinetic energy was absorbed during plastic deformation.

21.3.2 CFRP Panels

Laminates with three different lay-ups were tested and, the loading curve is approximated to linear. The experimental and finite element results are plotted in Fig. 21.3. The numerical results agreed quite well with the experimental results and the transverse stiffnesses were calculated as shown in Table 21.2.

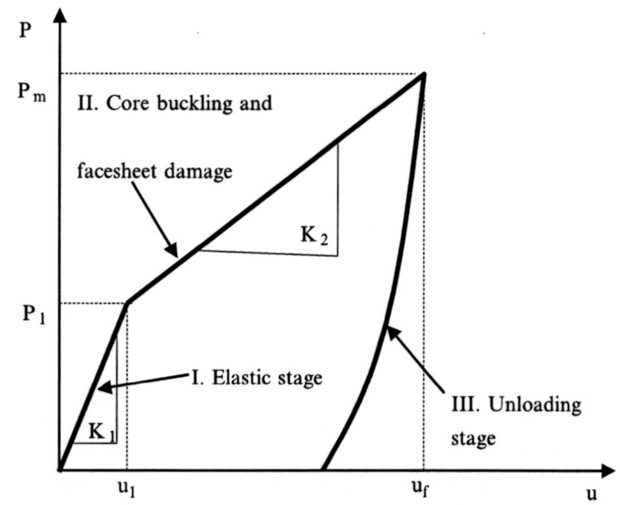

Fig. 21.1 Schematic diagram of the three stages for panels subjected to low velocity impact

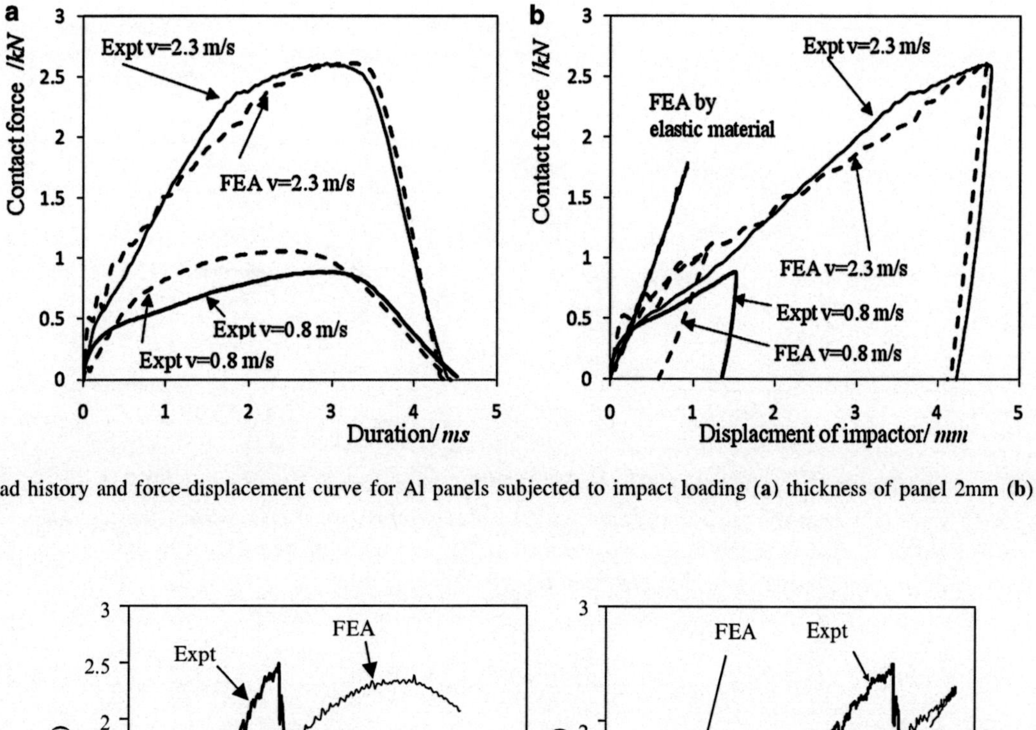

Fig. 21.2 Load history and force-displacement curve for Al panels subjected to impact loading (**a**) thickness of panel 2mm (**b**) thickness of panel 3mm

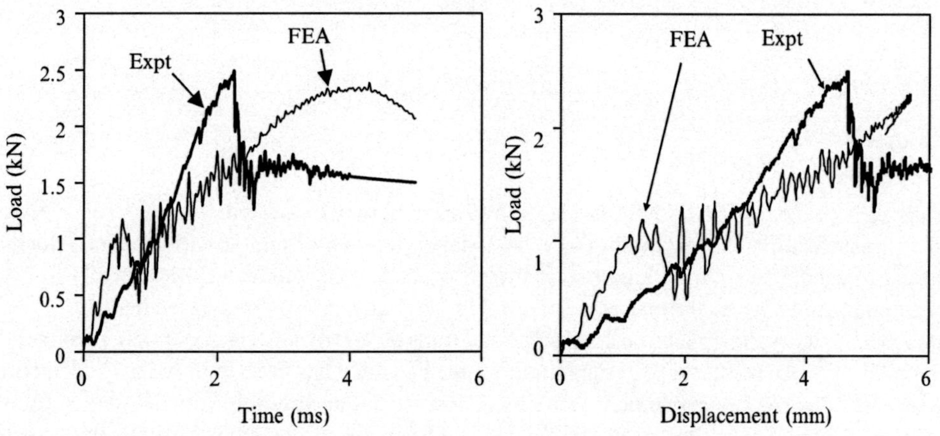

Fig. 21.3 Contact force history and force-displacement curve for layup [0/0/90/0/0/45/-45]s

Table 21.2 Transverse stiffness of laminates impacted by velocity of 2.31m/s

Layup	Maximum force (kN)	Displacement (mm)	Transverse stiffness (MN/m)
[0/0/90/0/0/45/-45]s	2.45	4.66	0.542
[45/0/0/90/0/0/-45]s	2.30	4.58	0.501
[45/0/0/90/0/-45/0]s	2.22	4.36	0.508

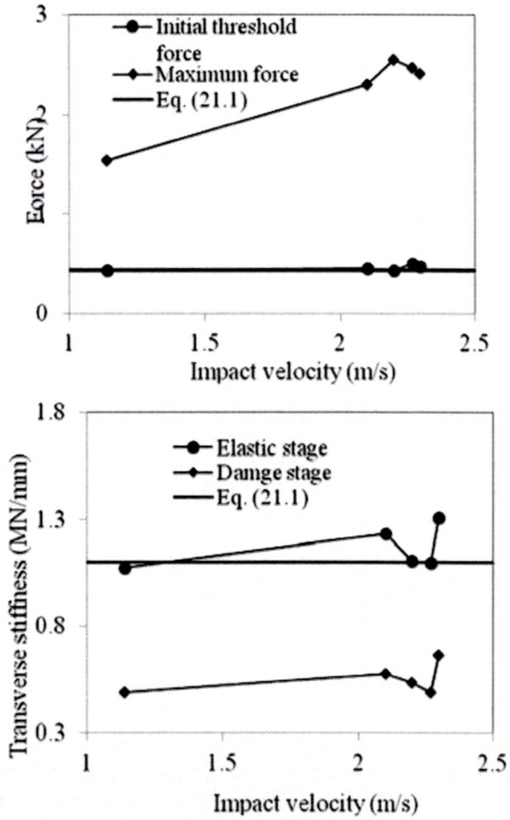

Fig. 21.4 Initial threshold force, maximum force and stiffness with impact velocity for sandwich panel $[0/90/0/90/0]_s$ 13:15

For laminates with different layup, the maximum forces and displacements are close to each other. Laminates with lay-up [0/0/90/0/0/45/-45]s not only bear the maximum force but also provide the largest transverse stiffness. Thus they have the best impact resistance compared with the other two laminates. The combination of +45/−45 plies in laminates has been known to improve the impact resistance, which has been reported by experimental results [5].

21.3.3 CSP Panels

For composite sandwich panels of [0/90/0/90/0]s 13:15, five specimens were tested with different impact velocities. The variation of initial threshold forceP_1, maximum force, and slopes for stage I and II with impact velocities are plotted in Fig. 21.4. The initial threshold force and transverse stiffness for stage I are predicted theoretically in [8]. The theoretical prediction has been compared with experimental results in Fig. 21.4 and shows good agreement.

The curves in Fig. 21.4 shows the initial threshold force and transverse stiffness are accurately predicted by the formula in Eq. 21.1. Larger impact velocity results in larger maximum force P_2 which has been reported both by theoretical method [8] and experimental result [9] while impact velocity has little effect on initial threshold force P_1 and stiffnessK_1 (Fig. 21.5).

Transverse stiffness K_1 and K_2 are illustrated in Table 21.3. The stiffness of sandwich panels in two stages were slightly increased with increasing impact velocities except the one specimen, which was also reported [9, 10].

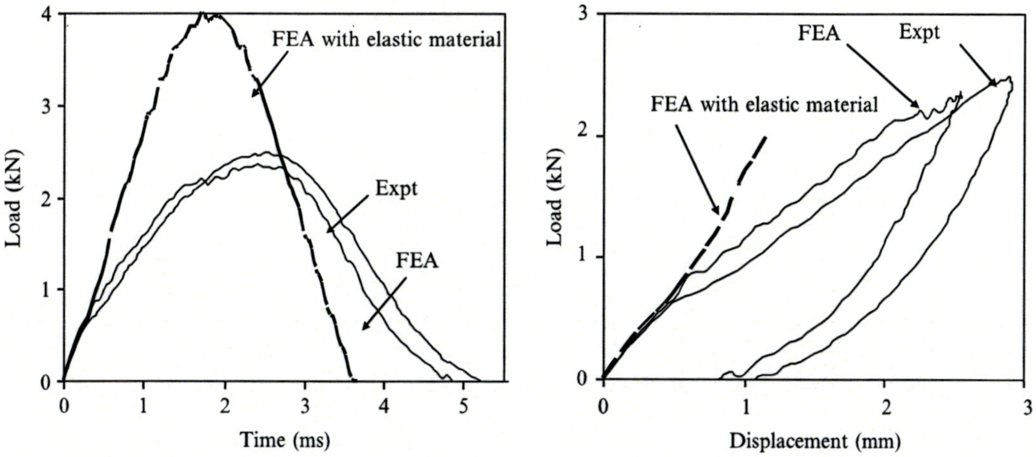

Fig. 21.5 Load history and load versus displacement for sandwich panel [+45/−45/0/90/0]s 3:20 impacted by 1.65 m/s

Table 21.3 Transverse stiffness of Al and CSP panels impacted with various velocities

Specimens	Impact velocity (m/s)	K_1(MN/m)	K_2 (MN/m)
Al	0.802	1.44	0.374
1100-H14	2.31	1.44	0.508
CSP	1.14	1.11	0.386
	2.11	1.26	0.565
	2.20	1.11	0.536
	2.27	1.12	0.586
	2.31	1.35	0.653

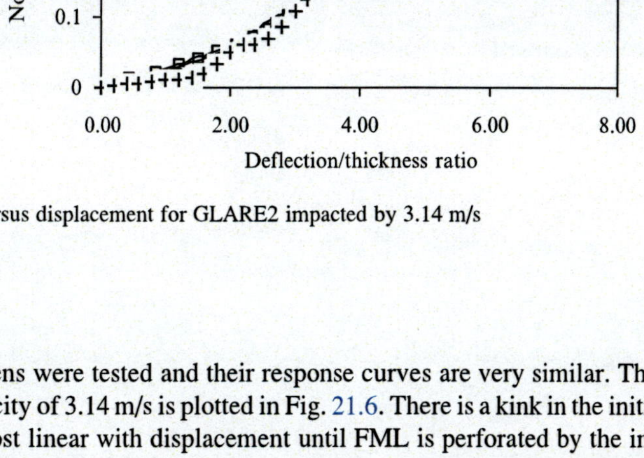

Fig. 21.6 Load history and load versus displacement for GLARE2 impacted by 3.14 m/s

21.3.4 FML Panels

In total, three GLARE specimens were tested and their response curves are very similar. The load-displacement history of GLARE2 impacted with a velocity of 3.14 m/s is plotted in Fig. 21.6. There is a kink in the initial stage of the curve; after that, the contact force increase almost linear with displacement until FML is perforated by the impactor. The response of FML subjected to impact loading has been predicted [11] by introduction of a non-dimensional force $p = P/(2\pi R h_0 \sigma_y)$, where P is the contact force, R is the radius of contact and $h_0, \sigma_y$are the thickness of one layer and yield stress of aluminum alloy.

The prediction of the contact force with displacement is also plotted in Fig. 21.6 and was compared with the experimental results. The failure modes of GLARE2 and GLARE3 are well-predicted and illustrated in Fig. 21.7. The deformation and damage evolution of GLARE2 are plotted graphically in Fig. 21.8. With increasing deflection and contact pressure by

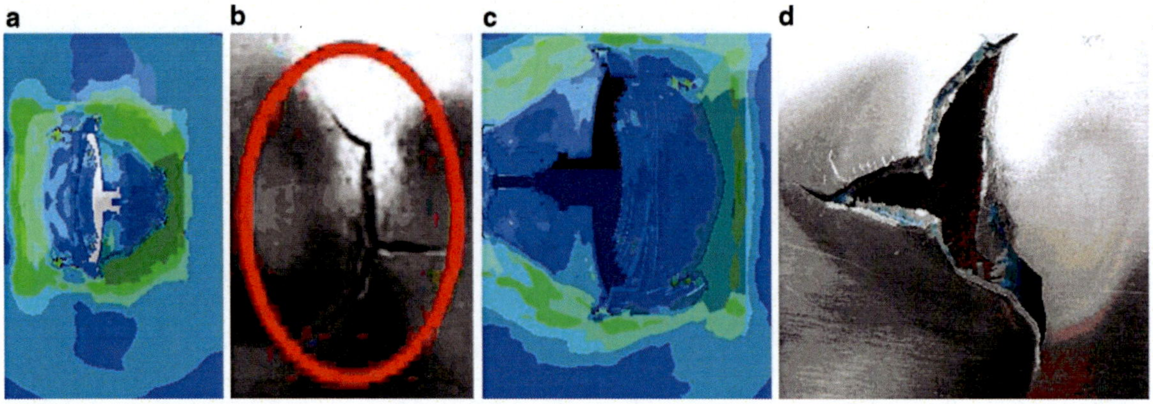

Fig. 21.7 Comparison of failure for specimens GLARE2 and GLARE3 FEA and experimental results: (**a**) FEA for B2; (**b**) Experiment for B2; (**c**) FEA for B3; (**d**) Experiment for B3

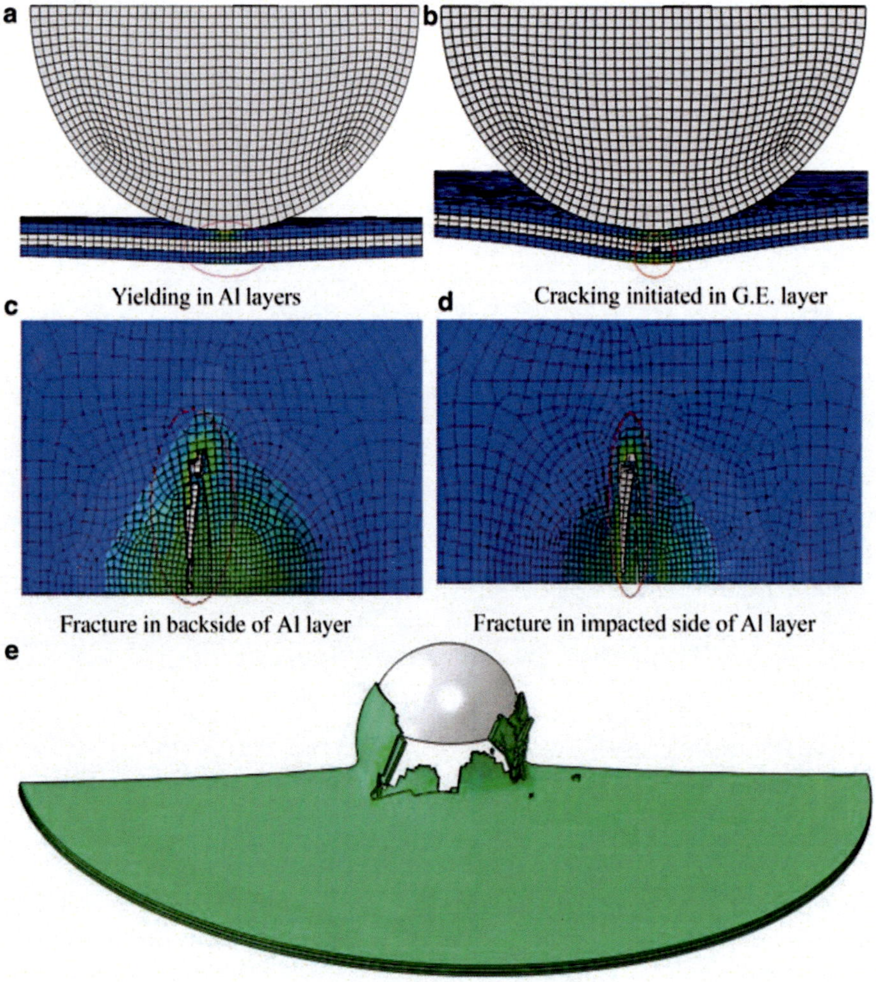

Fig. 21.8 Deformation and damage evolution of GLARE2 subjected to impact of velocity 3.14 m/s. (a) Yielding initiated in Al layer at 0.13 ms (*cut view*), (**b**) Cracking initiated in G.E. layer at 0.38 ms (*cut view*), (**c**) Cracking initiation in Al layer in back side at 1.7 ms(*bottom view*), (**d**) Cracking initiation in Al layer in top side at 1.8 ms(*top view*), (**e**) Graph of FML after perforation at 5 ms with half shown

impactor, the top AL layer in contact yields at first. This is followed by fracture in GRP layer. Then, the yield region increases and tensile fracture emerges in the bottom AL layer. In final stage, the impactor penetrates the aluminium layers and FML panel fails as a structure. During impact, the FML panel undergoes deflection up to six times of its thickness, thus having very good ductility [12].

21.4 Conclusion

A thorough experimental investigation of the impact properties of AL, CFRP, FML and CSP plates under transverse impact loading was carried out and their results were presented and discussed. These four different panels are commonly found in lightweight structures such as those found in the aircraft and aerospace application.

AL panel is very ductile and therefore also better at impact resistance. Most of the impact energy is absorbed by large plastic deformation in and around contact region. However, aluminum panel gives maximal area density. CFRP panel is the best in lightweight but inferior when impacted. The localized damage in and around contact region is critical and sometimes resulted in perforation. FML panels combined the advantages of ductility in aluminum alloy and the lightweight nature of fiber reinforced plastics. This combination dilutes the disadvantages of both materials. Damage in FRP is sometimes difficult to detect but it is not the case for FML since the first layer is made of metal with ductility. The damage of FML under impact is usually initiated by aluminum yielding at contact region followed by fracture in FRP layers due to large strain. Cracking in aluminum layers means the final perforation of FML panels.

In the tests, the bending stiffness of FML plates are found to be far less than that of the CS since the thickness of FML is far less than the height of CSP. The impacted area density and impact resistance of CSP are superior. CSP are lightweight and they can absorb energy not only from overall bending but also from indentation after core buckling. Therefore, CSP is the best choice to resist dynamic loading bearing in mind the constraint in height will be an issue.

FML and CSP responded with superior ductility as compare to CFRP panels and they have high strength to resist impact loading condition. The use of FML as facesheets in sandwich construction seems to be a promising combination for future research study.

References

1. Zhou G, Hill M (2007) Investigation of parameters governing the damage and energy absorption characteristics of honeycomb sandwich panels. J Sandw Struct Mater 9(4):309–342
2. Villanueva GR, Cantwell WJ (2004) The high velocity impact response of composite and FML-reinforced sandwich structures. Compos Sci Technol 64:35–54
3. Daniel IM, Wernera BT, Fennera JS (2011) Strain-rate-dependent failure criteria for composites. Compos Sci Technol 71(3):357–364
4. Stronge WJ, Matemilola SA (1994) Rate effect for impact damage initiation in CFRP laminates. In: 4th international conference on mechanical and physical behaviour of materials under dynamic loading, Paris, Journal de Physique IV, pp 225–230
5. Richardson MOW, Wisheart MJ (1996) Review of low-velocity impact properties of composite materials. Compos Part A 27:1123–1131
6. Hampson P, Moatamedi M (2007) A review of composite structures subjected to dynamic loading. Int J Crashworthiness 12(4):411–428
7. Jones N, and JK Paik (2012) Impact perforation of aluminium alloy plates. International Journal of Impact Engineering. 48:46–53.
8. Iqbal MA, Gupta NK (2008) Energy absorption characteristics of aluminum plates subjected to projectile impact. Lat Am J Solids Str 5(4):259–287
9. Foo CC (2008) Energy absorption characteristics of sandwich structures subjected to low-velocity impact. Nanyang Technological University, Singapore
10. Chai GB, Zhu S (2011) A review of low-velocity impact on sandwich structures. Proceedings of the Institution of Mechanical Engineers, Part L, Journal of Materials: Design and Applications, 225(4):207–230.
11. Zhu S, Chai GB (2012) Analytical investigation on low-velocity impact response of fibre-metal laminates. Composite structures
12. Tsartsaris N, Meo M, Dolce F (2011) Low-velocity impact behavior of fiber metal laminates. J Compos Mater 45(7):803–814

Chapter 22
Impact Monitoring in Aerospace Panels via Piezoelectric Rosettes

Francesco Lanza di Scalea, Hyonny Kim, Sara White, Zhi M. Chen, Salvatore Salamone, and Ivan Bartoli

Abstract This paper investigates the monitoring of impact damage in aircraft panels by the measurements of the resulting Acoustic Emissions (AEs) by Piezoelectric Rosettes. It has been recently demonstrated that the Piezoelectric Rosettes are essentially the dynamic equivalent of the Electrical Resistance Strain Gage Rosettes if certain conditions between the AE wavelength and the gage dimensions apply. Therefore, the Piezoelectric Rosettes can extract the principal strain angle of the AE wave, that is the wave propagation direction, without knowledge of the wave velocity. By intersecting two directions, the location of the AE source can be determined by a minimum of two Rosettes. Since no wave velocity is required, the AE source location is more accurate than conventional time-of-flight triangulation for complex structures where the wave velocity changes along different propagation directions (anisotropic materials) or along each propagation direction (tapered or layered geometries). The Piezoelectric Rosettes designed here are comprised of highly-flexible Macro-Fiber Composite (MFC) Piezotransducer patches that are conformable to curved surfaces and more durable than monolithic PZT sensors because they are less brittle. In this paper, the MFC Piezoelectric Rosettes are applied to monitoring the progression of damage during "blunt" impact tests of two curved carbon/epoxy composite panels with co-cured stringers and mechanically fastened shear-ties that are highly representative of aircraft fuselage structures. The damage progression was monitored satisfactorily without knowledge of the wave velocity in the panels and outside of certain time windows where the Rosette signals were saturated by the large impact loads. Additional impact tests are being planned to continue the development of this innovative impact monitoring technique.

Keywords Piezoelectric rosette • Lamb waves • Acoustic emission location • Fiber-reinforced composite materials

22.1 Introduction

Impacts (either at high velocity such as from hail ice, birds or projectiles, or at low velocity such as from ground equipment) are a realistic threat to exposed aircraft structures such as fuselage and wing panels [1]. The extensive use of laminated polymer composites in the design of aerospace structures has caused an increased attention to damage due to impacts because induced damage can often be barely visible in these materials.

Monitoring impact damage usually starts with locating the point of impact and/or the potential subsequent damage. In traditional Acoustic Emission (AE) monitoring, the impact or damage location is calculated from time-of-flight triangulation of elastic wave measurements taken at multiple receiving points. These waves are dispersive guided modes that are well supported by the plate-like geometries of aerospace fuselage and wing panels. One problem with time-of-flight triangulation

F. Lanza di Scalea (✉) • H. Kim • S. White • Z.M. Chen
Department of Structural Engineering, University of California San Diego, 9500 Gilman Drive, La Jolla, CA 92093-0085, USA
e-mail: flanza@ucsd.edu

S. Salamone
Department of Civil, Structural and Environmental Engineering, State University of New York at Buffalo,
212 Ketter Hall, Buffalo, NY 14260, USA

I. Bartoli
Department of Civil, Architectural and Environmental Engineering, Drexel University,
Curtis Hall, 3141 Chestnut St, Philadelphia, PA 19104, USA

E. Patterson et al. (eds.), *Composite Materials and Joining Technologies for Composites, Volume 7: Proceedings of the 2012 Annual Conference on Experimental and Applied Mechanics*, Conference Proceedings of the Society for Experimental Mechanics Series 44,
DOI 10.1007/978-1-4614-4553-1_22, © The Society for Experimental Mechanics, Inc. 2013

is the dispersive nature of the flexural guided wave mode (the dominant one from impacts) which makes it difficult to identify an accurate arrival time. Another, more severe problem with time-of-flight triangulation is the required knowledge of the wave velocity in the test material. The velocity is required to translate arrival time measurements into source location. This requirement is a fundamental limitation when monitoring anisotropic or geometrically-complex structures (e.g. aerospace panels), where the velocity is dependent upon propagation direction, $v = v(\theta)$. In addition, in the case of tapered or multilayered structures, the wave velocity can also change along a given propagation direction because of dispersion. Alternative AE source location approaches have therefore being proposed, including model-based approaches [2], artificial neural networks [3, 4], optimization routines [5, 6], and Piezoelectric Rosettes [7, 8].

Besides impact location, another desired outcome of an impact monitoring system is the determination of the time history of the impact force magnitude (impact force identification). Since impact force thresholds causing damage in composites are well known to the aircraft designers, knowledge of the impact force magnitude and duration (particularly the peak amplitude) can indicate whether an impact event is harmless or has caused damage to the structure [9]. The Piezoelectric Rosette technique is also being applied to the impact force identification.

The present paper aims at contributing to the area of impact monitoring in aerospace structures on the front of impact damage location. The first part of the paper quickly reviews the main concept of AE source location by Piezoelectric Rosettes. The second part of the paper summarizes the results of the monitoring of damage progression during "blunt" impact tests of two curved CFRP panels highly representative of aircraft fuselage structures.

22.2 Basics of Piezoelectric Rosettes for Acoustic Emission Source Location

It was previously demonstrated [7] that the voltage sensitivity of a rectangular ($l \times b$) MFC sensor, modeled as an orthotropic lamina in plane stress conditions, to a harmonic guided-wave strain field (such as an a_0 Lamb wave in a plate), can be written in a manner analogous to conventional Electrical Resistance Strain Gages (ERSGs). Accordingly, for $\lambda \geq l$(sensor length smaller than wavelength), the MFC sensor sensitivity can be decomposed into a longitudinal sensitivity factor, S_1, and a transverse sensitivity factor, S_2, as:

$$S \approx S_1 \cos^2\theta + S_2 \sin^2\theta \tag{22.1}$$

where θ is the wave propagation direction relative to the sensor's lengthwise direction, l, and S_1 and S_2 are given in Ref. [7]. By using strain transformation relations, the sensor's voltage response to strain becomes:

$$\overline{V} \approx S_1\,\bar{\varepsilon}_{11} + S_2\,\bar{\varepsilon}_{22} \tag{22.2}$$

where the wave strain components along the longitudinal and the transverse sensor directions, *1* and *2* respectively, are explicitly indicated. Equation 22.2 makes the MFC response formally equivalent to that of conventional ERSGs and hence rosette configurations can be considered.

Therefore, given the measurement of three independent strains from the response of three MFC sensors in a rosette configuration, it is possible to calculate the Cartesian strain components and the principal strain angle θ using well-known ERSG rosette reduction approaches. The principal strain angle of the wave, θ, measured from the global axis x (Fig. 22.1b), can be computed as:

$$\tan 2\phi = \frac{\bar{\gamma}_{xy}}{\bar{\varepsilon}_{xx} - \bar{\varepsilon}_{yy}} \tag{22.3}$$

Evaluation of the source location in a planar structure is readily achieved by the intersection of the principal directions determined by two rosettes (Fig. 22.1a).

22.3 Matlab© Acoustic Emission Source Location Script

A Matlab© script was written for the automatic location of AE sources from the MFC Rosettes. The script first loads the data collected by up to four Rosettes. It then applies time gates and passband filters the signals with a Butterworth filter of order 4. The upper frequency of the passband filter (HF) is determined based on the wavelength tuning behavior of the individual

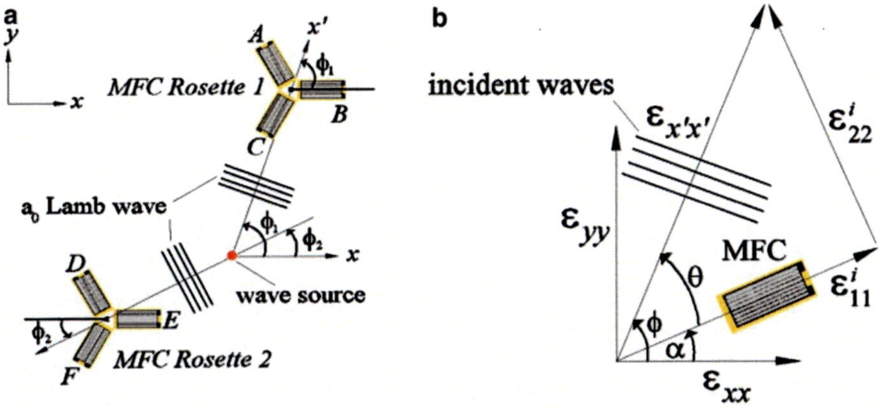

Fig. 22.1 (a) Concept of Piezoelectric Rosettes for wave source location in a plane, (b) reference system for the calculation of the Cartesian strains of the wave from the sensor's response

MFC sensors comprising the rosettes (with the main requirement being to use a frequency such that the smallest wavelength is equal to or larger than the largest dimension of the sensors). For the 14 mm × 7 mm MFC sensors used in the experiment tests discussed in the following section, the HF value was 100 kHz. The choice of the low frequency value (LF) is more flexible. The Matlab© program further features a rather sophisticated "event detection" scheme that uses two levels of controls to detect true damage events and discard artifacts such as secondary wave reflections from the panel edges. The first level of control imposes a minimum time delay (ΔT_{min}) between two successive AE events that prevents the detection of possible edge reflections following the damage events. The second level of control ensures that at least two Rosettes detect the same event, i.e. an AE hit with an amplitude exceeding a pre-determined threshold (THRSD) and arriving at the two Rosettes within a maximum time interval (ΔT_{max}).

The following parameters are among those given in input to the Matlab© program:

1. Number, position and orientation of up to four Rosettes;
2. Time gates for signals;
3. Low – and high-frequency cutoff values for passband Butterworth filter (LF and HF);
4. Amplitude threshold for signals to retain (THRSD);
5. Minimum time interval between two successive damage events (ΔT_{min});
6. Maximum time interval for same event detected by any two Rosettes (ΔT_{max});
7. Damage visualization parameters (size of panel, positions of shear ties and stringers).

Upon running, the program provides the user with a menu of choice to select different combinations of Rosettes to be used for the damage location. Once a selection is made, the program outputs graphs of the detected locations of damage, superimposed to a geometrical representation of the test panel, at four different time windows. Currently, the program takes approximately 45 s to run on a regular Personal Computer (Intel Core i5 CPU at 2.53 GHz) using 12 channels (four Rosettes of three sensors each) and 750,000 points in each channel acquired at a 2.5 MHz simultaneous sampling rate.

22.4 Experimental Tests and Results of Impact Damage Location

An extensive experimental testing program is being conducted to validate the AE source location technique on composite panels subjected to both projectile impacts and "blunt" impacts. This paper presents the results of the "blunt" impact tests conducted on two curved (carbon fiber reinforce plastic) CFRP panels. Specifically, "Test Panel 1" was a 1 m × 1 m curved panel with three hat-shaped co-cured stringers and L-shaped shear-ties (brackets) bolted using aerospace countersunk "shear head" fasteners (Hi-Lok) – Fig. 22.2a. The panel was instrumented with four Piezoelectric Rosettes, each comprised of three, d_{31} (type P2) MFC piezoelectric sensors (14 mm × 7 mm) arranged in a Delta configuration (relative angles of 120°)-Fig. 22.2b. A minimum of two Rosettes are needed for the damage location – the additional rosettes were meant to provide redundant data points.

The impactor (rubber bumper from ground equipment – Fig. 22.2c) contacted the specimen at the center of the panel, above the central stringer. The load history plot (Fig. 22.3) suggests that the impactor indented elastically for the first 200 ms,

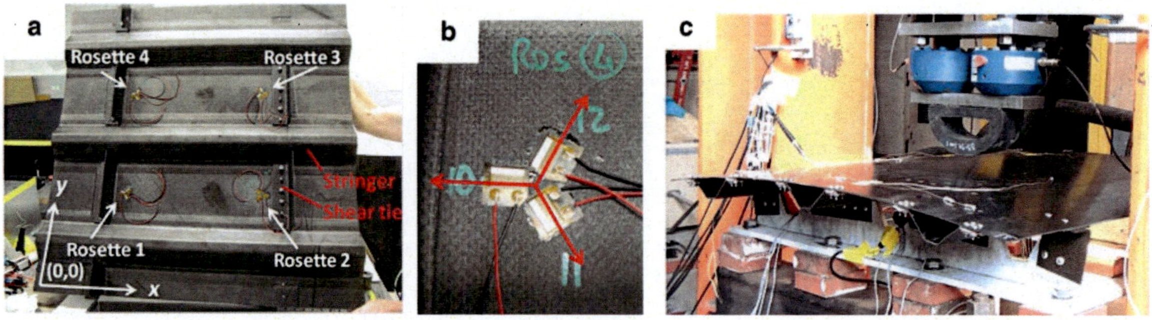

Fig. 22.2 (**a**) The curved CFRP Test Panel 1 with co-cured stiffeners and bolted shear-ties instrumented with the MFC Piezoelectric Rosettes. (**b**) Zoom of one of the Rosettes. (**c**) Setup of the "Blunt" Impact Test 1

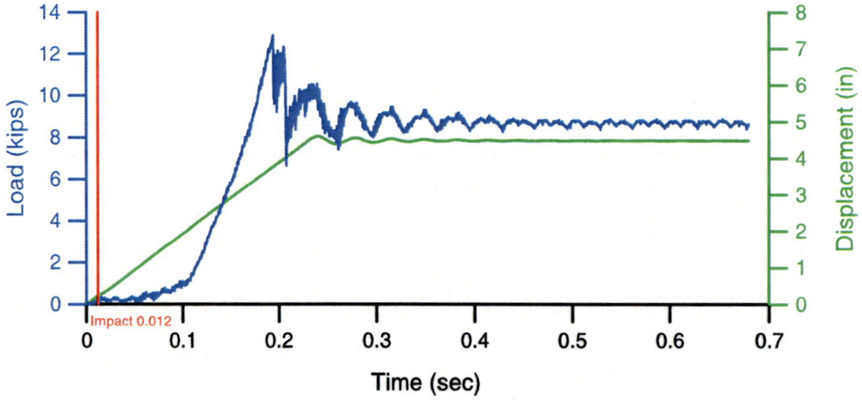

Fig. 22.3 Load and displacement time histories recorded from the impactor in Blunt Impact Test 1

followed by major damage (through-thickness cracking) with complicated rebounds. The large waves produced between 120 and 220 ms saturated the readings from the MFC sensors and made it impossible to detect damage progression in this time window. Nevertheless, several damage locations were detected outside of the saturated window.

The Matlab© results are shown in Fig. 22.4, where the damage locations ("+" symbols) determined from six rosette pairs (locations at triangle symbols) are plotted for four time windows (respectively: 150, 180, 280 and 300 ms from the beginning of the test – 0 s in Fig. 22.3). The Matlab© figure also shows the position of the stringers and shear-ties. The frequency band used for these results was 50–100 kHz. It can be seen that almost all of the damage occurred at the expected stress concentration areas, i.e., at the stringer-to-skin and shear-tie-to-skin interfaces. Within the time windows monitored, damage was seen to initiate at the central stringer, progressing to the top stringer, and then coalescing around the central area of the panel around the impactor location. Owing to the rosette technique, this result did not use any information on the material properties of the panel (e.g. wave speeds).

A second "blunt" impact test was conducted next. Shown in Fig. 22.5, "Test Panel 2" was the same material as Test Panel 1, but it had two, instead of three co-cured stringers. The other properties were the same as the previous panel (1 m × 1 m curved panel, bolted shear-ties, etc.). The four MFC Delta Rosettes were bonded at different locations than the previous test, as shown in the Fig. 22.5a.

The impact load and displacement time histories for this second test (Fig. 22.6) indicate that contact was maintained between 100 and 1,100 ms, with a peak impact load at around 200 ms. Figure 22.7 shows the Matlab© results for this test, obtained with the bandpass filter set to 40–100 kHz. The time windows considered in the plots are: 150, 180, 190, and 300 ms from the beginning of the test (0 s in Fig. 22.6). Some signal saturation occurred again between 200 and 230 ms hence this time window was not considered in the results by using appropriate time gates in the signals. Since one of the sensors in Rosette 4 lost a connection during the test, the results are shown using all pair combinations from three Rosettes. It can be seen that damage initiated at early times in the central region of the upper stringer. The damage then progressed horizontally along the upper stringer and into the top left shear-tie, with some isolated events detected in the bottom stringer. Again, this result did not use any information on the material properties of the panel.

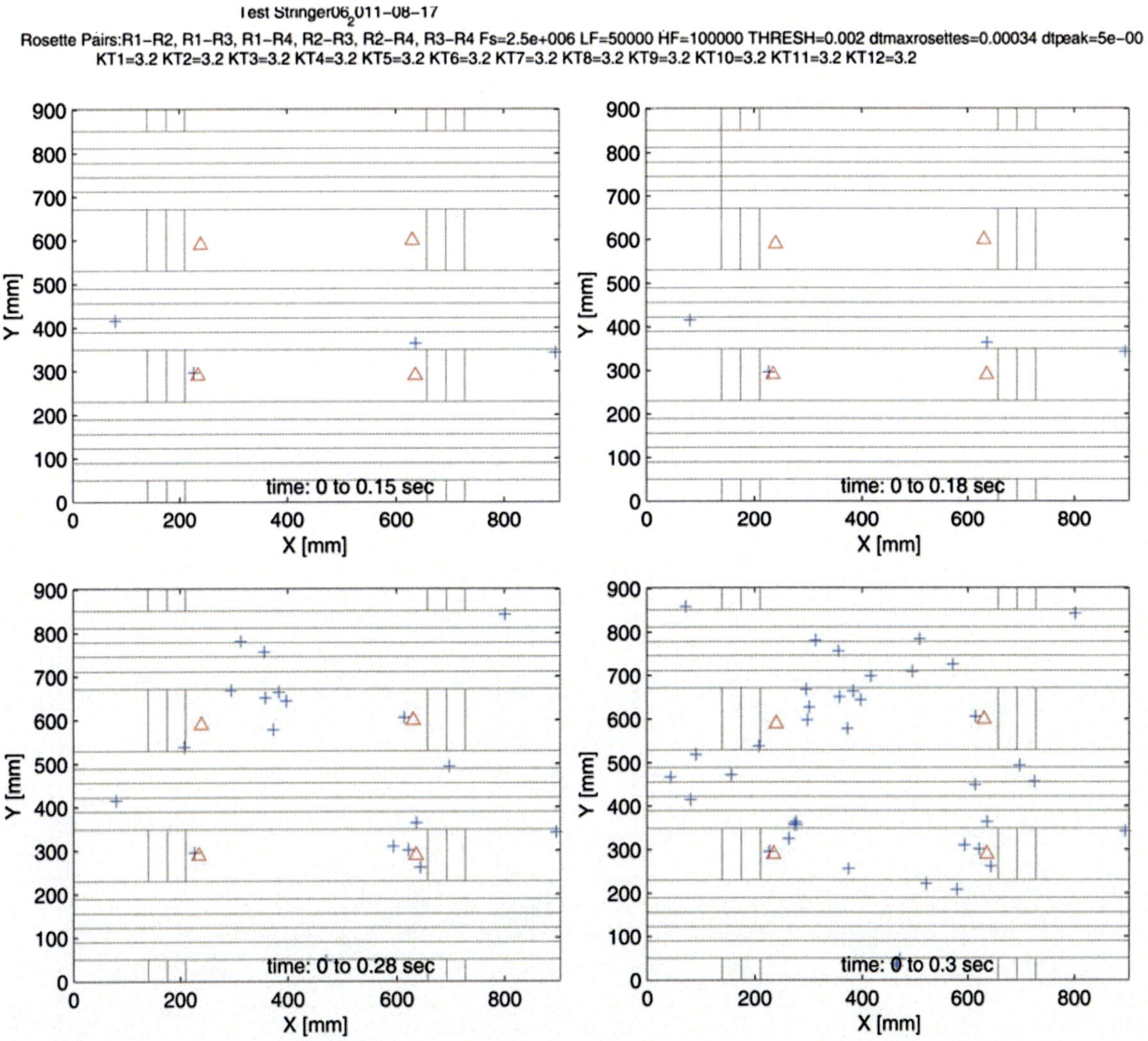

Fig. 22.4 Damage progression monitoring with MFC Rosettes in Test Panel 1 under "blunt" impact test. Output of Matlab© script for automatic damage visualization. The "+" symbols are damage locations

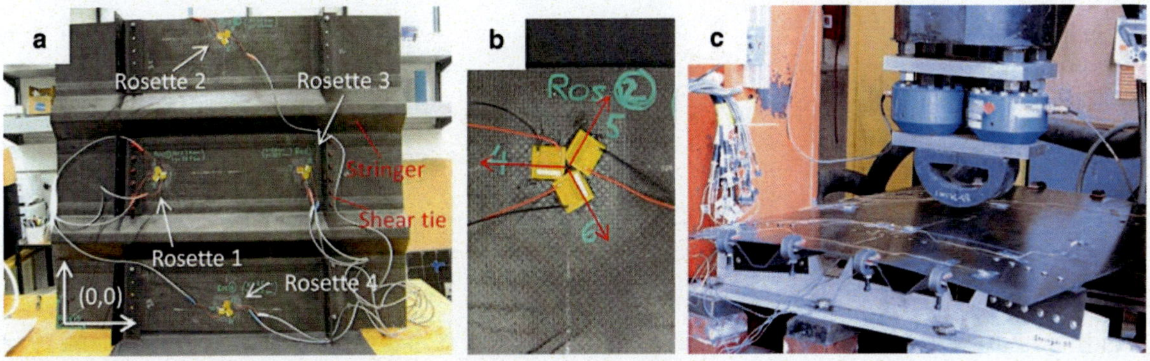

Fig. 22.5 (**a**) The curved CFRP Test Panel 2 with co-cured stiffeners and bolted shear-ties instrumented with the MFC Piezoelectric Rosettes. (**b**) Zoom in one of the rosettes. (**c**) Setup of the "Blunt" Impact Test 2

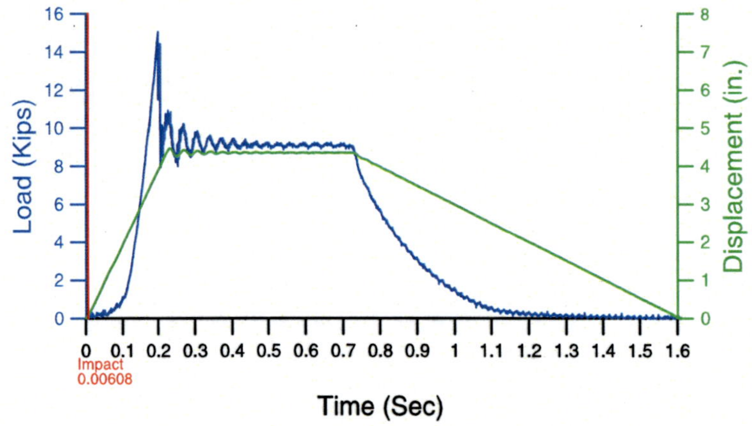

Fig. 22.6 Load and displacement time histories recorded from the impactor in Blunt Impact Test 2

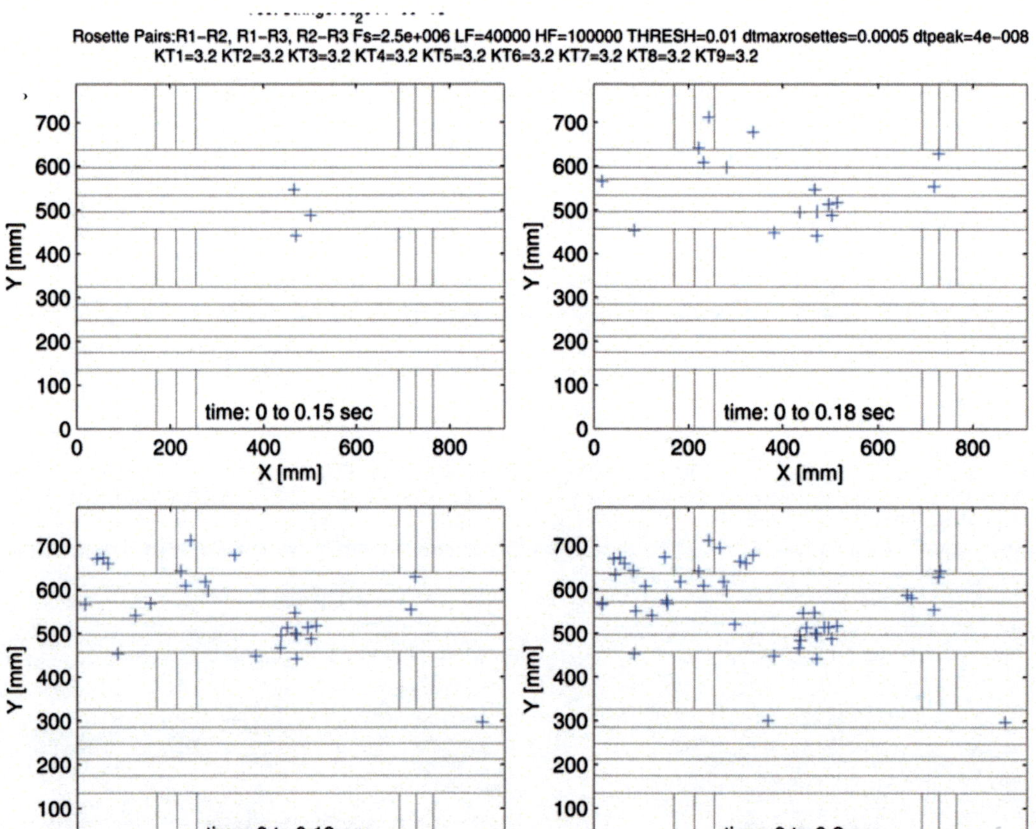

Fig. 22.7 Damage progression monitoring with MFC Rosettes in Test Panel 2 under "blunt" impact test. Output of Matlab© script for automatic damage visualization. The "+" symbols are damage locations

22.5 Conclusions

This paper has demonstrated the location of AE sources generated during "blunt" impact tests of aircraft composite panels by the Piezoelectric Rosette technique. The use of Piezoelectric Rosettes allows to determine the location of the AE source without knowledge of the material properties or wave speeds. This is particularly beneficial in the case of anisotropic structures, where the wave velocity changes with changing wave propagation direction, or tapered and multilayered structures, where the wave velocity changes along a given propagation direction. The Piezoelectric Rosettes used are comprised of highly-flexible Macro-Fiber Composite Piezotransducer patches that are conformable to curved surfaces and more durable than monolithic PZT sensors because they are less brittle. The Piezoelectric Rosette technique was here applied to monitoring the progression of damage in two, 1 m × 1 m curved CFRP panels with co-cured stringers and fastened shear-ties subjected to "blunt" impacts simulating those from aircraft ground maintenance equipment. The results indicated plausible progression of damage around the expected stress concentration areas, i.e., at the stringer-to-skin joints and at the shear ties-to-skin joints. The tests also showed some signal saturation at the highest load levels for which damage location was not possible. Additional tests are being planned on a larger curved panel where particular care will be taken to avoid signal saturation by maximizing the dynamic range of the data acquisition system.

In addition to impact/damage location, the Piezoelectric Rosette technique is being developed for the identification of the impact force, whose knowledge can lead to an informed decision-making following an impact event. The impact force identification approach is based on a Semi-Analytical Finite Element model of wave propagation [10].

Acknowledgments This work was funded in part by the US Air Force Office of Scientific Research under STTR Phase II project "Real-time In-situ Impact and Damage Locator in Anisotropic Aerospace Structures" with Dr. David Stargel as AFOSR Program Manager. Any opinions, findings and conclusions or recommendations expressed in this article are those of the authors and do not necessarily reflect the views of the US Air Force Office of Scientific Research.

References

1. Bartels G (1997) Aircraft structural health monitoring, prospects for smart solutions from a European viewpoint. In: Proceeding of international workshop on structural health monitoring, Stanford, pp 293–300
2. Chang F-K, Markmiller JFC (2006) A new look in design of intelligent structures with SHM. In: Proceeding of 3rd European workshop on SHM, Granada, pp 5–20
3. Staszewski WJ, Worden K, Wardle R, Tomlinson GR (2000) Fail-safe sensor distribution for impact detection in composite materials. Smart Mater Struct 9:298–303
4. Jones RT, Sirkis JS, Friebele EJ (1997) Detection of impact location and magnitude for isotropic plates using neural networks. J Intell Mater Syst Struct 7:90–99
5. Kundu T, Das S, Martine SA, Jata KV (2008) Locating point of impact in anisotropic fiber reinforced composite plates. Ultrasonics 48:193–201
6. Kundu T, Das S, Jata KV (2009) Detection of the point of impact on a stiffened plate by the acoustic emission technique. Smart Mater Struct 18:1–9
7. Matt HM, Lanza di Scalea F (2007) Macro fiber composite piezoelectric rosettes for acoustic source location in complex structures. Smart Mater Struct 16:1489–1499
8. Salamone S, Bartoli I, Di Leo P, Lanza di Scalea F, Ajovalasit A, D'Acquisto L, Rhymer J, Kim H (2010) High-velocity impact location on aircraft panels using macro-fiber composite piezoelectric rosettes. J Intell Mater Syst Struct 21:887–896
9. Kim H, Welch DA, Kedward KT (2003) Experimental investigation of high velocity ice impacts on woven carbon/epoxy composite panels. Compos A 34:25–41
10. Bartoli I, Marzani A, Lanza di Scalea F, Viola E (2006) Modeling wave propagation in damped waveguides of arbitrary cross-section. J Sound Vib 295(3–5):685–707

Chapter 23
Using Experimental Data to Improve Crash Modeling for Composite Materials

Morteza Kiani, Hirotaka Shiozaki, and Keiichi Motoyama

Abstract Accurate simulation of the composite material crash tubes subjected to axial impact is a challenging field of study in automotive or aerospace industry; however, analytical prediction of the crashworthiness behavior in composite materials is limited. In this paper, three different analytical approaches are presented which have been used to study the crashworthiness of a pultruded glass-polyester tube. The first model is established based on the single shell elements. This approach is very effective, when composite part is assembled in the full structure. However, this technique can be used when the experimental result is available. In the second approach, the crash tube is modeled by using multi-layered shell element (delamination model). Relying on coupon test information of the composite material, this modeling technique can provide reasonable result for the energy absorption of the tube. The third modeling approach is looking for crashworthiness prediction of the discussed tube by using the first model which its parameters are tuned based on the result of the second model. Finally, the sensitivity of the result is studied by changing the major parameters in the first model. This paper is looking for finding a method to reasonably estimate the crashworthiness behavior in the composite materials.

23.1 Introduction

Extensive research in the recent decades has shown that stable progressive crushing and high specific energy absorption can be achieved by fiber reinforced composite tubular structures. Most of the composite materials absorb crash energy through a combination of fracture and friction [1]. Indeed, the fracture due to compressive load extends in the composite material part through extensive micro cracks in the matrix and fibers in a short time. Moreover, fiber-matrix debonding, delamination and internal ply separation are resulted by compressive loading. These characteristics in composite materials increase energy absorption compared to the metalic structure [2, 3]. In automotive industry, to design auto body not only the strength to weight ratio but also high energy absorption should be considered. These two criteria motivate designers to replace metallic crash rail boxes with the composite material counterparts [4]. However, there is little doubt that the analytical methods are capable to predict the post-failure response of the composite material structures. Actually, as discussed before, composite material structure subjected to impact loading fail in several complex modes, so the energy-absorption characteristics in the composite material cannot be easily predicted [5]. In addition to failure mechanisms, the crash response of the most composite material parts is highly influenced by other parameters such as: geometry, fiber and matrix type, volume fraction, manufacturing process, lay-up, angle of impact, and the impact velocity. These factors have made a lot of limitations in the capability of the current software to predict or simulate the crushing behaviors of the composite materials. Today, the major commercial software that can be used in crash simulations include: LS-DYNA, ABAQUS Explicit, PAM-CRASH,

M. Kiani (✉)
Center for Advanced Vehicular System, Computational Engineering Program, Mississippi State University, Starkville, MS 39762, USA
e-mail: mk532@msstate.edu

H. Shiozaki
Mitsubishi Motors Corporation, 1 Nakashinkiri, Hashime-cho, Okazaki Aichi Prefecture 444-8501, Japan
e-mail: hirotaka.shiozaki@mitsubishi-motors.com

K. Motoyama
Center for Advanced Vehicular System, Mississippi State University, Starkville, MS 39762, USA
e-mail: keiichi@cavs.msstate.edu

E. Patterson et al. (eds.), *Composite Materials and Joining Technologies for Composites, Volume 7: Proceedings of the 2012 Annual Conference on Experimental and Applied Mechanics*, Conference Proceedings of the Society for Experimental Mechanics Series 44,
DOI 10.1007/978-1-4614-4553-1_23,

DYTRAN, and RADIOSS. Many studies show that these series of commercial software can simulate the crash response in the composite materials well [5–10] when the experimental data of crash response exist. To date, major studies have been assigned to compare the result of the crash simulation with the experimental data [6, 11–13]. However, by using some modeling techniques, a few researchers have started new studies to overcome the barriers of the crashworthiness prediction in composites. For example, Mamalis et al. [14, 15] studied the crash behavior of square composite tubes subjected to static and dynamic axial compression. In addition, Feraboli and his colleagues parametrically studied the crash simulation for a corrugated specimen Carbon Fiber Reinforced Plastic (CFRP) under quasi-static loading [16, 17]. They used LS-DYNA to simulate the crashworthiness of the composite materials for both quasi-static and dynamic impact. LS-DYNA developed by Livermore Software Technology Corp. (LSTC) is a multi-purpose explicit and implicit finite element program used to analyze the nonlinear response of both metallic and composite materials structures. This software by using fully automated contact algorithms, wide range of material models, and different element formulations enables users to solve impact problems especially in composite materials. Modeling the structure by using one single shell elements or multi-layered shell element are the two major approaches that the user can apply in the crashworthiness analysis in the composite materials. These two approaches and the capabilities which have been mentioned above give enough capacity to LS-DYNA to predict a reasonable behavior of the crash response for the composite material.

By using the techniques mentioned above, this paper studies the crashworthiness of the pultruded glass-polyester tube subjected to axial impact [7]. At the first step, the crash simulation has been performed based on the single shell elements. This model can be implemented for any complex geometry when the crashworthiness of the part or full structure is needed. However, using this method is limited because the crash results should exist to tune the parameters in the model. In the second approach, the tube has been modeled by using multi-layered shell element. This model results a reasonable crash response for the tube even the crash test data are not available. Finally, as the goal of this study, the paper presents a modeling technique to predict the crashworthiness behaviors of the composite tube. Actually, the third model has been constructed based on the mesh of the first model and applicable parameters of the second approach. In addition, as the first modeling approach needs the crash test data, the results of the second approach has been exploited as the target to tune the parameters in the third model. So, the third model can be used as the model that can predict the crashworthiness behavior of the discussed tube. The results of all approaches are compared to the experiment discussed in the literature [7]. Finally, the sensitivity of the major modeling parameters of the first approach is studied. This study realizes the effect of each parameter in the model, and helps to take a step in standardization of the crash simulation for the composite materials.

23.2 Material and Definition of the Impact

A circular cross section tube made of pultruded glass-polyester is subjected to impact by a mass of 68.85 kg and initial velocity of 9.3 m/s. The lay-up is $[0]_{6T}$ yielding an average laminate thickness of 3 mm, and the crash front of the tube has been chamfered by 45°. The outer diameter and the length of this tube are 50 mm and 220 mm, respectively. All material properties, taken from literature [18, 19], are listed in Table 23.1.

23.3 Finite Element Modeling Details

23.3.1 First Modeling Approach

In the first approach, the discussed tube is considered as a single shell, and the mesh is generated based on using element formulation of the Belyscko-Tsay quadrilateral. In this step, checking the direction of the elements is recommended. Most of the time, if the direction of the elements are aligned to the crushing direction, the accuracy of the simulation result will be enhanced. As Fig. 23.1a shows, two rows of elements are considered to be the trigger in this model. The thickness of the front

Table 23.1 Material properties for the pultruded glass-polyester

v_f	ρ (kg/m^3)	E_x(GPa)	E_y(GPa)	ν_{xy}	G_{xy} (GPa)
48.3	1,850	31.20	9.36	0.29	3.20
σ_{XX}^{T}**(MPa)**	σ_{YY}^{T} **(MPa)**	σ_{XX}^{C} **(MPa)**	σ_{YY}^{T}**(MPa)**	τ_{xy}^{u}**(MPa)**	**Interlaminar Shear Stress (MPa)**
483.0	34.9	409.0	92.2	73.3	55.0

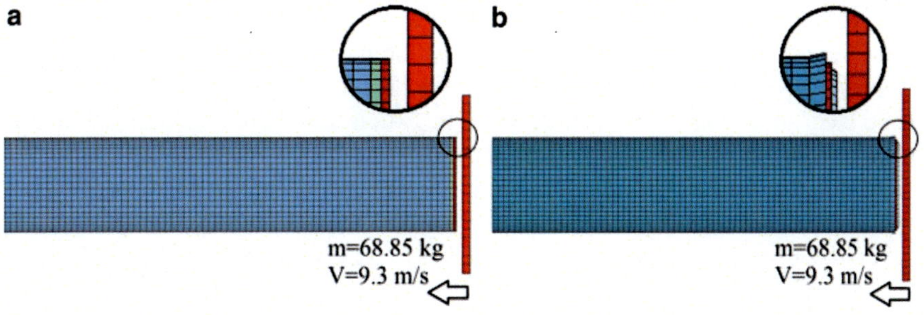

Fig. 23.1 (**a**) Finite Element (FE) model of the crash tube in first approach, (**b**) FE Model of the crash tube in the second approach

trigger is 0.5 mm, and the other trigger has the thickness of 2 mm. Usually, using two rows of elements as trigger portion prevents catastrophic breakage of the model. The mesh size of the model is another important issue in impact simulation. Many literatures show that the crash simulation result is highly influenced by the mesh size [20]. Improper mesh size not only brings instability but also induces errors in results. In this paper, the size of the elements for the FE model was considered equal to the thickness of the part, and this consideration yields 3,404 elements in the FE model. Usually, if the element size is considered to be equal to the thickness of the part, this issue yields improvement in the accuracy of the crash simulation for the composite materials.

Material model is another important parameter in the crash simulation. In LS-DYNA, there are several material models that can be used in the simulation of the crash in composite materials [21]. MAT54 is one of the best material models in LS-DYNA which is suitable for progressive failure analysis in the composite material [5, 22], of course, when the composite part is modeled by shell element. This material model can be used in simulation of the progressive failure in the composite materials. The parameters in this material model can be considered by two groups.

The first group includes the parameters which are related to the stiffness and strength of the composite materials. Implementation of coupon test provides enough information for the first group material parameters. The second group is related to the input parameters that cannot be acquired by implementing coupon test, but they can be found by trial and error fashion. SOFT, FBRT, and YCFAC are the major parameters which can be obtained just by trial and error. To find the proper value for these parameters, the user should have the crash test result and try to match the simulation result with the experiment by changing these parameters. This issue makes hard to predict the crashworthiness behavior for the composite material by using simulation techniques. Deleo et al. [23] investigated the effects of the major parameters in MAT54 on the crash response for a sinusoidal specimen made of carbon fiber reinforced plastic (CFRP). This investigation shows the parameter of compressive strength (σ_{XX}^{T}), Compressive Fiber Failure Strain (DFAILC), and SOFT which are the major parameters in the model that completely change the crashworthiness behavior of the composite. Whereas, other parameters in this material model cause instability or they do not have a great influence on crashworthiness behavior. Regardless of the first group of the parameters which can be found by test, this paper investigates the second group of the parameters.

Contact definition between the interacting parts is another important step which is necessary in the crash modeling. Two different contact algorithms are used in the first model. The first contact definition is Automatic_Nodes_To_Surface and it handles the interaction between the impactor and the composite tube. Also, to prevent the penetration of the deformed part boundary by its nodes, the second contact algorithm, Automatic_Single_Surface, is defined just for the elements of the tube.

Usually, finding all data for each input deck can be followed by using default values that are provided by LS-DYNA; however, the user can change most of the parameters until instability is observed in the simulation.

23.3.2 Second Modeling Approach

In the second modeling approach, the composite material tube is modeled using three layers of Belytschko-Tsay shell elements. The thickness for each layer is considered to be 1 mm, equally. As the Fig. 23.1b shows, trigger mechanism can be ignored, and the nodes in the first row of the inner and outer layer are bended toward inside and outside of the tube, respectively. This technique helps the crash simulation being stable and prevents catastrophic behavior.

In this kind of modeling, usually the number of the layers depends on the number of delamination occurred in the experimental crash test. However, the user can assume reasonable layers with considering the required calculation time. Similar to the first modeling approach, MAT54 has been considered as the material model and all the parameters in this card are filled out based on the coupon test; however, the SOFT parameter is the only parameter which its appropriate value can be obtained just by try and error and check the stability of the simulation. Also, Automatic_Nodes_To_Surface and Automatic_Single_Surface are defined for same the purposes which had been discussed previously in the first model. In addition to these two contact algorithms, two other contacts, Automatic_Surface_To_Surface_Tiebreak (option 8), were applied to tie the middle layer to the inner and outer layers. In option 8, three parameters are needed: NFLS and SFLS (normal and shear failure Stresses), and PARAM (critical distance at which the interface failure is complete). In this study, NFLS and SFLS are considered to be equal to interlaminar shear strength. Also, the time step in the second simulation is controlled to be less than 1E-6, similar to the first model.

By using the results of the coupon test and default options in LS-DYNA, most of the input cards have been filled out in the second model. Thus, the second model is constructed without any try and error on tuning the parameters.

23.3.3 A Model to Predict Crashworthiness Behavior of the Glass-Polyester Crash Tube

This model is supposed to predict the crashworthiness behavior of the glass-polyester tube discussed in [7]; moreover, it should be applicable in the full structure crash simulation. To construct this model, the parameters in the first model have been replaced by the similar parameters of the second model. In addition, to tune the parameters in the third modeling approach, the result of the second model has been considered as the reference. In this case, just the parameter of SOFT should be tuned by try and error to obtain the stability for the simulation. As the second model was established by using the coupon test results and the default values for the parameters in keywords, the third model can be considered as a model that can predict the crash response for the tube.

23.4 Results

Usually, the crashworthiness behavior of structures is evaluated by means of plotting Force-Displacement which gives the amount of energy absorbed by the structure. In Fig. 23.2a, Force-Displacement curves found by simulations are compared to the experiment [7], and the amount of energy absorption in each case is shown by Fig. 23.2b.

These results had been found by try and error on SOFT and conducting many simulations. The value of SOFT for each approach are 0.75, 0.905 and 0.72 corresponding to the first, second and third approach. Figure 23.3 shows the second approach provides reasonable failure patterns similar to the experiment [13], and Fig. 23.4 shows the final deformation for both experiment and second simulation.

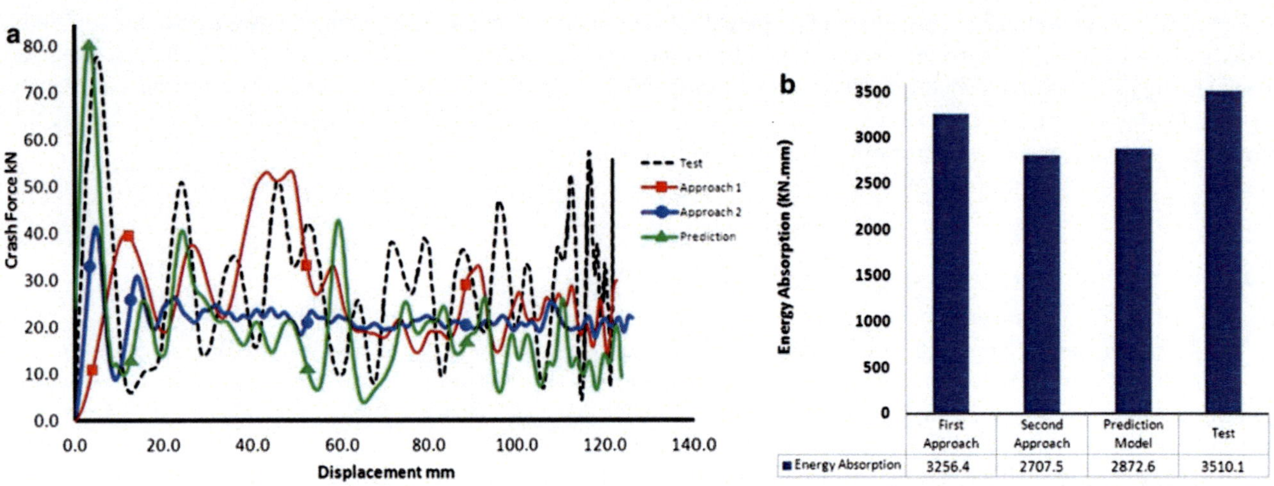

Fig. 23.2 (**a**) Comparison of Force-Displacement curves for experiment and simulations; (**b**) Comparison of energy absorption capacity

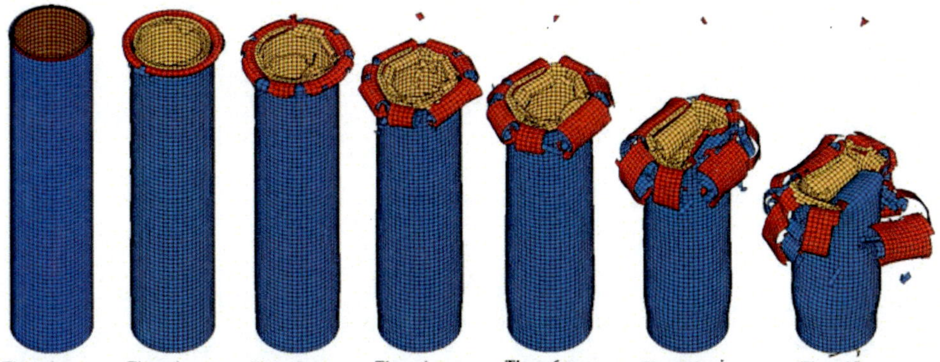

Fig. 23.3 Debonding between layers obtained from the second simulation

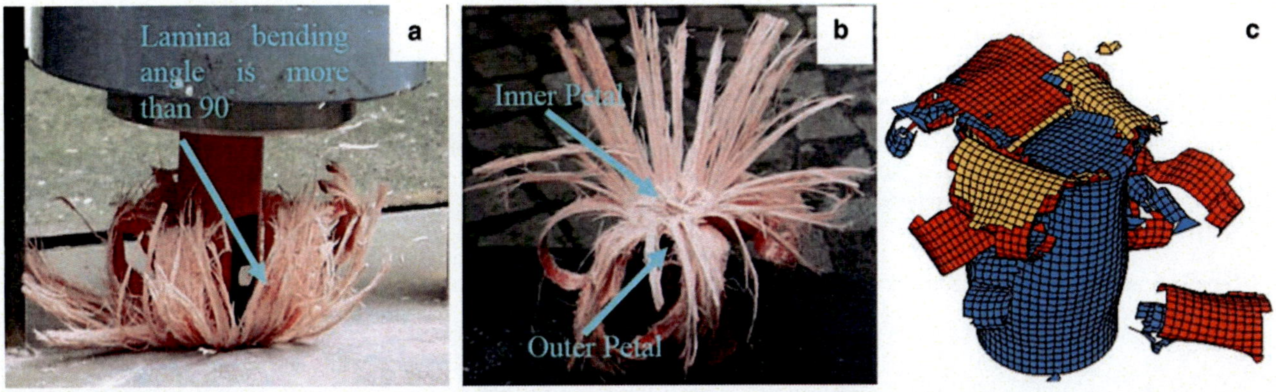

Fig. 23.4 (**a**, **b**) Comparison between the pattern of the deformation obtained by the experiment [13]; (**c**) final deformation resulted by the second modeling approach

23.5 Discussion

Deleo et al. [23] completely discussed about the effects of the major parameters in MAT54 on the crashworthiness behavior of the CFRP sinusoidal specimen under quasi-static impact loading. In this paper, the crashworthiness behavior of the glass-polyester crash tube is studied to investigate the sensitivity of the simulation results to the major parameters in the model. Here, the parameters that have more influence on the results of the first and second modeling approaches have been discussed. First of all, the parameters of MAT54 will be discussed by the following descriptions:

According to literature [21, 24], failure can occur if one of the following strength criteria is met:

Tensile fiber mode (Fiber rupture):

$$\sigma_{11} > 0 \quad then \quad e_f{}^2 = \left(\frac{\sigma_{11}}{\sigma_{XX}^T}\right)^2 + \beta\left(\frac{\sigma_{12}}{\tau_{XY}^U}\right) - 1\begin{cases} \geq 0 & Failed \\ < 0 & Elastic \end{cases},$$
$$E_{XX} = E_{YY} = G_{XY} = \upsilon_{YX} = \upsilon_{XY} = 0 \tag{23.1}$$

Compressive fiber mode (Fiber buckling and kinking):

$$\sigma_{11} < 0 \quad then \quad e_c{}^2 = \left(\frac{\sigma_{11}}{\sigma_{XX}^T}\right)^2 - 1\begin{cases} \geq 0 & Failed \\ < 0 & Elastic \end{cases},$$
$$E_{XX} = \upsilon_{YX} = \upsilon_{XY} = 0 \tag{23.2}$$

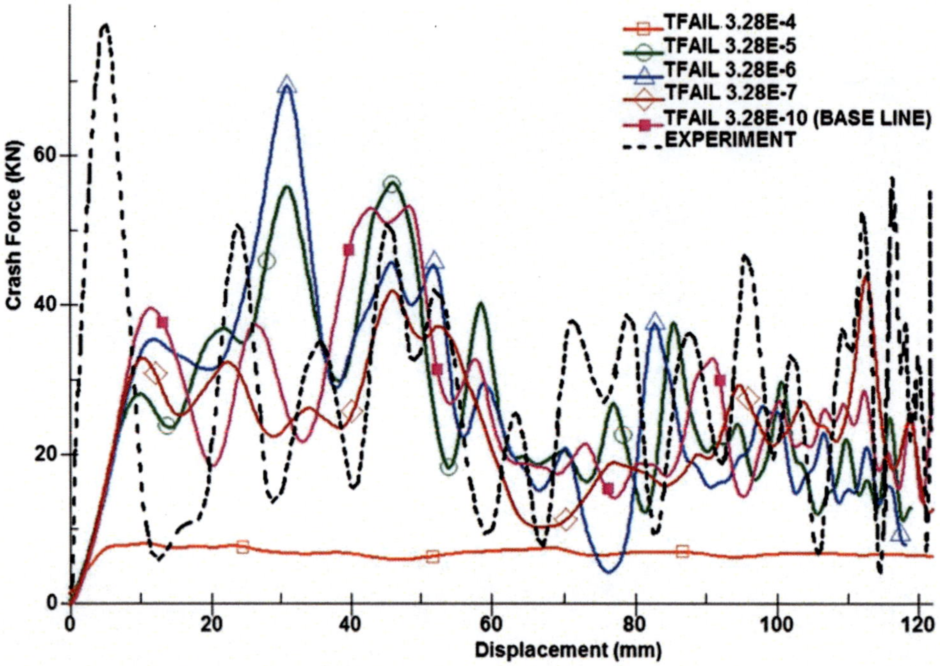

Fig. 23.5 Effect of TFAIL on crashworthiness

Tensile matrix mode (Matrix cracking under transverse tension and shearing):

$$\sigma_{22} > 0 \quad then \quad {e_m}^2 = \left(\frac{\sigma_{22}}{\sigma_{YY}^T}\right)^2 + \beta\left(\frac{\sigma_{12}}{\tau_{XY}^U}\right)^2 - 1 \begin{cases} \geq 0 & Failed \\ < 0 & Elastic \end{cases},$$
$$E_{YY} = \upsilon_{YX} = 0 \xrightarrow{\text{yields}} G_{XY} = 0, \tag{23.3}$$

Compressive matrix mode (Matrix cracking under transverse compression and shearing):

$$\sigma_{22} < 0 \quad then \quad {e_d}^2 = \left(\frac{\sigma_{22}}{2\tau_{XY}^U}\right)^2 + \left[\left(\frac{\sigma_{YY}^C}{2\tau_{XY}^U}\right)^2 - 1\right]\frac{\sigma_{22}}{\sigma_{YY}^C} + \left(\frac{\sigma_{12}}{\tau_{XY}^U}\right)^2 - 1 \begin{cases} \geq 0 & Failed \\ < 0 & Elastic \end{cases},$$
$$E_{YY} = \upsilon_{YX} = \upsilon_{XY} = 0 \xrightarrow{\text{yields}} G_{XY} = 0$$
$$\sigma_{XX}^C = 2\sigma_{YY}^C \quad for\ 50\%\ fiber\ volume. \tag{23.4}$$

In addition to strength characteristics, the most important modeling parameters affecting the crashworthiness behavior of the composite are related to the material model such as: TFAIL, ALPH, BETA, SOFT, FBRT, and YCFAC. These parameters have been changed in the first modeling approach, and the results are shown in Figs. 23.5, 23.6, 23.7, 23.8, 23.9, and 23.10. In the second modeling approach, two strength parameters NFLS and SFLS in the TIEBREAK contact algorithm are changed and their effect on the crashworthiness behavior of the crash tube is shown by the Fig. 23.11.

23.5.1 The Effect of TFAIL Parameter in MAT54

TFAIL is the time step size criterion for element deletion [21]. This parameter in a significant degree determines whether the overall crushing response of the compressed composite tube will be brittle or ductile [6]. In this simulation, the time step

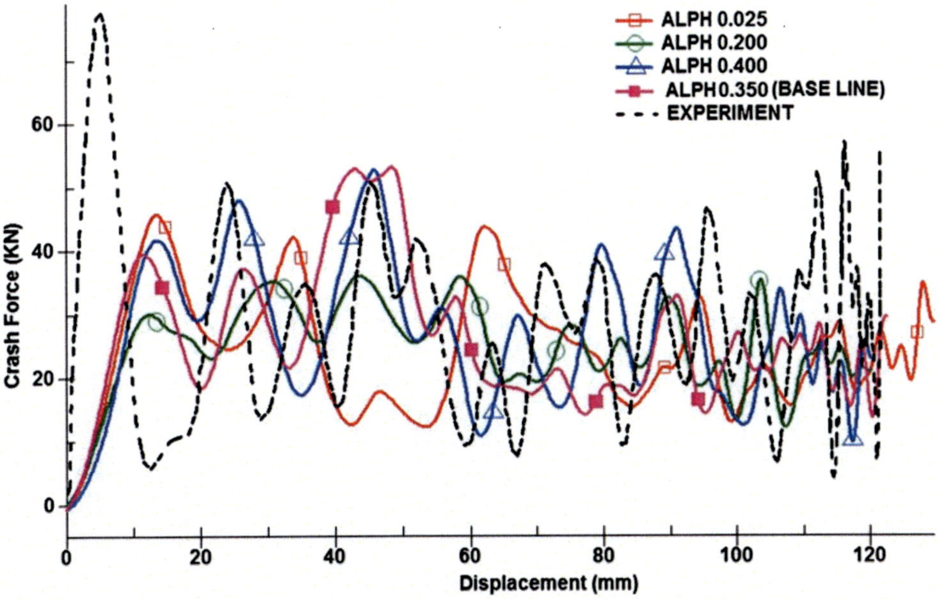

Fig. 23.6 Effect of ALPH on energy absorption of the glass-polyester tube

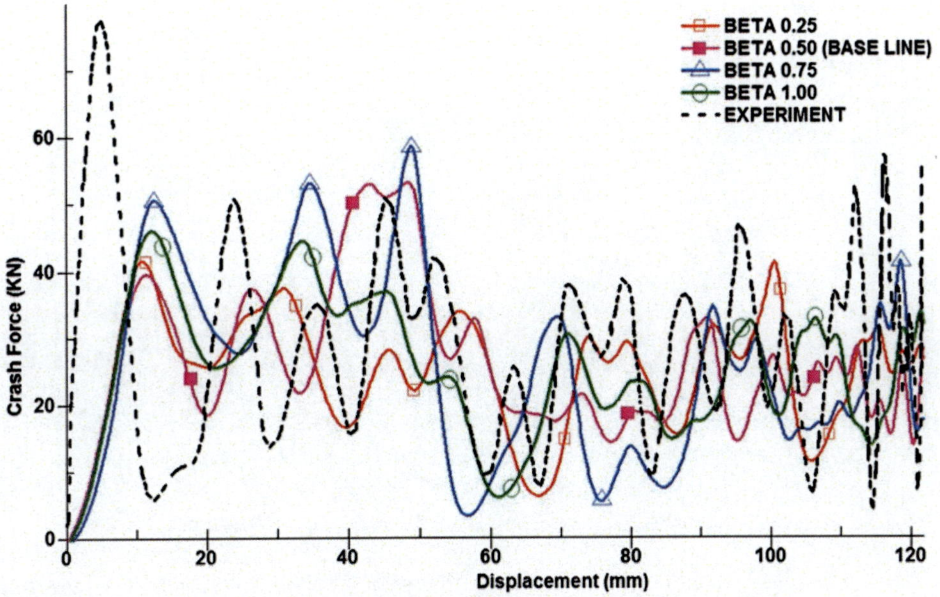

Fig. 23.7 Effect of BETA on energy absorption of the glass-polyester tube

calculated by LS-DYNA is 3.28E-6, and TFAIL for base line simulation is 3.28E-10. To find the effect of this parameter on crashworthiness, different values of TFAIL are applied and the results are compared to the base line model in Fig. 23.5. When TFAIL is larger than 3.28E-3, simulation will be stop due to violation of SOFT. Also, for TFAIL around 3.28E-4, the crash tube shows the behavior such as ductile material. If TFAIL reduces from 3.28E-5 to 3.28E-10, the energy absorption will grow and the stability of the simulation will enhance. This shows that MAT54 induces the behavior of ductility to the composite material when TFAIL is chosen by greater value than the time step calculated by LS-DYNA. The result of simulation does not change when TFAIL is less than 3.29E-10.

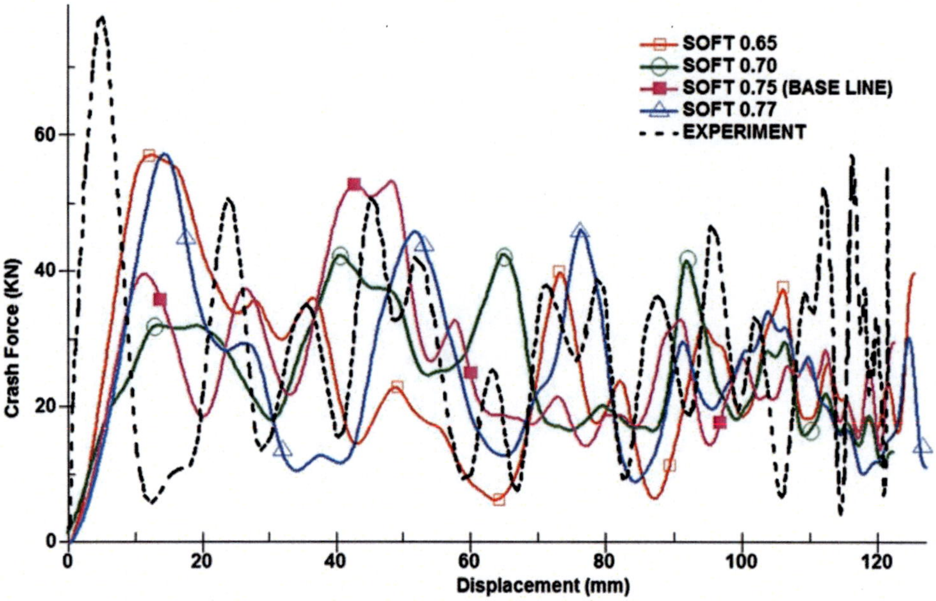

Fig. 23.8 The effect of SOFT on crashworthiness of the pultruded glass-polyester tube

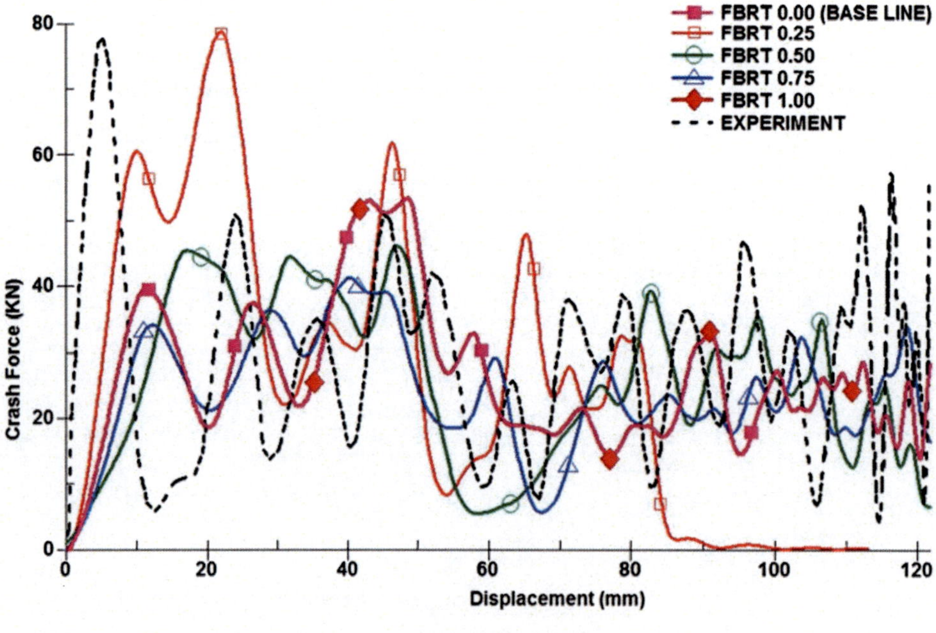

Fig. 23.9 The effect of FBRT on crashworthiness of the pultruded glass-polyester tube

23.5.2 *The Effect of ALPH Parameter in MAT54*

ALPH is the coefficient for nonlinear shear stress term in material model. Based on theory manual of LS-DYNA, in plane stress, the shear strain is given in terms of stress [21]:

$$2\varepsilon_{12} = \frac{1}{G_{XY}}\sigma_{12} + \alpha.\,\sigma_{12}^{2} \tag{23.5}$$

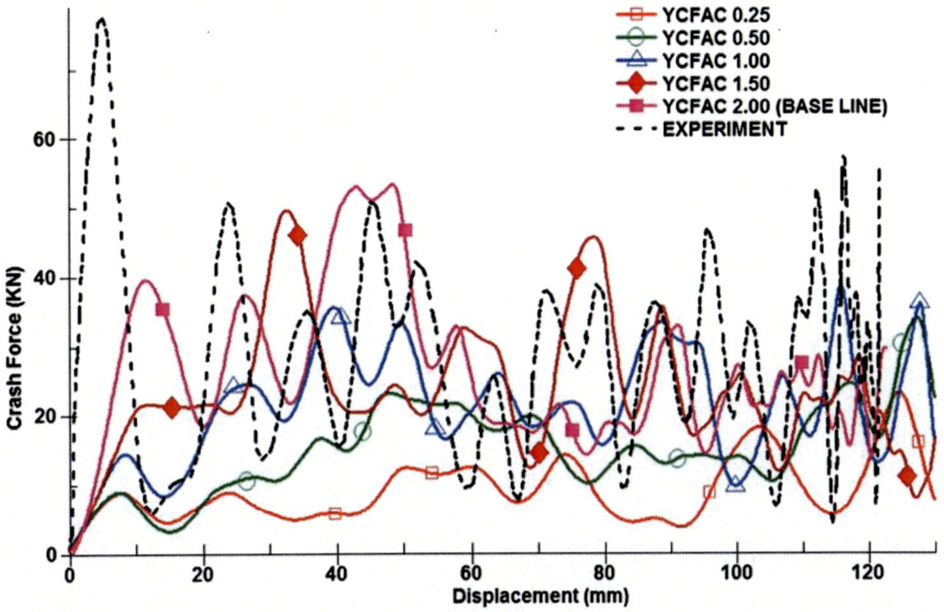

Fig. 23.10 The effect of YCFAC on crashworthiness of the pultruded glass-polyester tube

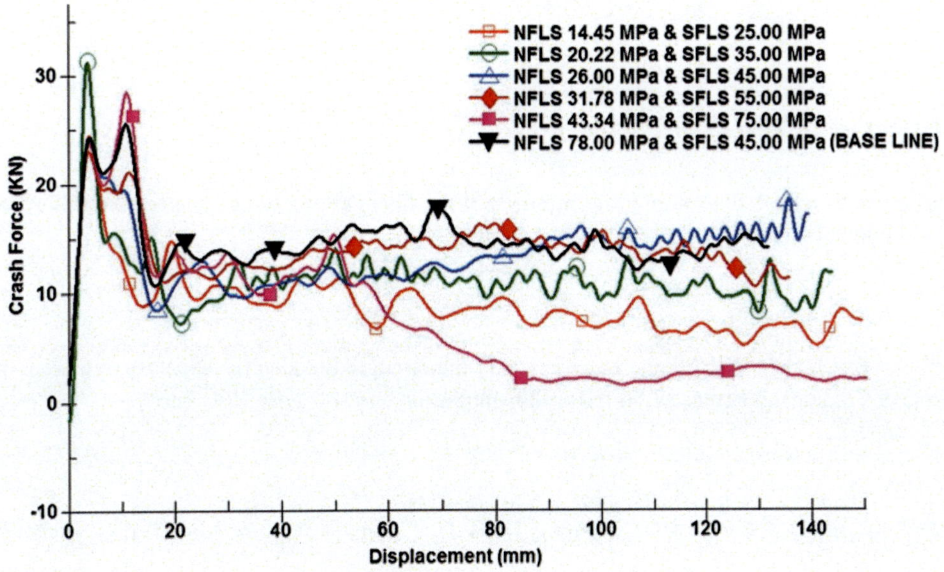

Fig. 23.11 Change in energy absorption due to change in NFLS and SFLS on crashworthiness

Some simulations have been executed by different values of ALPH. The results show that higher ALPH increases the strength of the tube and prevents the overall buckling in the tube. Figure 23.6 shows the effect of ALPH on energy absorption.

23.5.3 The Effect of BETA Parameter in MAT54

BETA is weighting factor for shear term in tensile mode, this parameter takes value between 0.0 and 1.0. In fiber tensile mode, the failure criterion will change to Hashin when BETA is 1.0. If BETA is considered to 0.0, the Eq. 23.1 induces the maximum stress criterion for the failure. Figure 23.7 shows the effect of BETA on the result. This figure shows Hashin criterion (BETA = 1.00) is not a good choice to simulate the pultruded glass-polyester crash tube in this paper. Also, considering maximum stress criterion (BETA = 0.00) cannot address a good result for the crash simulation.

23.5.4 The Effect of SOFT Parameter in MAT54

SOFT is the reduction factor for strength of the material in the crash front zone of the part. Among other parameters in MAT54, SOFT has the significant effect which not only induces instability but also causes overall buckling to the composite material part against impact loading. There is no mathematical way to find SOFT, yet the user performs a few simulations to figure out the maximum value of SOFT without observation of overall buckling or instability. In this study, SOFT has been found 0.750, 0.905, and 0.720 in the first, second and third modeling approaches, respectively. The effect of SOFT on the result of simulations can be seen in Fig. 23.8.

23.5.5 The Effect of FBRT Parameter in MAT54

FBRT reduces the tensile strength of the fiber in the remaining plies once matrix failure takes place. Figure 23.9 investigates the effect of FBRT on Force-Displacement result for the first approach. The study shows that the FBRT = 1.00 and FBRT = 0.00 are same while the lower value for FBRT reduces the energy absorption and brings overall buckling to the tube (FBRT 0.25).

23.5.6 The Effect of YCFAC Parameter in MAT54

YCFAC is the parameter to reduce the compressive strength of the fibers after matrix compressive failure occurs [21]. This reduction can be explained by Eq. 23.6 [21].

$$\sigma_{XX}^{C} = YCFAC.\sigma_{YY}^{C} \tag{23.6}$$

Figure 23.10 shows that higher YCFAC increases not only the strength but also the capacity of the energy absorption of the tube; moreover, this parameter brings stability for the simulation.

23.5.7 The Effect of NFLS and SFLS Parameter in Contact Algorithm Between the Layers

These two parameters define normal and shear failure stresses. In the second modeling approach, Automatic_Surface_To_Surface_Tiebreak (option = 8) is used to tie the layers. This contact algorithm allows the crack propagates over the interface between the layers based on the cohesive zone model, implemented in LS-DYNA as a delamination contact developed by Borg [25]. When damage is initiated, the energy released due to the failure of the interface is approximately defined by Eqs. 23.7 and 23.8 [21]:

$$G_{I_C} = \frac{1}{2}S.C_{Critical} \tag{23.7}$$

where

$$S = \sqrt{\max(\sigma_n, 0)^2 + |\sigma_s|^2} \tag{23.8}$$

By considering σ_n = NFLS and σ_s = SFLS, failure as a crack propagation will occur in the interface. Figure 23.11 shows the effect of the different values NFLS and SFLS on the crashworthiness behavior in the second modeling approach.

Figure 23.11 has been obtained by using Eqs. 23.7 and 23.8 and increasing the value of SFLS from 25 MPa to 75 MPa. These two equations give the value of NFLS and $C_{Critical}$ regarding to GIC and GIIC. The value of GIC and GIIC are 0.412 N/mm and 0.713 N/mm, respectively based on literature [26].

This investigation shows that by choosing SFLS higher than 55 MPa, overall buckling will occur for the tube. Moreover, the value of NFLS has less effect on the instability of the simulation.

23.6 Conclusion

Pultruded glass-polyester crash tube under impact loading has been simulated by using LS-DYNA. Two different approaches have been used to investigate the crashworthiness behavior of the crash tube. First approach, using single layer element, can be matched with the experiment, but this approach cannot be used to predict the crashworthiness behavior of the crash tube.

Second approach uses multi-layered shell element technique to simulate the composite tube under impact loading. In this study, by using multi-layered shell element technique, energy absorption has been calculated by 20 % error, yet the deformation pattern simulated by this method is similar to the experiment. Although the energy absorption calculated by this method is not very close to the experiment, it can be used to establish a prediction model based on the first approach. Actually, the result of the second approach can be assumed as the lowest bound of the energy absorption. Then, the result of the prediction model can be tuned based on the lowest bound created by the second approach. By considering the discussion section which is provided by this paper, the designer can tune the prediction model up to the point that the simulation reaches to instability.

References

1. Thornton PH (1979) Energy absorption in composite structures. J Compos Mater 13:247–263
2. Hamada H, Ramakrishna SA (1997) FEM method for prediction of energy absorption capacity of crashworthy polymer composite materials. J Reinforc Plast Compos 16:226–242
3. Carruthers JJ, Kettle AP, Robinson AM (1998) Energy absorption capability and crashworthiness of composite material structures: a review. Appl Mech Rev 51:635–649
4. Botkin M, Johnson N, Zywicz E, Simunovic S (1998) Crashworthiness simulation of composite automotive structures. In: 13th annual engineering society of detroit advanced composites technology conference and exposition, Detroit
5. Deleo F, Feraboli P (2011) Crashworthiness energy absorption of carbon fiber composites: experiment and simulation.In: SPE automotive composites conference, 98195–2400
6. Mamalis AG, Manolakos DE, Ioannidis MB, Papapostolou DP (2006) The static and dynamic axial collapse of CFRP square tubes: finite element modeling. Compos Struct 74:2213–2250
7. Palanivelu S, Paepegem WV, Degrieck J, Kakogiannis D, Ackeren JV, Hemelrijck DV, Wastiels J (2009) Numerical energy absorption study of composite tubes for axial impact loadings. In: 17th international conference on composite materials (ICCM-17), pp 27–31
8. Haug E, Fort O, Trameçon A, Watanabe M, Nakada I (1991) Numerical crashworthiness simulation of automotive structures and components made of continuous fiber reinforced composite and sandwich assemblies. SAE Technical paper series 910152
9. Fleming DC (2001) Modeling delamination growth in composites using MSC.DYTRAN. MSC User's conference proceedings
10. Belingardi G, Obradovic J (2010) Design of the impact attenuator for a formula student racing car: numerical simulation of the impact crash test. J Serbian Soc Comput Mech 4:52–65, ISSN:1820–6530, 2010
11. Haipeng H, Taheri F, Neil P, You L (2007) A numerical study on the axial crushing response of hybrid pultruded and +/−45[degree sign] braided tubes. Compos Struct 80:253–264
12. Hamidreza Z, Matthias K, Henrik A (2008) An experimental and numerical crashworthiness investigation of thermoplastic composite crash boxes. Compos Struct 85:245–257
13. Kakogiannis D, Hemelrijck DV, Wastiels J, Ackeren JV, Palanivelu S, Paepegem WV, Vantomme J, Nurick GN, Chung KS (2009) Experimental and numerical study of the energy absorption capacity of pultruded tubes under blast load. In: Proceedings of SEM annual conference
14. Mamalis AG, Manolakos DE, Demosthenous GA, Ioannidis MB (1997) The static and dynamic axial crumbling of thin-walled fiberglass composite square tubes. Compos Part B 28B(4):439–451
15. Mamalis AG, Manolakos DE, Ioannidis MB, Papapostolou DP (2007) On the response of thin-walled composite tubular components subjected to static and dynamic axial compressive loading: experimental. Compos Struct 69:407–420
16. Feraboli P, Deleo F, Garattoni F (2007) Efforts in the standardization of composite materials energy absorption. In: American society for composites 22nd technical conference, Seattle

17. Feraboli P (2006) Current efforts in standardization of composite materials testing for crashworthiness and energy absorption. In: 47th AIAA/ASME/ASCE/AHS/ASC structures, dynamics and materials conference, No. 2006–2217, Newport
18. Sims GD, Broughton WR (2000) Compr Comp Mater 2:151–197
19. Boukhili R, Hubert P, Gauvin R (1991) Loading rate effect as a function of the span-to-depth ratio in three-point bend testing of unidirectional pultruded composites. Composites 22:39–45
20. Heimbs S, Heller S, Middendorf P, Hähnel F, Weiße J (2009) Low velocity impact on CFRP plates with compressive preload: Test and modeling. Int J Impact Eng 36(10–11):1182–1193
21. LS-DYNA keyword user's manual (2010) Version 971 Rev. 5, Livermore Software Technology Corporation, Livermore
22. Chang FK, Chang KY (1987) A progressive damage model for laminated composites containing stress concentrations. J Compos Mater 21(9):834–855
23. Deleo F, Wade B, Feraboli P, Rassaian M (2010) Crashworthiness of composite structures: modeling of the crushing of UD tape sinusoidal specimens using a progressive failure model. In: Presented at AMTAS Fall Meeting, Seattle
24. Schweizerhof K, Weimar K, Rottner T (1998) Crashworthiness analysis with enhanced composite material models in LSDYNA: merits and limits. In Proceeding of fifth LSDYNA international user conference, Livermore Software Technology Corp., Livermore, Sept 1998
25. Borg R (2002) Simulation of delamination initiation and growth in fiber composite laminates Ph.D. thesis, Linköpings Universitet, Sweden
26. Szekrenyes A (2011) The influence of crack length and delamination width on the mode-III energy release rate of laminated composites. J Compos Mater 45(3):279–294

Chapter 24
Design of New Elastomeric Matrix Composites: Comparison of Mechanical Properties and Determining Viscoelastic Parameters via Continuous Micro Indentation

D. Zaimova, E. Bayraktar, G. Berthout, and N. Dishovsky

Abstract The aim of this study is to compare the influence of vulcanizates strategy on the mechanical properties and viscoelastic behaviour of the new elastomeric based composites based on NR/BR compound that we try actually to design the new compositions in order to improve the new family of elastomeric composites used extensively in engineering applications.

Essentially the basic mixture of NR/BR based elastomeric composites with different vulcanization temperatures and curing systems containing different filler s and reinforcements were characterized (for example in two vulcanizates: 140°C and 160°C) and to compare their tensile properties and viscoelastic behaviours were determined using of micro indentation technique that is a more reliable and efficient tool for practical applications.

Essentially, in detail with the elastic indentation load-depth model at the maximum indentation load describes the indentation depth versus time curves quite well for the four different compounds of elastomeric based composites. It is hoped that these parameters may be used in the evaluation and development of more strong and long lasting elastomeric composites. It was also observed that the model describes the indentation depth versus time curves very well for the mixture of NR/BR elastomeric composite sheets.

Keywords Vulcanization • Elastomeric composites • Curing • Tensile strength • Micro indentation • Viscoelasticity

24.1 Introduction

During the last two decades, very intense studies carried out on the vulcanization of elastomeric based composites containing different fillers-reinforcements have gained remarkable power with the introduction of a new class of compounds called new elastomeric based composites with different accelerators (even two or more). Because of the reduced time necessary for curing and production of improved conclusion stage utilize products, the study of new accelerator systems and reinforcements is still a hot topic of research. Even though the mechanism of vulcanization is still open for discussion, the use of different design of new elastomeric composites has gained consequence, depending on their performance.

As well known vulcanization temperature has a significant effect on crosslink structure. Optimum properties are when curing is done at the lowest possible temperature. However, to increase productivity, higher temperatures are frequently used. Additionally, increases in sulphur and accelerator concentrations give higher crosslink densities and, therefore, higher

D. Zaimova
Supmeca-LISMMA-Paris, Mechanical and Manufacturing Engineering School, St-Ouen, France

UTCM, University of Technology, Chemistry and Metallurgy, Sofia, Bulgaria

E. Bayraktar (✉)
Supmeca-LISMMA-Paris, Mechanical and Manufacturing Engineering School, St-Ouen, France
e-mail: bayraktar@supmeca.fr

G. Berthout
CSM Instruments SA, Rue de la Gare 4, CH-2034 Peseux, Switzerland

N. Dishovsky
UTCM, University of Technology, Chemistry and Metallurgy, Sofia, Bulgaria

E. Patterson et al. (eds.), *Composite Materials and Joining Technologies for Composites, Volume 7: Proceedings of the 2012 Annual Conference on Experimental and Applied Mechanics*, Conference Proceedings of the Society for Experimental Mechanics Series 44,
DOI 10.1007/978-1-4614-4553-1_24, © The Society for Experimental Mechanics, Inc. 2013

modulus, stiffness, hardness, etc. on the other hand, as the ratio of the concentration of accelerator to the concentration of sulphur increases, the proportion of monosulfidic crosslinks increases in natural rubber stocks (also called rubber compounds). Greater amounts of accelerator (with respect to sulphur) also give a plenty of groups of the type, which are attached to and "dangle" from the rubber molecular chains. Higher ratios of sulphur concentration to accelerator concentration give both more polysulphide crosslinks and more sulphur combined with the rubber chains to form sulphur-containing six-membered heterocyclic rings along the rubber molecular chains. The complexity of the vulcanization reactions, the types and relative concentration of crosslinks formed, and the nature and extend of main chain modification can vary considerably depending on the curing system itself and the conditions of vulcanization [1–13].

It is worth to give here a short information about two actual systems are discussed, conventional system and efficient system, to understand well the importance of the vulcanization strategy for improving of mechanical properties.

Conventional system (CV): These systems contain high sulphur levels, e.g. 1.5–3.5 phr and low level accelerator, e.g. 0.5–1 phr (S/A ratio from 1.5 to 70). At optimum cure the vulcanizates contains mostly polysulphide crosslink with relatively high level of chain modification. The higher the sulphur content creates the more polysulphide crosslink. The kind of crosslinks gives little resistance to reversion and oxidative aging and proper long-term flex life [9–11, 14]. Otherwise, these vulcanizates show good tensile and tear strengths, resistance to fatigue and good low temperature properties.

Efficient systems (EV): These systems contain low sulphur levels, e.g. 2.5–5 phr (S/A ratio from 0.05 to 0.3). They will form mainly mono – and disulphide crosslink and much less chain modification. Monosulphide crosslink are more stable than polysulphide crosslink. Short crosslink provides poor fatigue life and compression set [15, 16]; but improves resistance to reversion and oxidative aging. However the drawbacks include lower tear strength, fatigue resistance and abrasion resistance. These kinds of systems are appropriate for thick rubber items.

Addition of filler and/or reinforcements into elastomeric composites can lead an improvement in final mechanical properties and viscoelastic behaviours. Some authors [8–20] have studied the tensile strength and other physical properties of different conventional and efficient systems increasing the crosslink density in their papers. They concluded that physical properties are superior for conventional cure system at similar crosslink values [10–16, 21].

They explained these results by difference in the vulcanization mechanism of the two cure systems. The conventional cure system produces a homogeneous network while the efficient cure system produces a significant polymerization of double bonds of adjacent chains. This leads to a network of unevenly distributed crosslinks and results in impediment to NR crystallization and stress concentration that anticipates the compound failure. In other words, efficient system includes high crosslink areas, overcrosslinked domains, which inhibit the correct stress distribution, accumulating higher stress areas where micro-crack can be formed and crack propagation is increased.

The aim of this study is to compare the influence of vulcanizates strategy on the mechanical properties and viscoelastic behaviour of the new elastomeric based composites based on NR/BR compound that we try actually to design the new compositions in order to improve the new family of elastomeric composites used extensively in engineering applications.

Essentially the basic mixture of NR/BR based elastomeric composites with different vulcanization temperatures and curing systems containing different filler s and reinforcements were characterized (for example in two vulcanizates: 140°C and 160°C) and to compare their tensile properties and viscoelastic behaviours were determined using of micro indentation technique that is a more reliable and efficient tool for practical applications.

24.2 Experimental Conditions

The four compounds were mixed in laboratory using two rolling mixer. The moulding conditions were determined from torque data using moving die rheometer MDR 2000 (Alpha Technologies) for temperatures 140°C and 160°C. At the end, four compounds (A, B, C, D) based on sulphur vulcanized Natural rubber/Polybutadiene rubber blends were prepared containing certain fillers and/or reinforcements in order to investigate. The exact composition of the blends is kept in reserve because of confidentiality matters. The results discussed in the present paper give partial results carried out in the frame of the applied project supported by French – Bulgarian research cooperation

Cured sheets were prepared by compression moulding. The moulding took place at 140°C and 160°C and pressure 100 kg/cm^2. The thickness of the sheets is 2 mm. The measurements of physic-mechanical properties were carried out in accordance with ASTM D 412a$^{\varepsilon 2}$ (2010). The samples were cut from the moulded sheets in dumbbell shape. The sample length and thickness were measured. Tensile tests were performed on an Instron (model 4507). A minimum of three specimens was tested for each compound. Testing was done at room temperature with a cross head speed of 500 mm/min^{-1}.

Hardness was determined according to ASTM D 2240–05 (2010).

Micro indentation tests have been carried out on the CSM Indentation Tester with a Vickers diamond indenter. Nine tests were performed for each composition and then the average values were taken. The maximum indentation load (F_{max}) was 250mN, the rate charging – discharging was 500 mN/min. The load was held at maximum for 1950s.

24.3 Results and Discussion

Formulations of elastomeric composites are developed to meet specific requirements. Thus, each one of the various components taking part in a particular formulation has an important role as to give its contribution to the final properties [1–6, 8]. However, even with the wide choice of ingredients available, the whole set of characteristics may still not be reached and many times a combination of two or more rubbers has to be used in the preparation of compounds for special properties.

Thus, generally speaking, elastomeric materials are blended for properties improvement, better processing or lower cost [7]. Natural rubber (NR) is the adequate choice when good tensile and tear strengths are demanded, since these characteristics can be developed due to the capability this rubber has to crystallize under stress [8–13].

Polybutadiene rubber (BR) is characterized by its superior abrasion resistance, so that blends of NR and BR that combine the excellent processing and physical properties of the former with the superior abrasion resistance of the latter are largely used in the industry in the production of tyre treads and conveyor belts [1–6, 14].

Tensile strength and elongation values at failure are shown in Fig. 24.1a, b respectively, before each tensile test, the thickness of the specimens were measured with Mitutoyo device. Stress-stress curves were calculated as an average of three test measurements for each compound. All the tests have been carried out on the four different compounds from elastomeric composite sheets (thickness of the thin sheets, t = 2 mm);

In the case of the comparison of these four compounds, compound "C" has shown exceptionally different behaviour, higher tensile strength and elongation regarding the other compounds. This behaviour should be related to the low shear produced during the mixing process and it means that this behaviour is due to the less breakdown of the main chain during the mixing process;

In the frame of the present paper, more detail evaluation was given on the viscoelastic behaviour using micro indentation technique that is a very practical tool to evaluate typical parameters for each compound prepared in the program of the common research project going on. Thus, all of these results discussed here contain partial results of the general context of the research project going on.

At the first stage, for the micro indentation test, the indentation hardness (H_{IT}), indentation modulus (E_{IT}) and Vickers hardness (HV) were determined by using the Oliver and Pharr method transformed in the following relations:

$$A_p = f(h) \tag{24.1}$$

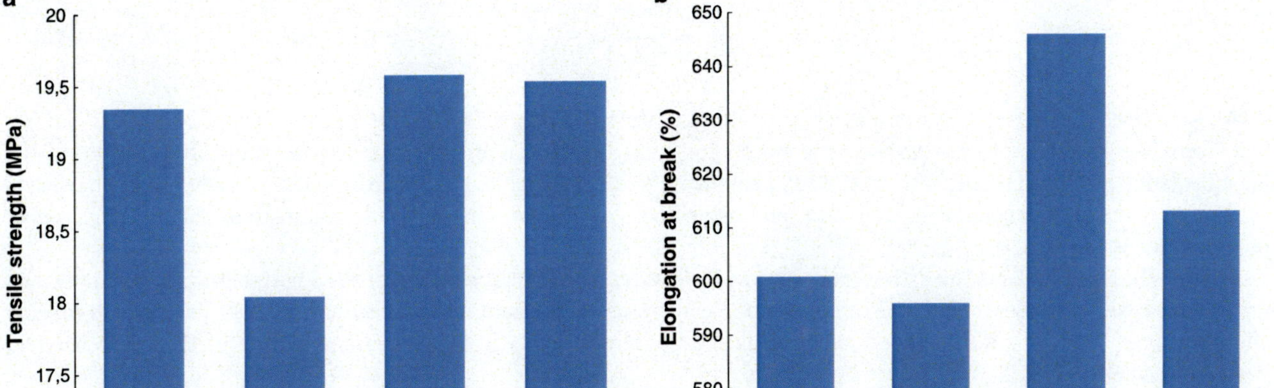

Fig. 24.1 (a) Tensile stress and (b) elongation values at failure obtained on the four different compounds from elastomeric composites sheets (t = 2 mm)

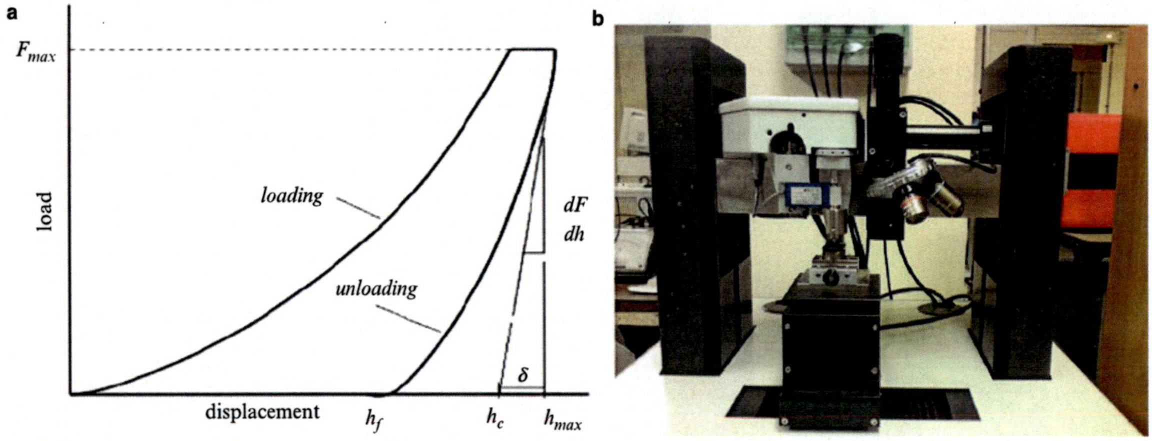

Fig. 24.2 Schematic representation of the force-depth curve for micro indentation procedure (**a**) and measurement device (**b**)

where A_p is projected contact area

$$H_{IT} = \frac{F_{max}}{A_p(h_c)} \tag{24.2}$$

$$E_{IT} = \frac{1 - v_s^2}{\frac{1}{E_r} - \frac{1 - v_i^2}{E_i}} \tag{24.3}$$

$$E_r = \frac{\sqrt{\pi}S}{2\beta\sqrt{A_p(h_c)}} \tag{24.4}$$

where

E_i – elastic modulus of the indenter (1,141 GPa)
ν_i – Poisson's ratio of the indenter (0.07)
E_r – reduced modulus of the indentation contact
ν_s – Poisson's ratio of the sample (0.35)

$$HV = \frac{F_{max}}{9.81A_p(h_c)} \tag{24.5}$$

As mentioned just before, the micro indentation test has become a popular technique due to its simplicity and to the fact that it provides valuable information about the morphology and mechanical properties of polymeric materials. Additionally, the indentation method has been also successfully employed to investigate the glass transition temperature of polymers. [15]

Recently, micro indentation appears as a promising tool for micromechanical and microstructural investigation of polymer blends [16, 21].

First of all, micro indentation differs from classical measurement of hardness, where the impressions are generated and then imaged with a microscope. Load and associated penetration depths are recorded simultaneously during both loading and unloading, producing a force-depth diagram as shown schematically in Fig. 24.2a and picture of the micro indenter device presented in Fig. 24.2b.

The Oliver and Pharr Method (Power Law Method) [22] recognizes the fact that the first portion of unloading curve may not be linear, and can be described by a simple power law relationship:

$$F = k(h - h_p)^m \tag{24.6}$$

Where k is a constant and m is an exponent which depends on indenter geometry. A power low function is used to describe the upper part of the unloading data.

$$F = F_{max}\left(\frac{h - h_p}{h_m - h_p}\right)^m \tag{24.7}$$

where the constants m and h_p are determined by a least of square fitting procedure. The contact stiffness S = (1/C) is given by the derivative at peak load:

$$S = \left(\frac{dF}{dh}\right)_{max} = mF_{max}\left[\frac{(h_m - h_p)^{m-1}}{(h_m - h_p)^m}\right] = mF_{max}(h_m - h_p)^{-1} \tag{24.8}$$

And the tangent depth, h_r, is thus given by:

$$h_r = h_m - \frac{F_{max}}{S} \tag{24.9}$$

The contact depth, h_c, is then:

$$h_c = h_m - \varepsilon(h_m - h_r) \tag{24.10}$$

where ε now depends on the power law exponent m.

The tangent is found by differentiating the unloading curve and evaluating at maximum load (F_{max}). The intercept of this tangent with the displacement axis yields h_r.

The viscoelastic properties of polymers are markedly dependent on the type of crosslinks and the degree of crosslinking. Crosslinking raises the glass-transition temperature (T_g) of a polymer by introducing restrictions on the molecular motions of the chain. Low degrees of crosslinking in normal vulcanized rubbers act in a similar way to entanglements and raise the Tg only slightly above that of the crosslinked polymer.

The effect of crosslinking is the most important and best understood in elastomers. Sulphur crosslinking of NR produces a variety of crosslinking types and crosslink lengths [10, 17–19, 23, 24].

It is well known that polysulfide linkages predominate with conventional sulphur vulcanization system whereas mono-sulfide and di-sulfide crosslinks are formed with an efficient vulcanization system, which has a higher accelerator/sulphur ratio. Sulphur also introduces main chain modifications either in the form of pendant groups or as cyclic sulfide linkages that have a large influence on the viscoelastic properties [10, 18–20, 23–24].

Networks containing high proportions of polysulfide crosslinks display different mechanical properties from those containing mono-sulfides crosslinks. It is shown in the literature that the increase in the accelerator level only results in small changes of the T_g. Hagen et al. studied the effect of different types of crosslinks on the viscoelastic properties of NR. They concluded that the rubbery tensile modulus was dependent on the crosslink density but almost independent of the crosslink type [17–20, 23–24].

In the frame of the present paper, elastomeric based composite specimens (basically contain were tested extensively using micro indentation device that a very useful tool to evaluate the viscoelastic behaviours of these composites).

Figures 24.3, 24.4, 24.5, and 24.6 describe evolution of the applied load depending on the indentation depth during the loading – unloading stages, indentation depth versus time data at maximum indentation load and evolution of the applied load and indentation depth depending on the time during the loading – unloading stages respectively. In comparison of the viscoelastic behaviour of these four compounds, one may evaluate many different parameters of viscoelasticity of the elastomeric composites.

It is revealed that samples with compound "C" have a higher viscoelastic deformation than the corresponding samples with compounds A, B and D due to improved chain mobility (moderately due to subdivision and allow more relaxation after loading). For mechanical properties, the values of the compound "C" give higher indentations after loading, as compared to the other three compounds, "A, B and D" that is related directly vulcanization strategy and crosslink densities (Fig. 24.7).

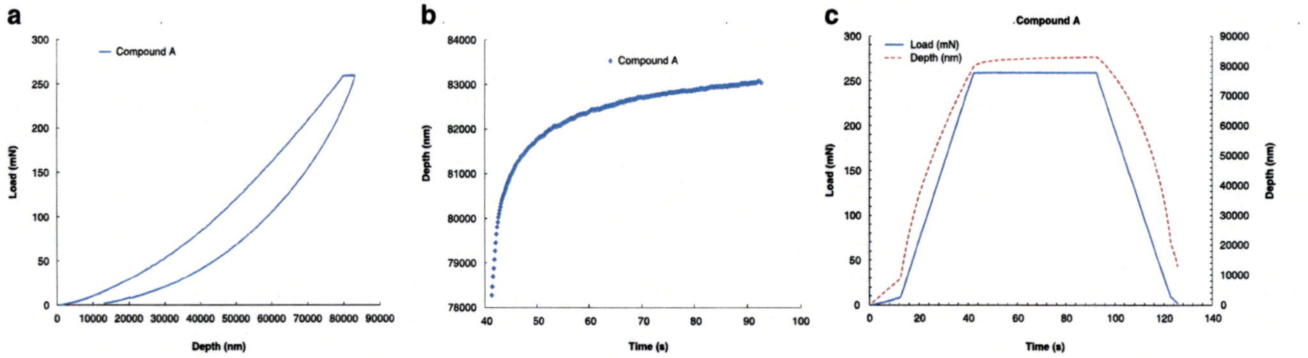

Fig. 24.3 (a) Compound "A": Evolution of the applied load depending on the indentation depth during the loading – unloading stages, (b) Indentation depth versus time data at maximum indentation load and (c) Evolution of the applied load and indentation depth depending on the time during the loading – unloading stages

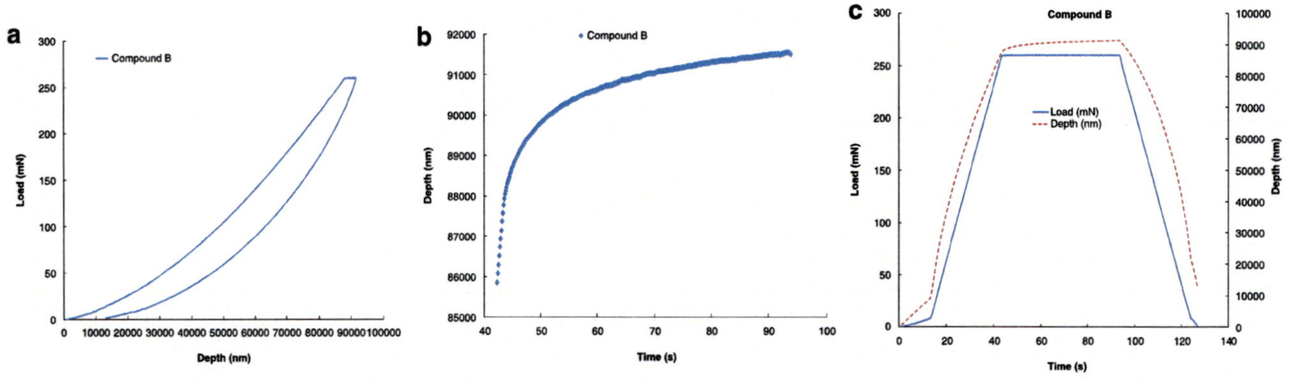

Fig. 24.4 (a) Compound "B": Evolution of the applied load depending on the indentation depth during the loading – unloading stages, (b) Indentation depth versus time data at maximum indentation load and (c) Evolution of the applied load and indentation depth depending on the time during the loading – unloading stages

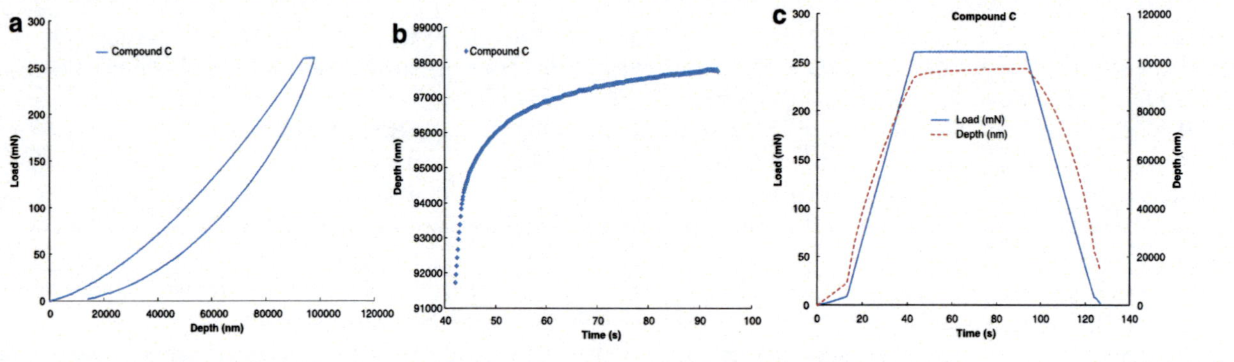

Fig. 24.5 (a) Compound "C": Evolution of the applied load depending on the indentation depth during the loading – unloading stages, (b) Indentation depth versus time data at maximum indentation load and (c) Evolution of the applied load and indentation depth depending on the time during the loading – unloading stages

In the same way, a detail comparison of four compounds was given in Fig. 24.8 for the measurements of indentation hardness, indentation modulus and also Vickers hardness presented as real – actual data. All of these values will be used for modelling the viscoelastic behaviour of the elastomeric composites that we design actually, in the frame of the next paper.

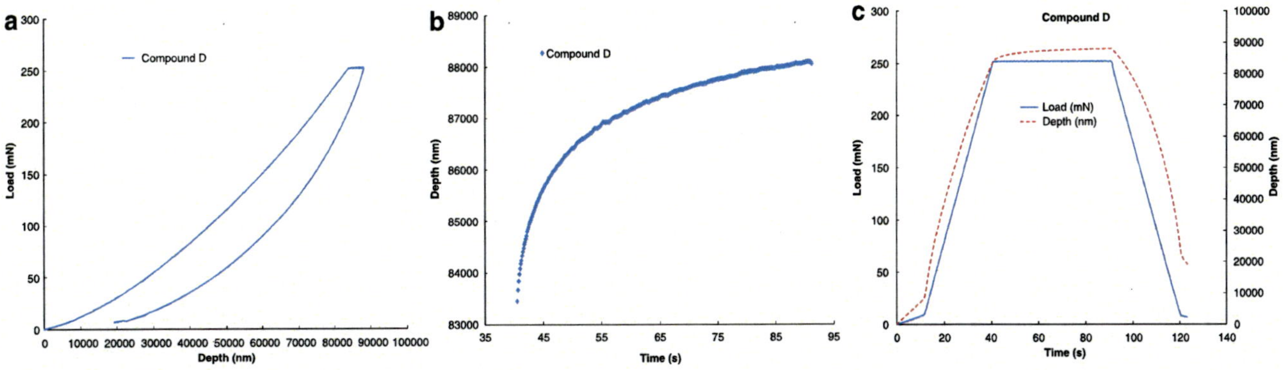

Fig. 24.6 (**a**) Compound "D": Evolution of the applied load depending on the indentation depth during the loading – unloading stages, (**b**) Indentation depth versus time data at maximum indentation load and (**c**) Evolution of the applied load and indentation depth depending on the time during the loading – unloading stages

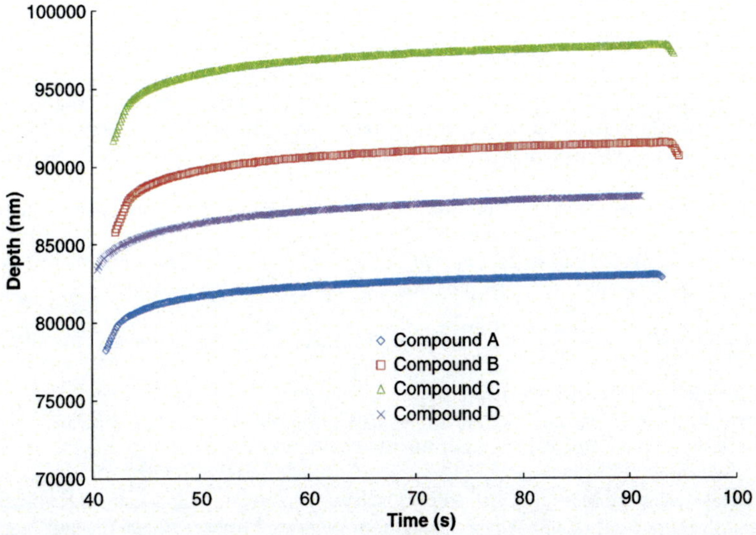

Fig. 24.7 Comparison of the evolution of indentation depth values depending on the time at maximum indentation load (Actual data presented for the compounds A, B, C and D)

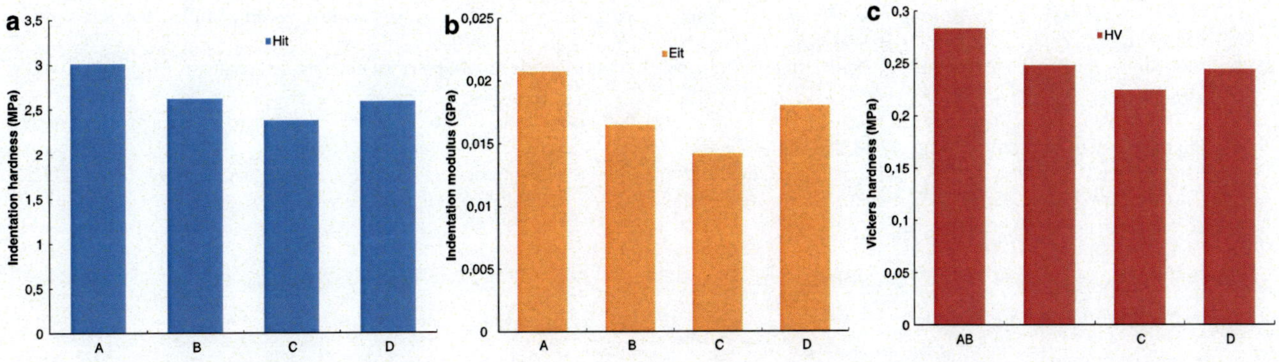

Fig. 24.8 Comparison of four compounds sheets: (**a**) Measurements of indentation hardness (**b**) indentation Modulus and (**c**) Vickers hardness as real – actual data was presented

24.4 Conclusion Remarks

In the frame of the research project, the processing, mechanical and viscoelastic parameters of new elastomeric composites containing basically the mixture of NR-BR vulcanizates with different addition-reinforcements have been evaluated.

Essentially, in detail with the elastic indentation load-depth model at the maximum indentation load describes the indentation depth versus time curves quite well for the four different compounds of elastomeric based composites. It is hoped that these parameters may be used in the evaluation and development of more strong and long lasting elastomeric composites. It was also observed that the model describes the indentation depth versus time curves very well for the mixture of NR/BR elastomeric composite sheets.

All of these measurements from tensile results for the observed mechanical properties, essentially by using micro indentation method were compared and found to follow the same tendency among the two basic mixtures. Based on the processing, mechanical characteristics and viscoelastic behaviours, optimum composition between four compounds studied here, compound "C" gives optimistic choice that will be useful to make a detail suggestion after results of the research project that is going on.

References

1. Siqueira Filho AS, BoÂ MC (1985) Controle de Qualidade na InduÂ stria de Artefatos de Borracha, Manuais CNI
2. Klei B, Koenig JL (1997) The competitive vulcanization of natural rubber and polybutadiene blends. Rubber Chem Technol 70:231–238
3. Thorn A.D., Robinson R.A. (1994) Compound design. In: Rubber products manufacturing technology. Marcel Dekker, New York, p 1, Chap. 1
4. Brydson J.A. (1988) Natural rubber. In: Rubber materials and their compounds. Elsevier Science, New York, p 69, Chap. 4
5. Kohl JG, Singer IL, Simonson DL (2008) Determining the viscoelastic parameters of thin elastomer based materials using continuous micro indentation. Polym Test 27(6):679–682
6. Eng A.H., Ong E.L. (2001) Hevea natural rubber. In: Bhowmick AK, Stephens HL (eds) Handbook of elastomers, 2nd edn. Marcel dekker, New York
7. Bloomfield GF (1967) New forms of natural rubber. In: Brydson JA (ed) Developments with natural rubber. Maclaren & Sons, London
8. Ciesielski A. (1999) Types of rubber and their essential properties. In: An introduction to rubber technology, Rapra Technology Limited, Shawbury, Shrewsbury, Shropshire
9. Andrews EH, Gent AN (1963) Crystallization in natural rubber. In: Bateman L (ed) The chemistry and physics of rubber like substances. Wiley, New York
10. Fischer-Cripps AC (2004) A simple phenomenological approach to nanoindentation creep. Mater Sci Eng A 385:74–82
11. Nyburg SC (1945) X-ray determination of crystallinity in deformed natural rubber. Br J Appl Phys 5:321
12. Kohl JG, Bolstes RN (2001) A study on the elastic modulus of silicone duplex or bi-layer coatings using micro-indentation. Prog Org Coating 41:135–141
13. Subramaniam A (1987) Natural rubber. In: Morton M (ed) Rubber technology, 3rd edn. Van Nostrand Reinhold Company, New York
14. Gent AN, Razzaghi-Kashani M, Hamed GR (2003) Why do cracks turn sideways? Rubber Chem Technol 76:122
15. Hamed GR, Al-Sheneper AA (2003) Effect of carbon black concentration on cut growth in NR Vulcanizates. Rubber Chem Technol 76:436
16. Kohl JG, Singer IL (1999) Pull-off behavior of epoxy bonded to silicone duplex coatings. Prog Org Coating 36:15–20
17. Hofmann A. (1988) Natural rubber. In: Rubber technology handbook. Hanser Publishers, New York
18. Nielsen LE, Landel LE (1994) Mechanical properties of polymers and composites, 2nd edn. Marcel Dekker, New York
19. Chapman AV, Porter M (1988) In: Roberts AD (ed) Natural rubber science and technology. Oxford Scientific, New York, Chap. 12
20. Singer IL, Kohl JG, Patterson M (2000) Mechanical aspects of silicone coatings for hard foulant control. Biofouling 16:301–309
21. Brady RF, Singer IL (2000) Mechanical factors favoring release from fouling release coatings. Biofouling 15:73–82
22. Oliver WC, Pharr GM (1992) An improved technique for determining hardness and elastic modulus using load and displacement sensing indentation experiments. J Mater Res 7:1564–1583
23. Kohl JG, Singer IL, Schwarzer N, Yu VY (2006) Effect of bond coat modulus on the durability of silicone duplex coatings. Progr Org Coating 56:220–226
24. Ngan AHW, Wang HT, Tang B, Sze KY (2005) Correcting power-law viscoelastic effects in elastic modulus measurement using depth-sensing indentation. Int J Solids Struct 42:1831–1846
25. Kohl JG, Singer IL, Griffith JR (2000) Evaluating the durability of silicone duplex coatings using a scratch tester. Rubber Chem Technol 73:607–618

Chapter 25
Polyurea-Based Composites: Ultrasonic Testing and Dynamic Mechanical Properties Modeling

Wiroj Nantasetphong, Alireza V. Amirkhizi, Zhanzhan Jia, and Sia Nemat-Nasser

Abstract Many scientists and researchers study polyurea due to its excellent blast-mitigating properties. In this work, we have studied two polyurea composite systems with filler materials intended to improve dynamic mechanical properties. The two filler materials are milled glass and fly ash. The shape and quantity of filler significantly affect the dynamic mechanical properties of the composite. Ultrasonic tests were conducted on samples with both fillers. The volume fraction of the inclusions was varied to study the effect of filler quantity on mechanical properties. Moreover, computational models based on the methods of dilute-randomly-distributed inclusions and periodically-distributed inclusions were created to improve our understanding of polyurea-based composites and serve as tools for estimating the dynamic mechanical properties of similar composite material systems. The experimental and computational results were compared and show good agreement. The experiments and modeling have been conducted to facilitate the design of new elastomeric composites with desirable impact- and blast-mitigating properties.

Keywords Polyurea • Fly ash • Milled glass • Dynamic mechanical properties • Dilute-randomly-distributed inclusions • Periodically-distributed inclusions

25.1 Introduction

The failure of structures and materials in response to blast loadings is an active research area. Over the past 40 years, polyurea has been investigated as a potential coating material to help absorb and dissipate blast energy. Polyurea is the generic name for a block copolymer formed from a chemical reaction of diisocyanates with polyamines [1]. The reaction is generally very fast and insensitive to humidity and low temperatures [2]. This viscoelastic material is stable and incredibly tough, making it a popular material for coating applications. Additionally, researches have shown that polyurea also has an excellent blast-mitigating capability. Structures selectively coated with polyurea could potentially absorb more impact energy before failure and enhance the dynamic performance of structure [3–6].

Recent researches have focused on modifying polyurea with filler materials to create composites with enhanced properties. Qiao et al. [7, 8] reduced the density but increased the strength of polyurea by mixing it with fly ash. As the volume of fly ash increases, the density of the composite decreases, while the storage modulus as measured by dynamic mechanical analysis (DMA) increases. Alternatively, glass fibers can be blended with a polymer to greatly improve the strength of the polymer [9]. Tests on milled-glass-fiber-filled polyethylene terephthalate-co-isophthalate composites were conducted by Velasco et al. [10]. Their results showed a trend of increasing modulus with increasing fiber volume fraction. Increasing wear resistance is another advantage of polymer composites containing milled glass fibers [11].

In this work, two kinds of composites with different inclusion shapes were studied with two objectives in mind: (1) to investigate the effect of inclusion content on dynamic mechanical properties and (2) to present the models which effectively

W. Nantasetphong (✉) • A.V. Amirkhizi • Z. Jia • S. Nemat-Nasser
Department of Mechanical and Aerospace Engineering, Center of Excellence for Advanced Materials, University of California, San Diego, 9500 Gilman Drive, La Jolla, CA 92093-0416, USA
e-mail: wnantase@ucsd.edu

E. Patterson et al. (eds.), *Composite Materials and Joining Technologies for Composites, Volume 7: Proceedings of the 2012 Annual Conference on Experimental and Applied Mechanics*, Conference Proceedings of the Society for Experimental Mechanics Series 44,
DOI 10.1007/978-1-4614-4553-1_25,

estimate the mechanical properties of the composites. Polyurea filled with fly ash or milled glass composites were fabricated, with 5%, 10%, and 20% volume fraction of inclusions. The morphology of composites was inspected under Scanning Electron Microscopy (SEM). Their dynamic mechanical behavior was determined by ultrasonic measurements. The experimental results were compared with the computational models.

25.2 Experimental Details

25.2.1 Material Characteristics

Polyurea serves as the matrix component in the composite material. It was synthesized from the reaction of Isonate 143 L a polycarbodiimide-modified diphenylmethane diisocyanate [12], and Versalink P-1000, an oligomeric diamine [13]. Theoretically, Isonate 143 L and Versalink P-1000 should be mixed in a reaction ratio of 1:1, i.e., the total number of isocyanate groups must equal the total number of hydroxyl groups in order to obtain complete chemical reaction. However, a slight excess of Isonate 143 L was used to ensure that the reaction was completed and produced some cross-linking between the hard domains formed from semi-crystallization of the diamine molecules. The amount of excess Isonate 143 L was estimated through weight measurements of containers before and after the processing.

Fly ash particles originating from coal combustion were applied as the filler. Fly ash particle is hollow and spherical with a porous shell. Our samples were sieved to an average outside diameter of 105 μm. The average inner diameter was estimated to be around 91 μm. The density of fly ash particle is about 0.78 g/cm^3.

Milled glass particles are in the short fiber form, approximately 200 μm in length and 16 μm in diameter. The particles have very extreme aspect ratio (~12.5:1) and its density is about 2.2–2.7 g/cm^3.

25.2.2 Preparation of Composites

25.2.2.1 Polyurea with Fly Ash Composite

Due to the short gel time of polyurea at room temperature, fly ash was added to the Versalink P-1000 prior to the polymerization process in order to have time to achieve a homogenous distribution. The fly ash particles were first preheated to 110°C for 1 h and cooled down under dry conditions before being added to the Versalink P-1000 in a predetermined proportion and thoroughly mixed. Using a magnetic stirrer, the components were mixed for another 2 h while being degassed to eliminate trapped air bubbles. The Isonate 143 L was degassed separately to remove trapped air bubbles. The Versalink-fly-ash mixture and Isonate were then combined and mixed rapidly for 5 min while degassing. This prepolymer was transferred to a teflon mold at room temperature and allowed to cure for 1 week before being tested. To control the ambient humidity, the mold with curing material was placed in an environmental chamber that maintained a relative humidity (RH) level of 10%. The fly ash volume fractions in the final composites were 5%, 10%, and 20%.

25.2.2.2 Polyurea with Milled Glass Composite

The polyurea matrix was produced by mixing Isonate 143 L and Versalink P-1000, which are diisocyanate and polyamine components respectively. These two components of polyurea were prepared separately in two flasks, stirred by magnetic stirrer, and degassed at 1 Torr for 2 h. The milled glass was added to the Versalink P-1000 and mixed by hands before degassing. The degassed Isonate 143 L was then poured into the Versalink flask. The components were stirred under vacuum for another 5 min. The mixture was transferred to teflon molds and cured overnight in the 10% RH environmental chamber. The following day, the samples were taken out of the mold and allowed to fully cure in the chamber for 2 weeks before testing. The milled glass volume fractions of the samples were 5%, 10% and 20%.

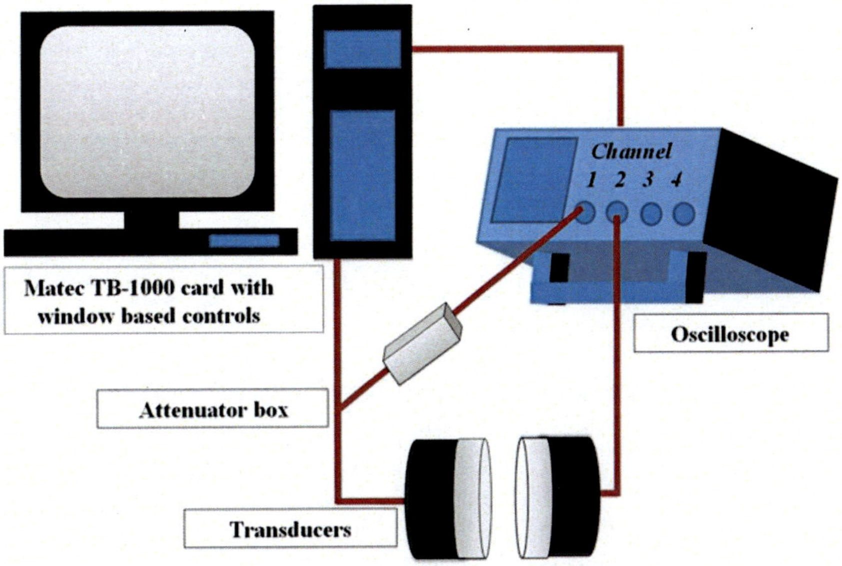

Fig. 25.1 Experimental setup

25.2.3 *Scanning Electron Microscopy (SEM)*

The morphology of the composites was observed under SEM using a Philips XL30 ESEM scanning electron microscope. During SEM sample preparation process, composite specimens were immersed in liquid nitrogen until thermal equilibrium was achieved, at which point they were removed and immediately fractured with a hammer. The fragments were coated with a thin layer of iridium (75 nm thick) in an automatic sputter coater and then the fracture surfaces were observed using an acceleration potential of 15KV.

25.2.4 *Ultrasonic Measurement*

25.2.4.1 Experimental Setup for Longitudinal Waves

Direct contact measurement was used to measure the speed of longitudinal wave in the composites. The experimental setup consisted of a desktop computer containing a Matec TB-1000 Toneburst Card, two Panametrics videoscan longitudinal transducers (V103 Panametrics-NDT OLYMPUS), an attenuator box, and a digital Oscilloscope. As shown in Fig. 25.1, toneburst signals of various frequencies were sent from the card to the attenuator box and were fed through BM-174-3 cables to the generating transducer. The received signal was sent directly to the oscilloscope and displayed on the oscilloscope where the amplitude and travel time were measured. To find wave speed and attenuation of longitudinal wave, two tests were performed using two different sample thicknesses. Samples were sandwiched between a pair of longitudinal transducers. The longitudinal wave couplant was applied to all contact surfaces. Tests were conducted at 1 MHz and -50°C to 30°C with 10°C increments. The speed of longitudinal wave was determined by measuring the difference between the times of travel through two samples of different thicknesses. The attenuation was measured using the transmitted wave amplitudes of these two tests. The results were used to calculate storage and loss longitudinal moduli for each material type.

25.2.4.2 Experimental Setup for Shear Waves

The previously described setup for wave measurement was used with shear transducers (V153 Panametrics-NDT OLYMPUS) instead of longitudinal transducers. Due to an unavoidable longitudinal wave signal generated by the shear transducers, an aluminum rod was inserted between the generating transducer and the sample to delay the shear wave, separating it from the longitudinal wave. Wave speed and attenuation of the shear wave were calculated after two tests (with samples of different thicknesses) were conducted. Shear wave couplant was applied to all interfaces to get better surface contact. Similar to longitudinal wave testing, shear wave tests were conducted at 1 MHz and −50°C to +30°C with 10°C increments. The speed of shear wave and its attenuation were used to calculate storage and loss shear moduli for each material.

25.3 Modelling Details

Two models were developed to represent two structures of composites: spheroids and ellipsoids distributed in matrix.

25.3.1 Composite with Dilute and Randomly Distributed Hollow Spherical Inclusions

The moduli of the matrix and particles are known. Two assumptions were made: (1) The particles do not interact with each other (Fig. 25.2) and (2) the actual particles are small and have a uniform dilute distribution, so that they can be considered as spherical (Fig. 25.3). The effective moduli of the composite are calculated by considering the change in strain energy in a loaded homogeneous body due to the insertion of inhomogeneities. With appropriate choice of admissible stress or strain field, combined with the minimum complementary energy and minimum potential energy theorems, the bounds for the moduli can be obtained [14, 15]. The structure of the model is shown in Fig. 25.4.

25.3.2 Composite with Periodically Distributed Cylindrical Inclusions

This model depicts a polymer with circular-cylindrical inclusions that can be long or short fibers. The approach was first introduced by S. Nemat-Nasser et al. [16]. The unit cell of the model is shown in Fig. 25.5. To estimate the overall moduli of the composite, the total elastic energy per unit cell, which depends on the properties of its constituents, is equated with the corresponding elastic energy of a homogenized model with uniform properties throughout the entire volume. The model represents the uniaxially-oriented cylindrical inclusions embedded in the matrix (Fig. 25.6). In general, the moduli are anisotropic. However, the model can be modified to represent the moduli of composite with randomly oriented cylindrical

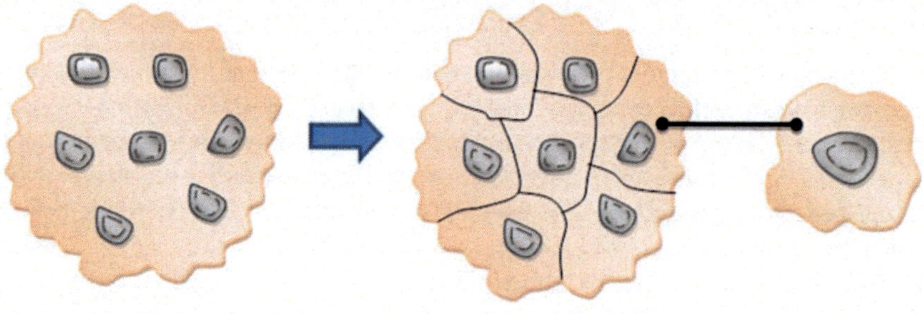

Fig. 25.2 Since the particles do not interact with each other, they can each be treated as a particle embedded in an infinitely large matrix

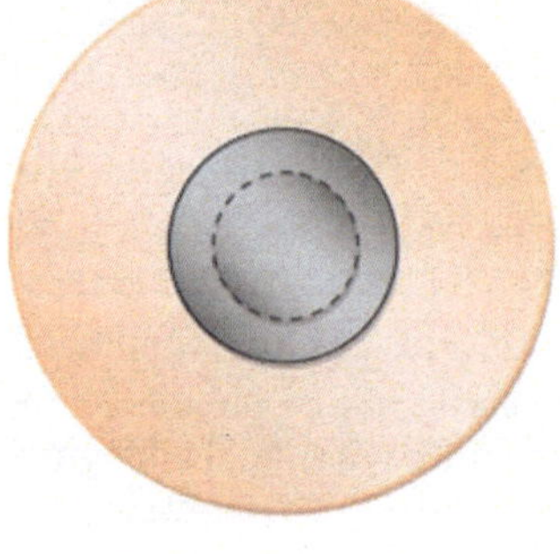

Fig. 25.3 A spherical shape inclusion in an infinitely large matrix may be modeled in a fully spherically symmetric geometry

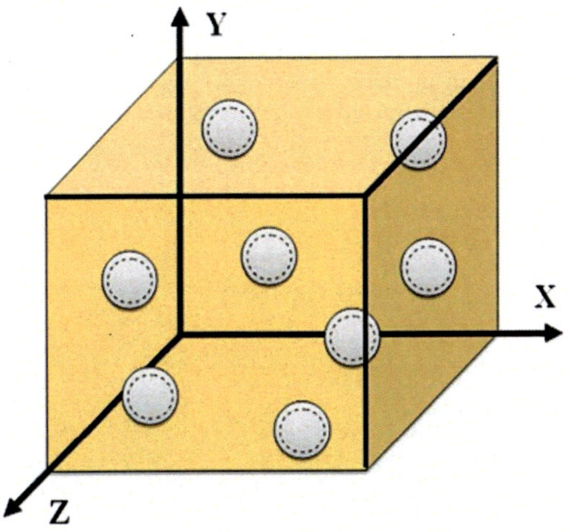

Fig. 25.4 The composite with dilute and randomly distributed hollow spherical inclusions

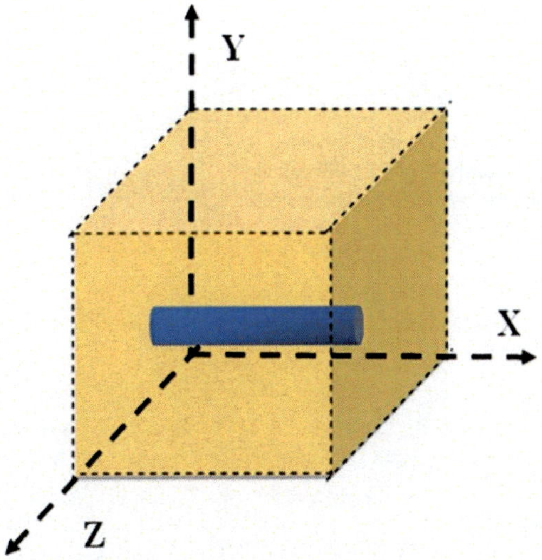

Fig. 25.5 A unit cell of composite with periodically distributed cylindrical inclusions model

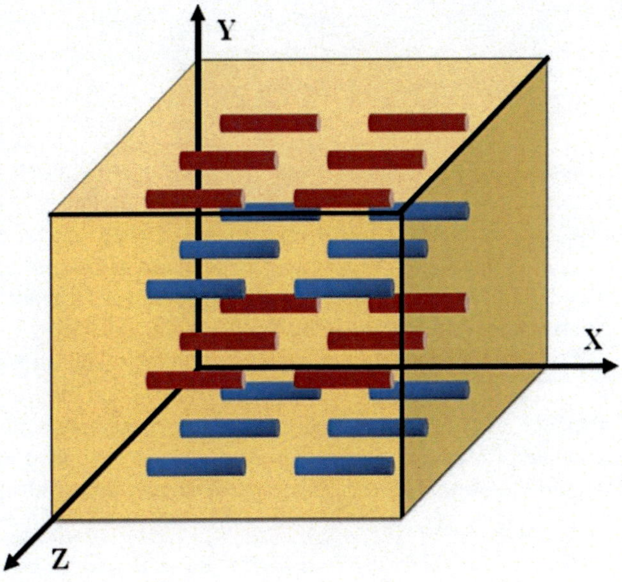

Fig. 25.6 Composite with periodically distributed cylindrical inclusions model

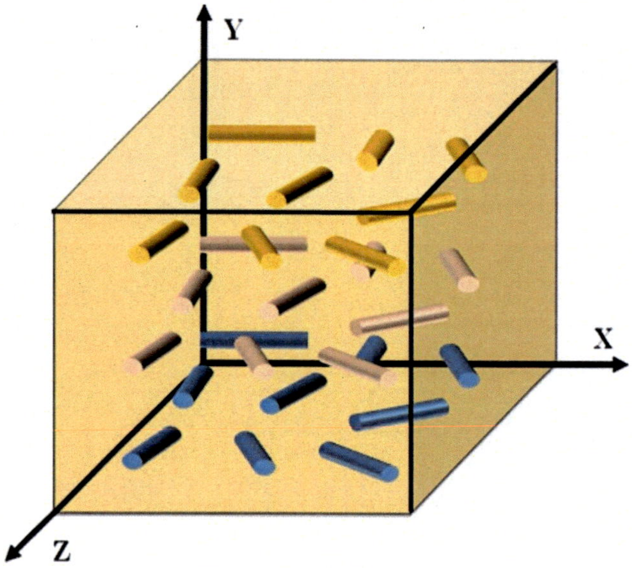

Fig. 25.7 Composite with randomly oriented cylindrical inclusion in a particular plane (xz-plane in this picture)

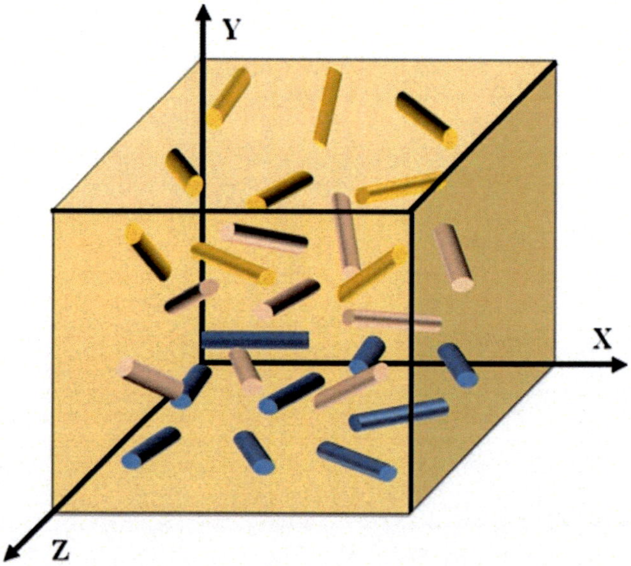

Fig. 25.8 Composite with randomly oriented cylindrical inclusion in three-dimensional space

inclusion in a particular 2D plane (Fig. 25.7) or randomly oriented in 3D (Fig. 25.8). The modification can be made by calculating various modulus tensors using periodically-distributed uniaxially-oriented cylindrical inclusions model (Fig. 25.6) at many different fiber-angle orientations and averaging the modulus tensors of all orientations. In this work, the 2D model (Fig. 25.7) was used to compare with experimental data due to the assumption that during the curing process milled glass fibers tended to lie parallel to the mold surface because of their relative mass and being pressed into the mold. This assumption was proved by SEM picture.

The actual composite, for example, polyurea with milled glass composite does not have periodic structures. In reality, inclusions are essentially randomly distributed and may, in fact, form clusters and other anomalous structures. However, when the density of the inclusions is large, the interaction effect between inclusions becomes dominant and must be accounted for in a more direct manner than is contained in the dilute and randomly distributed model. In this sense, the assumption of periodicity has been proved very powerful in representing the inclusion interaction effects.

25.4 Results and Discussion

25.4.1 Composite Morphology Under Scanning Electron Microscopy (SEM)

SEM was used to investigate the alignment of milled glass fibers in polyurea matrix. From the SEM pictures (Fig. 25.9, 25.10, and 25.11), it appears that most of milled glass fibers tended to lie parallel to the sample surfaces similar to the model in Fig. 25.7. The milled glass fiber, which has extreme aspect ratio and large density compared to fly ash, can be used to create anisotropy in the thin samples fabricated by pressing a small amount of the pre-composite mixture during curing.

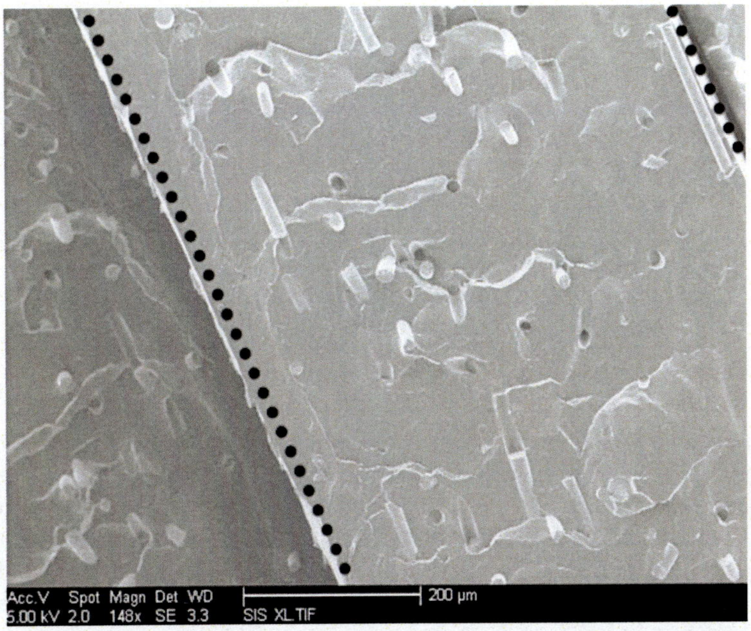

Fig. 25.9 SEM picture of 5% volume fraction polyurea with milled glass composite. The *black dotted lines* show the surfaces of the sample

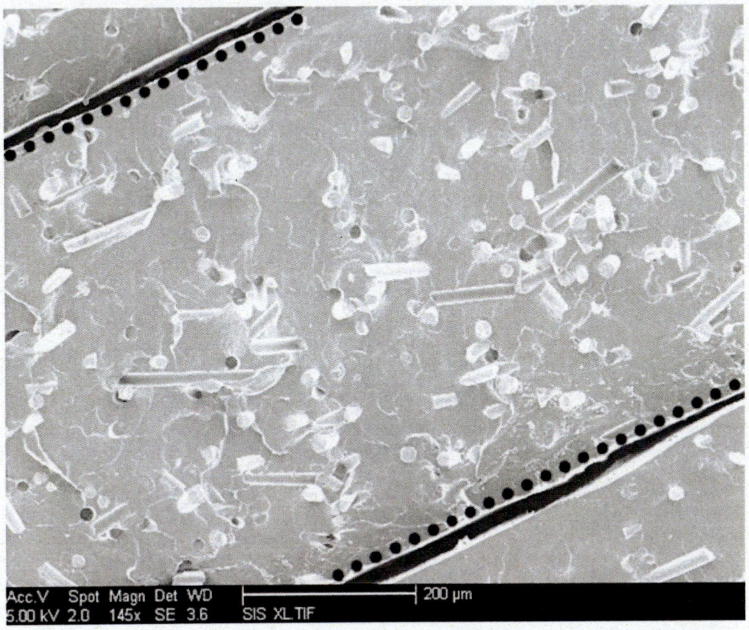

Fig. 25.10 SEM picture of 10 % volume fraction polyurea with milled glass composite. The *black dotted lines* show the surfaces of the sample

Fig. 25.11 SEM picture of 20% volume fraction polyurea with milled glass composite. The *black dotted lines* show the surfaces of the sample

25.4.2 Longitudinal Wave Ultrasonic Testing

Fly ash and milled glass particles polyurea composites have similar trend in longitudinal wave speed. As temperature increases, both composites get softer and their longitudinal wave speed decreases. Similarly, the attenuation per wavelength decreases continuously as temperature increases. Both composites get stiffer as the particle volume fraction increases. In other words, the storage and loss longitudinal moduli increase when more inclusions are added to both composites. Interestingly, from 10% to 20% volume fraction, polyurea with milled glass particles exhibits a steeper increase in storage and loss longitudinal moduli, whereas the polyurea with fly ash shows a moderate increase. We believe this is due to the extreme aspect ratio of the milled glass particles. At high volume fractions (e.g. 20%), the length of the milled glass is likely to lead to more interactions between particles than in the spherical fly ash particle composites. The storage and loss longitudinal moduli from experiment agree very well with the computational results using the models previously discussed.

25.4.3 Shear Wave Ultrasonic Testing

Polyurea composites containing fly ash or milled glass particles show changes in shear wave speed similar to those observed in the longitudinal wave speed. As the temperature linearly increases, the shear wave speed nearly linearly decreases but not as smoothly as in the longitudinal tests. This is due to the shear wave signal being quite dissipative compared to the longitudinal wave signal. The amplitudes of shear waves are not uniform and also smaller than those of the longitudinal waves, making the shear wave data more difficult to analyze. Unlike the longitudinal wave, the shear wave attenuation per wavelength tends to increase as the temperature increases. Both composites get stiffer in shear as the particle volume fraction increases. In other words, storage and loss shear moduli increase when more inclusions are added, in a similar manner to the changes in the longitudinal storage and loss moduli. Like the longitudinal moduli, polyurea with milled glass exhibits a steeper increase in storage and loss shear moduli at 20% volume fraction, whereas the polyurea with fly ash has a nearly uniform increase. The reason for this difference was discussed in Sect. 25.4.2. The storage and loss shear moduli from experiment correspond very well with the computational results of the models previously discussed.

Acknowledgements This research has been conducted at the Center of Excellence for Advanced Materials (CEAM) at the University of California, San Diego. This work was partially supported through the Office of Naval Research (ONR) grant N00014-09-1-1126 to University of California, San Diego.

References

1. Roland CM, Casalini R (2007) Effect of hydrostatic pressure on the viscoelastic response of polyurea. Polymer 48:5747–5752
2. Broekaert M, Pille-Wolf W (2000) The influence of isomer composition and functionality on the final properties of aromatic polyurea spray coatings. In: Proceedings of the Utech 2000 conference, Netherlands Congress Centre, The Hague
3. Mock W, Balizer E (2005) Penetration protection of steel plates with polyurea layer, Presented at Polyurea properties and enhancement of structures under dynamic loads, Airlie
4. Amini MR, Isaacs JB, Nemat-Nasser S (2006) Effect of polyurea on the dynamic response of steel plates. In: Proceedings of the 2006 SEM annual conference and exposition on experimental and applied mechanics, St Louis, 4–7 June 2006
5. Bahei-el-din YA, Dvorak GJ (2006) A blast-tolerant sandwich plate design with a polyurea interlayer. Int J Solids Struct 43(25–26):7644–7658
6. Tekalur SA, Shukla A, Shivakumar K (2008) Blast resistance of polyurea based layered composite materials. Compos Struct 84:271–281
7. Qiao J, Amirkhizi AV, Schaaf K, Nemat-Nasser S (2010) Properties of elastomer-based particulate composites. In: Proceedings of the SEM annual conference, Indianapolis, 7–10 June 2010
8. Qiao J, Amirkhizi AV, Schaaf K, Nemat-Nasser S (2011) Dynamic mechanical analysis of fly ash filled polyurea elastome. J Eng Mater Technol 133(1):011016, 7 pages
9. Kinsella M, Murray D, Crane D, Mancinelli J, Kranjc M (2001) Mechanical properties of polymeric composites reinforced with high strength glass fibers, Technical paper of 33rd international SAMPE technical conference, Seattle, 5–8 Nov 2001
10. Velasco JI, Arencón D, Sánchez-Soto M, Gordillo A, Maspoch MLI (2003) Milled glass fiber filled-poly(ethylene terephthalate-co-isophthalate) composites – thermal and mechanical properties. J Thermoplast Compos Mater 16:365–380
11. Vasconcelos PV, Lino FJ, Magalhaes A, Neto RJL (2005) Impact fracture study of epoxy-based composites with aluminium particles and milled fibres. J Mater Process Technol 170:277–283
12. The Dow Chemical Company (2004) Isonate 143L; Modified MDI (Dow Chemical, Midland, MI)
13. Air Products Chemicals, Inc. (2003) Polyurethane specialty products (Air products and chemicals. Allentown)
14. Hashin Z (1962) The elastic moduli of heterogeneous materials. J Appl Mech 29:143–150
15. Lee KJ, Westmann RA (1970) Elastic properties of hollow-sphere-reinforced composite. J Compos Mater 4:242–252
16. Nemat-Nasser S, Iwakuma T, Hejazi M (1982) On composites with periodic structure. Mech Mater 1:239–267